P9-DUB-662

MOBILE INTER-NETWORKING WITH IPv6

THE WILEY BICENTENNIAL–KNOWLEDGE FOR GENERATIONS

Each generation has its unique needs and aspirations. When Charles Wiley first opened his small printing shop in lower Manhattan in 1807, it was a generation of boundless potential searching for an identity. And we were there, helping to define a new American literary tradition. Over half a century later, in the midst of the Second Industrial Revolution, it was a generation focused on building the future. Once again, we were there, supplying the critical scientific, technical, and engineering knowledge that helped frame the world. Throughout the 20th Century, and into the new millennium, nations began to reach out beyond their own borders and a new international community was born. Wiley was there, expanding its operations around the world to enable a global exchange of ideas, opinions, and know-how.

For 200 years, Wiley has been an integral part of each generation's journey, enabling the flow of information and understanding necessary to meet their needs and fulfill their aspirations. Today, bold new technologies are changing the way we live and learn. Wiley will be there, providing you the must-have knowledge you need to imagine new worlds, new possibilities, and new opportunities.

Generations come and go, but you can always count on Wiley to provide you the knowledge you need, when and where you need it!

William J. Pesce
WILLIAM J. PESCE
PRESIDENT AND CHIEF EXECUTIVE OFFICER

Peter Booth Wiley
PETER BOOTH WILEY
CHAIRMAN OF THE BOARD

MOBILE INTER-NETWORKING WITH IPv6

Concepts, Principles, and Practices

RAJEEV S. KOODLI and CHARLES E. PERKINS

WILEY-INTERSCIENCE
A John Wiley & Sons, Inc., Publication

Published by John Wiley & Sons, Inc., Hoboken, New Jersey.
Published simultaneously in Canada.

Library of Congress Cataloging-in-Publication Data:

Koodli, Rajeev S., 1968–
Mobile inter-networking with IPv6 : concepts, principles, and practices / Rajeev S. Koodli and Charles E. Perkins. — 1st ed.
p. cm.
"A Wiley-Interscience publication."
Includes bibliographical references and index.
ISBN 978-0-471-68165-6
1. Mobile communication systems. 2. TCP/IP (Computer network protocol) 3. Internetworking (Telecommunication) I. Perkins, Charles E., 1951– II. Title.
TK6570.M6K66 2007
621.3845'6—dc22 2006052808

Printed in the United States of America.

10 9 8 7 6 5 4 3 2 1

Contents

List of Figures

List of Tables

Preface

Internet access and mobility continue to change our perception of communications and computing. We are becoming more dependent on both, as can be seen by the continued proliferation of mobile telephones, personal digital assistants (PDAs), and laptop computers. Improvements in miniaturization and higher-speed communications media are creating an ever-increasing population of users for these electronic gadgets. To a certain extent, the basic applications remain very familiar: messaging, voice communications, e-mail, and Web access. More recently, we have seen that imaging and music are becoming important features in mobile communication. Image processing and rich video are not far behind.

Consumers want access to these applications wherever they go, and they naturally want to interconnect their gadgets for any of myriad reasons. The image should be easily tagged and shared with a group, and perhaps should be backed up locally on a home disk. The message should be transferred into an e-mail folder. The e-mail headers should be read aloud into the earphones hooked up to the mobile telephone. These are very natural requirements, yet far from universally satisfied. The inevitability of mobile computing and mobile Internet access will soon cause these features to be taken for granted.

Underneath the shiny personalized faceplates and the high-resolution LCD displays, there is a wholly different world of engineering to make these separate, mobile devices able to communicate on behalf of the users' applications. The genius of the Internet allows every device in the world to be separately addressable, and thus separately managed for the purpose of establishing communications pathways for the applications. In recent years, the Internet protocols for basic packet delivery have been devised to handle the requirements of global mobility and application transparency. After all, the user doesn't want to worry about whether things will break just because the computer or telephone or communications device is operating in a different location. Up until now, unfortunately, we have had to worry about this far too much. The purpose of *Mobile IP* is to eliminate this source of bewilderment and worry.

This book provides the reader with a thorough understanding of how Mobile IP works, particularly in the context of the new Internet Protocol standard known as *IP version 6* (IPv6). Mobile IPv6, although functionally similar to Mobile IPv4 (which was explained in a previous publication), is quite different in detail due in part to conformance with IPv6 specifications and to experience gained over the years. For instance, "Route-Optimized" communication between mobile nodes and their correspondents is now an integral part of the Mobile IPv6 specification, not a separate protocol. Mobile IPv6 is specified as the mobility protocol for the *CDMA* cellular packet systems and is being specified as the mobility protocol for the *WiMax* networks.

While Mobile IPv6 is designed to provide basic mobility support on the Internet, there are some newer protocols known as *Fast Handover*, *Context Transfer*, and *Hierarchical Mobile IP* or *Regionalized Registration.* These protocols enable better performance for real-time applications such as voice and video, for which the connectionless aspect of base IP operation is typically insufficient for an enjoyable user experience. These protocols are gaining increasing importance as more and more communication, spearheaded by Voice over IP (VoIP), becomes dependent on the Internet Protocol.

Part IV of the book investigates how IP mobility is used in practice. We look at the adoption of Mobile IPv6 in CDMA cellular systems. We also investigate mobility in enterprises, where Virtual Private Networks (VPNs) and access enforcement devices such as firewalls end up interacting with mobility protocols.

Part V describes some experimental work, such as performance of VoIP over WLAN during handovers, multi-access networks and VoIP handovers, as well as emerging topics such as *Location Privacy.* This part is intended to expose the readers to experimentation involving mobility and to provide a glimpse of research one could envision.

Many chapters in this book have *Exercises* to encourage readers to pursue further the topics covered in the book. All chapters contain *References*, some of which may still be "works in progress" reflecting the nature of standardization in the *Internet Engineering Task Force* (IETF), the organization that specifies Internet standards such as TCP/IP, http, and SIP.

In order to take full advantage of the information in this book, you should be familiar with TCP/IP protocols, including IP, the *Internet Protocol*, and TCP, the *Transmission Control Protocol*). Rich Stevens' book "TCP/IP Illustrated Vol 1: The Protocols" and Douglas Comer's "Internetworking with TCP/IP" both provide excellent introductions to the topic. The book is otherwise self-contained, so that familiarity with Mobile IPv4 is not a requirement.

This book is designed to be useful for students interested in mobile networking with insights into communication protocols and details of operations. Engineers reading this book will be able to implement Mobile IPv6 more effectively, having a clear understanding of the system impact of mobility. Experts may find it as a useful reference.

As you read the book, you will notice many italicized terms, some of which have conventional meanings which may be different from the expected ones (for example, *home agent*). These terms are defined in the Glossary, so please be sure that you understand their meaning before continuing.

We used LaTeX to write this book. It is a great tool, but it has a mind of its own! At times, it generously introduces extra space in exactly those places where you least wish for it. We have tried our best to "hand craft" around such tantrums, but there are a few places where we just had to let it go. We hope that our readers will be kind and understanding if there is any inconvenience.

Finally, a few comments on the *packet header*, which we will frequently encounter in this book. Any communication protocol must be able to define the syntax and semantics for communication. Such definitions must be clear and unambiguous. This is especially important for Internet communication protocols that traverse individual subnetworks. IP is a packet-based protocol, which basically means that communication takes place without dedicating a "circuit" for the sole use of a particular communication, but instead by using small chunks of *datagrams* or *packets*, a stream of which could be used to constitute a *session* or a *connection*. A header in IP defines fields which are essential for the IP software in the communicating end-points to understand how to process each packet. A packet header is shown in Internet Standards using ASCII pictures! In this book, we use the same ASCII notation (while admittedly sacrificing the aesthetics) to be consistent with the protocol specifications that readers will encounter in these standards.

R. S. KOODLI AND C. E. PERKINS

Mountain View, California

Acknowledgments

Many intelligent people have made major contributions to Mobile IPv6 and its companion protocols including Network Mobility, Fast Handovers, Hierarchical Mobility and so on. Thanks are due to all of them, especially to the long-suffering participants of the Mobile IP working group of the IETF. In addition to learning a lot and enjoying many good conversations during the years of development, we appreciate the many benefits and improvements that have resulted from the group effort. In these acknowledgments, we would like to mention specifically some people whose personal interactions have enriched our experience of developing the protocols and writing this book.

Our research colleagues in Nokia Research Center, in Mountain View, California played a very important part in helping to shape the way we understood the promise and potential of Mobile IPv6 by way of numerous prototype implementations from a very early version all the way to the eventual standard. We shared the joys and the pain of Connectathons, TAHI test suites, ETSI interoperability events, and countless internal demonstrations for various product managers and research associates. Our hearty thanks go to Vijay Devarapalli, Jari Malinen, T.J. Kniveton, and Meghana Patil nee Sahasrabudhe, as well as our fond hopes of working together again in the future. Thanks also to Hannu Flinck, who worked tirelessly and endured great stress to obtain funding for our many projects, apart from contributing to implementation. We also acknowledge the support of the head of our laboratory, Reijo Juvonen, especially for bearing with us on this seemingly never-ending (book) project. By extension, we wish also to give thanks to Nokia for supporting our research efforts. Our team at Mountain View was responsible for a long stream of Internet Drafts, protocol enhancements, testbeds, demonstrations, RFCs, and published papers on the subject. From an outsider's point of view, it may have appeared like we had a team of twenty top researchers to have accomplished all we did.

Thanks to Pekka Nikander, who collaborated in the creation of the BAKE (Binding Authorization Key Establishment), which was one of the first protocols that could scale to the numbers needed for IPv6 and essentially offered us the confidence that the job could in fact really be done. Thanks to Basavaraj Patil, who as co-chair of the Mobile IPv6 working group and valued colleague within Nokia, never failed in his encouragement for us and assistance at numerous times. Thanks also to Francis Dupont, a long-time contributor to Mobile IPv6, who also supported the idea of inline mobility management signaling and produced a draft on the topic.

This book has benefitted immensely from the meticulous proofreading by the Wiley production staff, and Cedric Westphal, our longtime trusted colleague at Nokia. We are grateful to both of them.

No list of acknowledgments would be complete without mentioning the kind understanding and support we have derived from our families. Charlie would like to acknowledge his love and thanks to Diane, Greg and Phil! It was not so long ago that the boys were Gregory Russell and Philip Austin. Some day, you two might read and understand this book, at least as a historical artifact and maybe as a step along the way to a much better technological future.

Rajeev would like to dearly thank Vidya for her endurance and astute sense of knowing when to ask "How is it going?", and mother Ramaa Koodli for her persistence in asking "When are you finishing?". Rohan and Krithi have brought unfailing joy and cheer in life. You two already understand mobility, beginning with all the photos and videos you have taken on the phone, so perhaps you will take this book for granted!

K. P.

Acronyms

AAA	Authentication Authorization and Accounting
AH	Authentication Header
AP	Access Point
ARP	Address Resolution Protocol
BAck	Biding Acknowledgement
BCE	Binding Cache Entry
BS	Base Station
BSS	Basic Service Set (in Wireless LAN)
BU	Binding Update
CoA	Care-of Address
CN	Correspondent Node
CDMA	Code Division Multiple Access
DAD	Duplicate Address Detection
DHCP	Dynamic Host Configuration Protocol
DNS	Domain Name System
EAP	Extensible Authentication Protocol
ESP	Encapsulating Security Payload

FBU	Fast Binding Update
GGSN	GPRS Gateway Support Node
GPRS	General Packet Radio System
GSM	Global System for Mobile Communication
HA	Home Agent
HoA	Home Address
ICMP	Internet Control Message Protocol
IETF	Internet Engineering Task Force
IID	Interface Identifier
IKE	Internet Key Exchange (protocol)
IMS	IP Multimedia Subsystem
IP	Internet Protocol
IPsec	IP security
IPv6	Internet Protocol version 6
MAC	Media Access Control
MAP	Mobility Anchor Point
MN	Mobile Node
MH	Mobility Header
MR	Mobile Router
MS	Mobile Station
NAR	New Access Router
NAT	Network Address Translator
NEMO	Network Mobility
ND	Neighbor Discovery
PAR	Previous Access Router
PDSN	Packet Data Serving Node
QoS	Quality of Service
RFC	Request For Comments (an IETF protocol specification)
RO	Route Optimization
RTP	Real-time Transport Protocol
SA	Security Association
SAD	Security Association Database
SCTP	Stream Control Transmission Protocol
SGSN	Serving Gateway Support Node

SIP	Session Initiation Protocol
SPD	Security Policy Database
SPI	Security Parameter Index
SSID	Service Set Identifier
SSL	Secure Socket Layer
TCP	Transmission Control Protocol
TLS	Transport Layer Security
UDP	Unreliable Datagram Protocol
URI	Universal Resource Identifier
URL	Universal Resource Locator
VPN	Virtual Private Network
VoIP	Voice over IP
WLAN	Wireless LAN

Part I

Introduction and Background

As the Internet continues to shape and reshape our lives, there is a parallel phenomenon taking place in wireless (cellular) communication which is equally (if not more) influential. Personal and mobile voice communication has proved to be a major paradigm in the history of human communication. As both the Internet and mobile telephony continue their dramatic advances, their confluence is already emerging in image sharing, instant messaging and even some rudimentary forms of voice communication. However, it is only fair to recognize that this trend is in its infancy, and also perhaps has been affected by the existing Internet and cellular deployment limitations and deficiencies. The following few functionalities are considered necessary in the Mobile Internet:

- An ability to readily provide addressing support for billions of devices without severely restricting the spectrum of applications (and without burdening the network infrastructure itself)
- Easily support movement of users on the Internet from one network (e.g., Wireless Local Area Network (WLAN)) to another (e.g., 3G cellular)
- Provide adequate support for security

Perhaps there are others, but the above are arguably essential for a Mobile Internet.

In this part of the book, we provide a preliminary introduction to our main subject of Internet Protocol (IP) Mobility. We introduce the reader to the underlying problems in mobility on the Internet. We recognize that users value *persistence* of their communication, as well as the ability of their peers to reliably *reach* them in spite of mobility on the Internet. We will walk the reader through the events that take place as a device connected to the Internet disconnects and plugs back into the Internet.

We will provide background on *IPv6*, the protocol designed primarily to provide an abundance of IP addresses for billions of devices, a large portion of which are going to mobile devices accessing the Internet. IPv6 is crucial for the evolution of the current Internet into a Mobile Internet. We also provide background on Internet Protocol security, or *IPsec*, which is a suite of protocols designed to provide comprehensive security at the IP layer. IP mobility protocols are obliged to make use of IPsec in order to protect messages that require adequate authorization. Chapters 2 and 3 present information which is immediately relevant for the rest of this book. We suggest that readers requiring additional information on these topics consult other references. On the other hand, those already familiar with these topics may still find the chapters refreshing.

1

Mobility on the Internet: Introduction

The best way to predict the future is to invent it. –Alan Kay

In order to understand what mobility on the Internet means, let us consider Figure 1.1. Bob, a user connected to the Internet by means of an access network (such as a cable modem or a DSL or a dial-up network), is "talking" to Alice, who is another user connected to the Internet by a different access network (e.g., WLAN). We do not show Bob and Alice themselves, but only their devices. Now, consider that Alice "moves" from her current access network to another access network (e.g., the Code Division Multiple Access (CDMA) cellular network). There are two basic problems we can see. First, how can Bob continue the existing conversation with Alice? Second, how can Bob reliably reach Alice once she has moved? The first problem can be broadly considered to be the *handover* problem. It is quite similar to that of users continuing their calls on their cell phones in spite of movement (such as in a moving train). The second is the *reachability* problem. It is quite similar to being able to reach users on their cell phones even when they are out of town (i.e., roaming). These two problems, which appear to be very straightforward, create many technical challenges associated with mobility on the Internet.

Perhaps it is tempting to assume that Mobile Internet is a given. Even so, it might be worthwhile to consider why the Internet Protocol is the best fit for a Mobile Internet. As we know, IP has been an unquestionable success in "gluing" disparate networks the world over. As the Internet becomes increasingly mobile, spearheaded by the

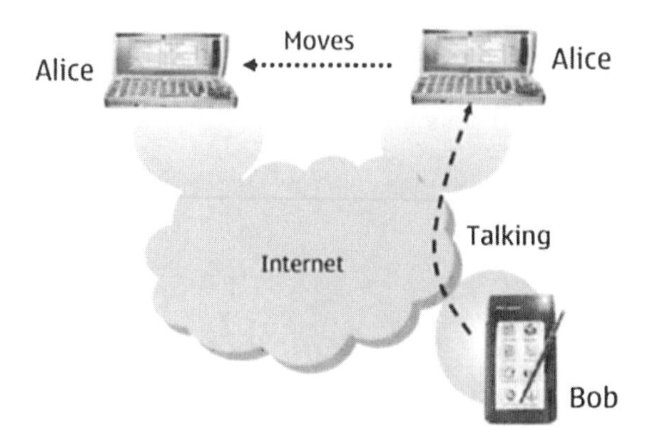

Fig. 1.1 Mobility on the Internet

mobile communication devices, users are bound to roam freely and attach to a variety of networks. In fact, this is already beginning to take place. Mobile Smartphones and Personal Digital Assistants equipped with cellular packet radio and WLAN are becoming all too common. In a Mobile Internet composed of numerous wireless (and wired) technologies, no single access network itself can provide mobility support across all kinds of networks. IP is best suited to address this imminent problem. More specifically, IP mobility is primarily a routing problem, but it includes problems involving fast switching of connections, smooth handovers, transport protocol optimizations, and many others.

The general problem of mobility is quite complicated, and no one protocol can be expected to provide a complete solution. In fact, just considering the problems introduced by mobility at the network layer (i.e., IP mobility), we still cannot expect to find a single protocol solution for all of them. These problems include the following:

1. Authentication for access rights to the new network

2. Obtaining an IP address at the new point of attachment

3. Routing packets to each new point of attachment

4. During transitions between points of attachment, taking appropriate actions to minimize handover delay and data loss not directly attributable to unavailability of the underlying communications medium

5. When packet is to be transmitted, selecting the appropriate source IP address

Solving the first two of these problems without attempting to solve the other problems can be said to provide a solution for portable computing. In other words, one can establish a link and start to work at a new point of attachment, but the user experience more closely resembles rebooting than relocating. Whatever activities were in progress before the new link is established may be aborted or start behaving unpredictably. In this book, making use of this kind of portable computing will be called *roaming*.

Solving the third problem is often related to network-layer protocol considerations because the act of supplying a new IP address depends on being able to prove that one is authorized to receive it. There are many different preexisting solutions to the problem of access control and authorizations. Many or even most of them are challenging to adapt to the needs of network-layer mobility. To authorize roaming is not so demanding, because, by the above definitions, the mobile user is already expecting significant disruption of whatever activities (if any) might have been in progress on the mobile device. For roaming scenarios, then, we can expect the access control to proceed without any dependency on mobility management, if indeed there is any mobility management. Once the access control is granted, the IP address can be allocated and the routing (or bridging, even) enabled for the particular device.

Roaming in this way is particularly easy for devices that are not named within the global *Domain Name System* or DNS [1, 2]. This is because there is no need to update the DNS entry for such devices, even when they receive new IP addresses [1]. On the other hand, for devices that do have a persistent name that is published by DNS, getting a new IP address immediately invalidates the DNS entry and effectively makes the device unreachable for incoming transmissions that are based on resolving the device's persistent name. In order to solve this problem, methods have been specified (see, for instance, [3]).

It is worthwhile to emphasize that updating DNS carries with it an obligation to provide strong proof that the update is properly authorized. Otherwise, if DNS were to accept such update requests without sufficient assurance about the identity of the requesting node, havoc would ensue. In fact, in today's Internet, these secure update procedures are not so widely deployed, and effectively must be considered to be unavailable to the majority of roaming nodes.

In this way, we can see already that the restrictions imposed by focusing only on the roaming problem have the effect of relegating roaming nodes to the status of "second-class citizens" of the Internet. They can operate as clients of well-known services (typically including e-mail services and browsing web pages that are statically addressable). But, for instance, such devices cannot by themselves receive telephone calls by way of the Internet, or publish web pages, or operate in true peer-to-peer fashion. For instance, Bob (in Figure 1.1) can neither reach Alice reliably nor maintain his call once she moves to a different network. Many people are content with these restrictions, because right now web browsing is the dominant application, and e-mail is not closely tied to real-time interactions.

[1] A DNS entry is required in order for a node to determine the IP address of its peer with which it wishes to communicate. The DNS contains the IP addresses for entries such as www.yahoo.com or any user record.

We can distinguish between roaming and handover by observing that handover usually means arranging for transfer of control or responsibility. In the case of handover for a mobile network device, this means that a new network entity has to begin taking over some of the functions currently operating at some other network entity that is interacting with the mobile device. When the handover completes, all interactions are handled by the new entity, and the previous one can deallocate whatever resources were assigned for use with the mobile device.

That's all very abstract. If we think of network entities as access routers, then they are supposed to be delivering packets to the mobile device and forwarding traffic from the mobile device to its peers. From the standpoint of IP networking, the basic responsibility of the access routers is to forward packets to their destination, as indicated by the IP address of the destination. However, the access routers only have this responsibility for the addresses that are *topologically-correct* for their place in the network, and when the access routers are in different places, their customers will be expected to have different IP addresses. From this perspective, the access routers do not really offer any handover features to the mobile device unless they have been augmented with extra functions, such as those described in later chapters.

Using the same idea, we can think about designing other sorts of mobility-related handovers. For instance, one could handover a session from one display device to another. A new display controller would have to take over responsibility for showing the pixels and a new path created for transmitting the picture information to the display controller. If the devices are on different hardware platforms, some translation has to occur so that the picture information is presented to the display controller in a way that is suitable for the hardware. Perhaps the size of the screen is different, or perhaps one screen has fewer colors than the other. If the displays are attached to different points of the network, then some communications protocol has to be defined as a way of transferring the picture information to the new network node. If this is done carefully, one could well imagine a smooth handover with no frames lost. Perhaps with a lot of buffering, there could be a time when the same images are streaming to both display controllers, so that the viewer would not miss a single dot.

Perhaps it is useful to discuss topological correctness of IP addresses a little bit. The Internet addressing is often termed as *hierarchical* meaning that organizations typically get chunks of IP addresses which are valid for all the hosts attached to *that* particular segment of the Internet. In order to maintain its (amazing) scalability, the Internet routing works based on the network prefixes and not on individual host IP addresses; only when a packet reaches the destination subnet does the host-based forwarding take place. This means when a host with an IP address IP_A valid on network A moves to network B with its own set of addresses on the Internet topology, the Internet routing cannot ensure forwarding of packets to the host's IP address IP_A. The host must obtain another IP address on network B, and then everything moves like clockwork. In other words, a host address must be *topologically-consistent* or *topologically-correct* for packets to reach it.

In order to tackle the problem of change of IP address, Mobile IP provides a way for a mobile device to maintain a persistent IP address at the same time that it acquires a new IP address from every new access network that it visits. The mobile device may be viewed as taking some of the responsibility for overcoming the effects that result from changing its IP address by shielding the change from its applications. Since the applications don't see any change of address, they continue without major disruption when the mobile device acquires a new IP address at its new point of attachment.

Mobile IP allows the mobile device to remain addressable at its persistent IP address, known as the device's *Home Address.* This is the address used by the applications running on the mobile device, and it is the address registered in the global DNS. Since the address doesn't change, the need for secure updates to DNS is sidestepped completely. That is already a major benefit. By the same token, existing Transmission Control Protocol (TCP) connections can survive relocation to a new point of attachment, since Mobile IP allows the device to continue using its persistent address even when the device is only reachable at an address other than its persistent address. Such a "roaming address" is called as *Care-of Address* in Mobile IP. Both of these addresses are known to the mobile node as well as its trusted partner called the *Home Agent*, which can be considered as a router on the mobile node's *home network*, the IP subnet that the mobile node normally resides in when it is not roaming. In a way, quite a bit of the Mobile IP protocol is between the mobile node and its home agent. The rest of the protocol is between the mobile node and its *Correspondent Node*, which is any Internet-enabled device that communicates with the mobile node.

A good portion of this book describes the essentials of the Mobile IP protocol in the context of IPv6. The other parts are dedicated to protocols that provide the necessary performance for real-time applications such as Voice over IP (VoIP). In order to fully understand these protocols, it is necessary to understand the basics of IPv6 and some elements of the IP security (IPsec) architecture. In the next two chapters, we provide a brief introduction to these two topics, expecting the users to consult references elsewhere for a much more detailed discussion. We will return to the detailed description of the Mobile IP protocol in Part II.

The *Request For Comments* (RFCs) in References in this chapter and the rest of the book can be accessed via http://ietf.org/rfcs/html.

REFERENCES

1. P. Mockapetris. "Domain Names - Concepts and Facilities," RFC 1034, Internet Engineering Task Force, November 1987.

2. P. Mockapetris. "Domain Names - Implementation and Specification," RFC 1035, Internet Engineering Task Force, November 1987.

3. B. Wellington. "Secure Domain Name System Dynamic Update," RFC 3007, Internet Engineering Task Force, November 2000.

2

IP Version 6

640K ought to be enough for anybody. –Bill Gates, 1981

2.1 MOTIVATION

As the Internet began to grow at a dramatic rate during the late 1980s and early 1990s, engineers realized that the current version of the IP protocol would not be adequate to meet the demands of the Internet's growth. Of particular concern was the availability of adequate IP addresses for all kinds of devices accessing the Internet. With 32 bits, IPv4 can theoretically provide up to 4.2 billion addresses. However, even with uniform allocation, this is insufficient given the magnitude of growth seen in wireless and mobile communication. Allocation policies in practice make this worse; a few organizations have an overabundance of addresses, whereas many others have ended up with too few. These and other considerations led to a design that could support every imaginable device that would need an IP address in the foreseeable future. With this design, one estimate is that every atom of each person on earth could be assigned seven unique addresses!

IPv6 also provides additional features. Address auto-configuration allows hosts to autoconfigure their IP addresses without the need for a centralized server. The *header*, which contains the essential fields for the IP protocol to work, has been simplified for more efficient processing; even with 40 bytes, the IPv6 header is more amenable to header compression than the IPv4 header. Extensions to the base IPv6 header provide better support for security and mobility; the *Mobility Header* in particular has been designed to support the Mobile IP protocol and its enhancements. We consider the header format in detail in the following section.

With the basic addressing support and additional support, especially for security and mobility, the new IPv6 protocol can form the basis of the next-generation Internet. It is, hence, important to understand some of its basic operations. The following sections provide this information. Detailed description of the protocol are found in multiple references dedicated to the subject.

In this chapter, we will describe the motivation, the IPv6 addresses, the IPv6 protocol basics, the Neighbor Discovery protocol, and Address auto-configuration. These are essential for an understanding of the Mobile IPv6 protocol operation. Those already familiar with the topic may skip most of this chapter, or find it a good review. In any case, readers can revisit this chapter when reading the subsequent chapters to refresh their memories. We begin with a set of definitions.

2.2 DEFINITIONS

We define some commonly used terms in the literature. These definitions are adapted from standards documents and from common usage in the field.

- A *link* is the physical medium over which communication can take place. In the IP terminology, this implies a layer which can carry IP packets and provide demultiplexing at the receiver. Examples are Ethernet, ATM, Frame Relay, and IP itself! [1].

- An *interface* refers to an attachment point to a link. An interface can be physical, such as a WLAN card, or logical, such as a tunnel interface.

- A *link-layer address* is an identifier associated with an interface. The IEEE 802 MAC address is an example of a link-layer address.

- An *interface identifier* (IID) is a bit string of known length that uniquely identifies the interface on the link. In many cases, the IID is the same as the link-layer address of the interface. Sometimes, a pseudo-random number may be used as an IID.

- A *prefix* is a bit string that denotes the topological position on the Internet. The prefix is part of an IP address.

- If two or more nodes are attached to the same link, they are considered *neighbors.* They share at least one common prefix.

[1] IP packets can be carried inside IP packets, providing a "tunnel," which itself is carried over one or more physical media.

- When sending a packet to a destination, the sender checks if the destination address is *on-link*, based on the advertised prefix on the link and the prefix part of the destination's address.

- Often the term *tunnel* is used in IP literature. It means encapsulating an IP datagram in another IP datagram so that a certain level of indirection in forwarding can be achieved. For instance, the datagram can be sent to an agent to effect specialized processing before the original datagram is delivered to the eventual destination.

2.3 IPV6 FORMAT

The Internet protocols often define standards for hosts and routers to communicate. IPv6 is one such important protocol. In order for a standard to be implementable as well as interoperable, it needs to define unambiguous messages that the protocol uses. The messages carry information for the IP software to interpret, as well as a user-generated "payload," such web pages, streamed content and VoIP. The information carried for the IP software can be generally referred to as the "header", which specifies how the message must be interpreted and processed. The header definition and semantics must be unambiguous so that the sender and the receiver can communicate. A header "format" does just that. It allows a sender to construct and fill the fields in the header in a way that the receiver can understand and interpret. The following section provides a description of the IPv6 header.

2.3.1 IPv6 Header Format

The header in an IPv6 packet is shown in Figure 2.1.

The description of each of the header fields follows. Each "+-+" corresponds to 1 bit. Hence, the *Version* field consists of 4 bits. As we discussed in the Preface, all the IETF standards use ASCII notation for specifying the message formats. We will also use the same notation throughout this book.

The *Version* field is set to 6 to indicate the protocol version.

The *Traffic Class* field actually consists of two subfields. The first 6 bits of this field constitute the Differentiated Services Code Point (DSCP), and are used for providing different forwarding treatment (Quality of Service [QoS]) to traffic. A source node may mark these bits to request certain "differentiated services" [1], and a router may provide the corresponding forwarding behavior. These bits may be remarked by intermediate routers to ensure that the traffic is compliant to what is agreed upon. The remaining two bits of the field are reserved for "Explicit Congestion Notification (ECN)" [9], which is used to inform in advance the transport protocol of congestion along the path a packet takes.

The *Flow Label* is used to identify a flow, which is a sequence of packets from a source to a destination. Presumably, the sequence of packets in a flow can be characterized by some common needs and behavior. Traditionally, the flow classification

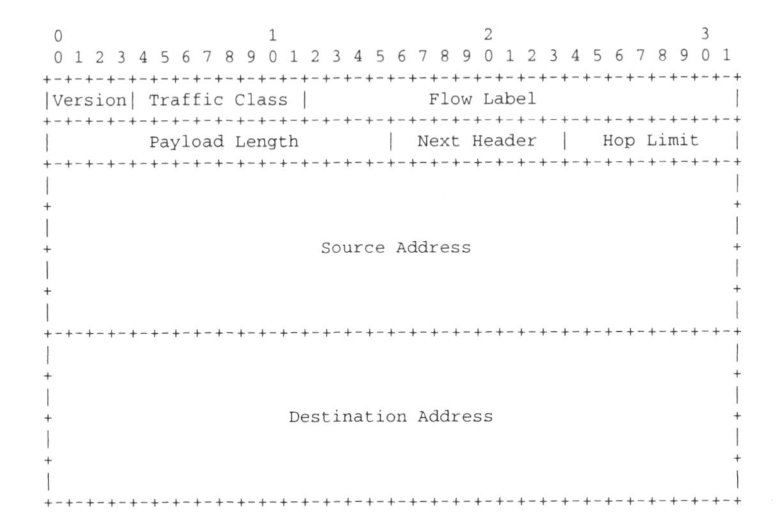

Fig. 2.1 IPv6 Header

has been based on the 5-tuple of source and destination IP addresses, source and destination ports, and the transport protocol type. The 5-tuple may be unavailable, for instance due to encryption, or locating the transport header fields may be inefficient. Flow Label serves as an IP header field that can assist in flow classification at the intermediate nodes.

The length of the rest of the packet following the IPv6 header is denoted by *Payload Length.*

The type of the header immediately following the IPv6 header is identified by *Next Header.* For data, this is typically TCP or Unreliable Datagram Protocol (UDP). However, IPv6 defines multiple extension headers which may be present. For instance, packets using Mobile IP may use *Routing Header* and *Destination Option* headers, which we discuss further below and in the following chapters.

Hop Limit determines how far a packet should traverse on the Internet. Its value is decremented by one by each node that forwards the packet. When the value reaches zero, the packet is decremented.

The Source and Destination Addresses specify the originator and the receiver of the packet, respectively. The Destination Address does not specify the ultimate receiver if the Routing Header is present.

2.3.2 IPv6 Extension Headers

Additional IP layer information may be carried in extension headers which appear between the IPv6 header and an upper-layer header such as a TCP header. The Next Header field in the IPv6 header specifies the presence of extension headers each of which itself contains the Next Header field. When multiple extension headers are present, the receiver must process them sequentially, beginning from the first extension header following the IPv6 header. This requirement exists because contents and semantics of an extension header determine whether to process the next extension header or not. There is also a strict ordering requirement when constructing multiple extension headers (See Section 4.1 in [2]).

We describe two such extension headers which are used in mobility protocols.

2.3.2.1 Routing Header When a host wishes a packet to be processed by a set of intermediate nodes before it is processed by the eventual destination, it uses the Routing Header extension. The Routing Header has a value of 43 in the Next Header field. Different types of routing headers may be defined. The IPv6 protocol itself defines a Type 0 routing header. As we shall see later, Mobile IP defines a Type 2 routing header.

The format of a Type 0 Routing Header is as shown in Figure 2.2.

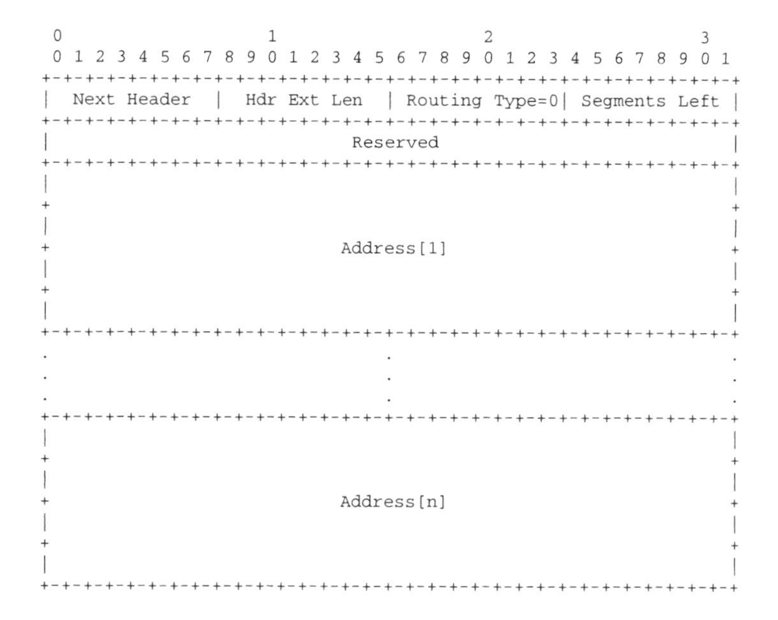

Fig. 2.2 Routing Header Extension Header

A host needs to specify the IPv6 addresses of all the intermediate nodes the packet needs to visit and specify the Segments Left field accordingly. For instance, if there are two intermediate nodes which a packet needs to traverse, the Segments Left field is set to 2 at the source. When this value reaches zero, the receiver proceeds to the Next Header in the chain. When it is non-zero, the intermediate node swaps the destination IP address in the IPv6 header with the next address in the list of addresses in the routing header chain. In accordance with this operation, the eventual destination appears last in the routing header chain and the destination IP address in the IPv6 header always points to the next node to be visited.

2.3.2.2 Destination Options Header The Destination Options extension header is meant to be processed by the destination specified in the IPv6 header. It is identified by a value of 60 in the Next Header field. When it appears before the Routing Header, it is also meant to be processed by the nodes specified in the Routing Header chain. The processing actions are specific to each option itself; protocols which make use of this extension header need to specify the processing semantics and actions as part of the "Options" in this extension header, whose format is shown in Figure 2.3. In other words, this extension header does not define its own Type field, unlike the routing header.

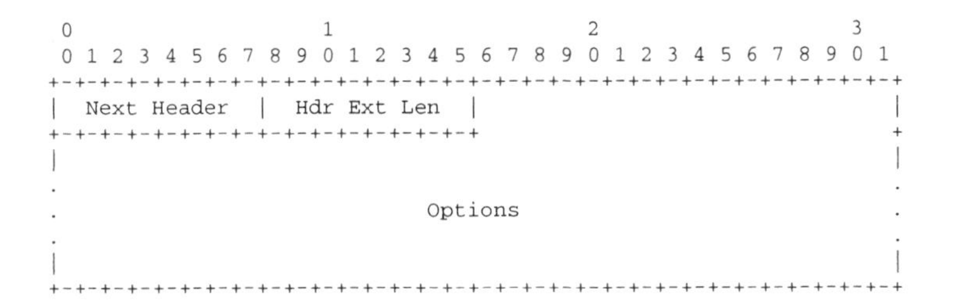

Fig. 2.3 Destination Options Extension Header

When designing the option format for Destination Options (and also for another extension header called *Hop-by-Hop Options*), the option fields may need to be aligned so that multiple octet fields fall on their natural boundaries. The alignment requirement is often stated in the notation "xn+y", which indicates that the Option Type field must appear at any multiple of 'x', plus 'y' octets. RFC 2460 [2] provides a guideline on how to construct the fields in the options, suggesting that fields be arranged in ascending order and that alignment be derived based on the largest field (up to 8 octets). For instance, if an option contains two fields which are 4 and 8 octets long, the alignment requirement is 8n+2, since the Option Type appears at an offset of 2 octets from the start of the header based on the alignment for the 8-octet field. This is shown in Figure 2.4. See Appendix B in [2] for additional examples.

2.4 IPV6 ADDRESSES

An IPv6 address is assigned to a network interface [2]. If a node has multiple network interfaces, each can have one or more IPv6 addresses of its own, and any address

[2]It is worthwhile to note that an IP address is always assigned to a network interface and not to a host.

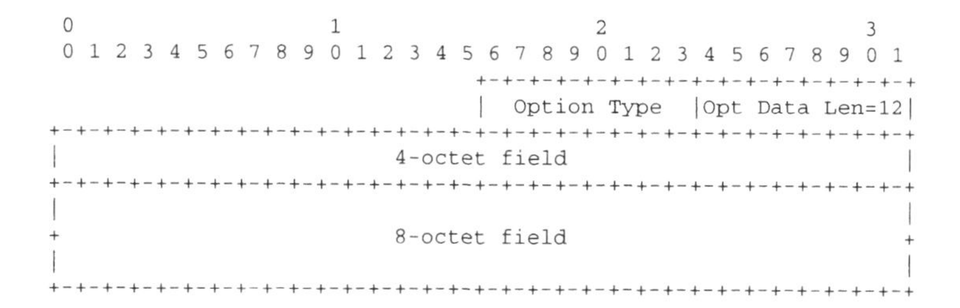

Fig. 2.4 Option Alignment

can be used to reach the node subject to address scope and routing availability. Like IPv4, IPv6 continues to associate a subnet prefix and a host part to an address. A single interface can also be assigned multiple IPv6 addresses, either based on different prefixes or using different host identifiers.

There are three types of addresses in IPv6: unicast, anycast, and multicast. Each of these addresses can have different scopes, which limits their applicability and usage. There are no broadcast addresses in IPv6.

The addresses are represented in text form using the notation "X:X:X:X:X:X:X:X", where each X is the hexadecimal value corresponding to 16 bits of the overall address. An example is ABCD:1234:FEDC:0123:0:0:0:1. A contiguous string of zeros can be compressed using the notation '::'. For example, the above address can be represented as ABCD:1234:FEDC:0123::1. Sometimes, the address prefix is also represented using the notation "IPv6-Address/Prefix-Length", where Prefix-Length is the number of leftmost bits in the IPv6-Address that make up the prefix. The remaining bits in the IPv6-Address are assumed to form the host part, the IID.

IPv6 predefines numerous addresses for nodes. In the following, we describe some commonly used addresses we will encounter in this book.

A unicast address uniquely identifies a network interface attached to an IPv6 network. A Unicast Address is formed by using a network prefix and an IID. The network prefix can determine the scope of the unicast address. The IID comes from the host. It is typically, but not always, also the link-layer address associated with the interface.

A *link-local address* is formed by prepending the network prefix FE00::0 to an IID. It is valid only on a link, and can be used for communication between the hosts without requiring a router. A link-local address is required on an interface supporting IPv6.

A *Unique Local address* is for use within a site. It is unique with a high probability across the Internet, but is not used for global communication. It is formed using a well-known prefix, a "Global ID" (for uniqueness) and the subnet prefix which provides hierarchy.

A *global address* is formed by prepending a globally unique prefix to an interface identifier. This address is valid for communication over the Internet.

An *unspecified address* is denoted by 0:0:0:0:0:0:0:0. It is not assigned to any interface, but is used when a node is attempting to formulate an address for its interface. It can be used as a source IP address, not as a destination address. The recipient responds typically to the all-nodes multicast address (see below).

A *loopback address* is denoted by 0:0:0:0:0:0:0:1. It is not assigned to any physical interface. A node may use it to send a packet to itself, perhaps to confirm that the IP stack is forwarding packets correctly. The loopback address is never used as a source IP address.

An IPv6 anycast address is an address assigned to more than one interface, but only one interface receives the packet sent to that address. The routing protocol chooses the "nearest" interface according to its measure of distance. Each router is expected to support a subnet-router anycast address, which is formed by appending to the unicast prefix an interface identifier set to zero.

An IPv6 multicast address is assigned to a set of nodes. They begin with the prefix "FF", followed by 4 bits of flags, a 4-bit scope field, and finally the group ID itself of 112 bits. The scope field specifies the limit of the multicast group. Some common values used for the scope field are:

1: node-local, which means the scope is limited to a single node

2: link-local, which means the scope is limited to the link

5: site-local, which means the scope is limited to a site

E: global scope

The following reserved multicast addresses are noteworthy. They are used for a variety of purposes including address configuration, address resolution, and router discovery.

The *all-nodes multicast addresses* are FF01::1 and FF02::1. The former has node-local scope and the latter link-local scope.

The *all-routers multicast addresses* are FF01::2, FF02::2 and FF05::2 meant for node-local, link-local, and site-local, respectively.

The *solicited-node multicast address* is formed by appending the low-order 24 bits of the unicast or anycast address to the prefix FF02:0:0:0:0:1:FF00/104.

There are many other addresses that IPv6 defines (see [5]). However, a host is required to understand the following addresses:

- a link-local address for each of its interfaces
- all assigned unicast addresses
- the loopback address
- all-nodes multicast addresses
- a solicited-node multicast address for each of its assigned unicast and anycast addresses

- subscribed multicast addresses

In addition to the above addresses, a router is required to recognize the following addresses:

- the subnet-router anycast addresses for the interfaces for which it is acting as a router
- all-routers multicast addresses
- all other anycast addresses, if any have been configured

With this introduction to IPv6 and IPv6 addresses, we will review the Neighbor Discovery protocol, which is used to deliver packets on a link.

2.5 NEIGHBOR DISCOVERY PROTOCOL

In a physical network such as the Ethernet or the token ring, nodes need to know the physical or Media Access Control (MAC) addresses of their peers in order to communicate using the Internet Protocol. If Bob and Alice were on the same wireless subnetwork, how would Bob's VoIP packets reach Alice? An IP packet needs to be encapsulated within a MAC layer frame for transport over the physical medium. The MAC layer frame itself needs to have the destination's address. The process of resolving a destination's IP address to that destination's MAC address is referred to as, unsurprisingly, *address resolution*. In IPv4, this task is performed by the *Address Resolution Protocol* (ARP) [7]. In IPv6, this function is performed by the Neighbor Discovery protocol (ND) [6] [3].

The ND protocol also allows hosts to discover routers willing to forward packets for them. The protocol provides mechanisms to assess the reachability of a previously resolved neighbor. All these functions are necessary for user communication over a specific physical medium.

2.5.1 Router Discovery

Routers send *Router Advertisements* to announce their presence on a link. These advertisements allow hosts to learn the router's MAC address, as well as many other important parameters necessary for communication, including the network prefixes supported, the type of address configuration to use, the hop limit to use when sending packets, and optionally the Maximum Transmission Unit (MTU) for the link. The Router Advertisements are sent either unsolicited at regular intervals or as replies to *Router Solicitations*. The frequency of unsolicited advertisements, as well as the response time associated with solicited advertisements, is important for IP mobility purposes, as we will see in later chapters.

[3]The task of resolving a MAC layer address to an IP address is performed by Reverse ARP in IPv4, and by Inverse Neighbor Discovery in IPv6.

The format of a Router Advertisement message is shown in Figure 2.5.

```
 0                   1                   2                   3
 0 1 2 3 4 5 6 7 8 9 0 1 2 3 4 5 6 7 8 9 0 1 2 3 4 5 6 7 8 9 0 1
+-+-+-+-+-+-+-+-+-+-+-+-+-+-+-+-+-+-+-+-+-+-+-+-+-+-+-+-+-+-+-+-+
|     Type      |     Code      |          Checksum             |
+-+-+-+-+-+-+-+-+-+-+-+-+-+-+-+-+-+-+-+-+-+-+-+-+-+-+-+-+-+-+-+-+
| Cur Hop Limit |M|O|  Reserved |       Router Lifetime         |
+-+-+-+-+-+-+-+-+-+-+-+-+-+-+-+-+-+-+-+-+-+-+-+-+-+-+-+-+-+-+-+-+
|                         Reachable Time                        |
+-+-+-+-+-+-+-+-+-+-+-+-+-+-+-+-+-+-+-+-+-+-+-+-+-+-+-+-+-+-+-+-+
|                          Retrans Timer                        |
+-+-+-+-+-+-+-+-+-+-+-+-+-+-+-+-+-+-+-+-+-+-+-+-+-+-+-+-+-+-+-+-+
|   Options ...
+-+-+-+-+-+-+-+-+-+-+-+-
```

Fig. 2.5 Router Advertisement Message Format

The format uses the Internet Control Message Protocol (ICMP) structure [8]. The *Type* and *Code* fields specify the binary values that identify the message type and the specific "status" value associated with the message [4]. The *Checksum* field specifies computation (one's complement of the one's complement sum of the ICMP message starting with the ICMP Type) over the ICMP header.

The *Cur Hop Limit* field specifies the default value that should be used in the Hop Count field in the IP header. The 'M' and 'O' flags specify that a host should use stateful configuration of address and other parameters. We will consider how hosts can use address auto-configuration in a later section. Use of such flags is common in message formats to indicate a specific purpose. Indeed, Mobile IPv6 defines its own ('H') flag in the Router Advertisement to indicate that the router is also Mobile IPv6-capable.

The *Router Lifetime* specifies how long the router sending the Router Advertisement is willing to act as a default router. This value only applies to the router's ability to act as a default router and not to any other values supplied in the Router Advertisement. The *Reachable Time* and the *Retransmit Time* specify how long a node should consider a neighbor to be reachable once the reachability is confirmed and the time between retransmitted solicitations to resolve a neighbor, respectively.

The Router Advertisements contain the network prefix(es) that allow hosts to determine which destination addresses are on the same link and which are not. Such an *on-link* determination allows hosts to either communicate directly with a peer or use the router when sending packets to a destination. The prefix information is crucial for mobile nodes to determine that they have roamed across IP subnetworks.

The format of the prefix information option carried in the Router Advertisement is shown in Figure 2.6.

We see again the use of flags in the option format. When set to one (often simply stated as "when set"), the 'L' flag indicates that the supplied prefix can be used to

[4]The Code value is similar to the Unix *errno*.

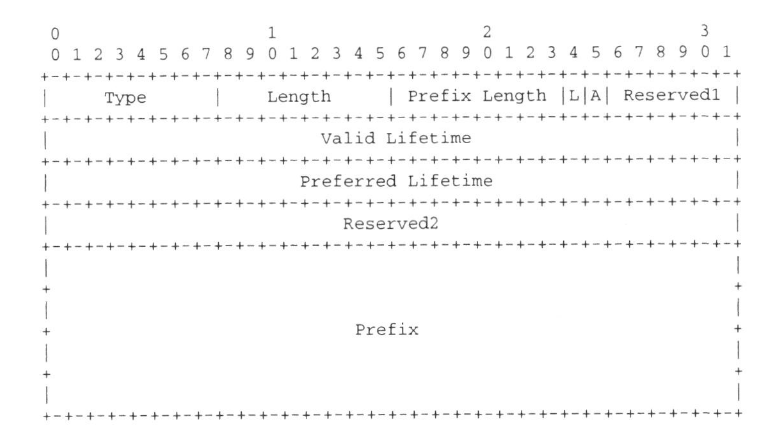

Fig. 2.6 Prefix Information Option Format

determine that any other node using the same prefix is also on-link. Interestingly, when it is not set, the 'L' flag makes no statement as to whether the supplied prefix can be used for on-link determination or not. The 'A' flag specifies that the supplied prefix can be used for configuring an IP address without the need for a server. The *Valid Lifetime* and *Preferred Lifetime* indicate how long (in seconds relative to when the packet is sent) the prefix is valid and how long the address is considered "preferred" [10].

The *Prefix* field has a size of 16 bytes because it can be an IPv6 address itself. The *Prefix Length* field indicates the number of leading bits which constitute the prefix; the rest are assumed to be initialized to zero. Having the Prefix field contain the actual IP address is important when the router's global address needs to be known, as in Mobile IP.

Once a node receives a Router Advertisement, it has sufficient information to construct a globally routable IP address: Either use a stateful mechanism such as the Dynamic Host Configuration Protocol (DHCP) [4] or use the stateless address auto-configuration. In the following section, we describe how a node can construct an address without the need for a server as in IPv4.

2.6 STATELESS ADDRESS AUTOCONFIGURATION

IPv6 defines a mechanism whereby hosts can generate unique link-local and globally routable unicast addresses simply by using a unique identifier local to the host and by using the prefix advertised by the routers. This is often referred to as *stateless address autoconfiguration* since it requires no state maintenance on the network in order to perform address assignment. The overall design goals are to allow hosts to communicate with each other even in the absence of a centralized address management server and also in the absence of a router (by means of autoconfigured link-local addresses). The design also aims to avoid manual configuration on the part of system administrators as much as possible.

The protocol works on multicast-capable links and is performed when an interface becomes enabled, such as at the time of power-up, when an interface joins a link etc. Hosts first generate a link-local address by prepending the well-known prefix of FE80::0 to the interface identifier. This address is "tentative" until confirmed to be unique. The process of confirming the address uniqueness is called *Duplicate Address Detection* (DAD). Since DAD uses the Neighbor Solicitation and Neighbor Advertisement messages defined in [6], we briefly discuss them first.

2.6.1 Looking for a Neighbor

When a node has a packet to send to a neighbor, the node needs to have the link-layer address of the neighbor. The binding of IP address to link-layer address is maintained in a *neighbor cache*, which also specifies in which state an entry is in the cache. So, the node first consults its neighbor cache for an entry that confirms that the neighbor is "Reachable." If there is no entry, the node creates an entry in "Incomplete" state and sends a *Neighbor Solicitation* message to the solicited-node multicast address (see 2.4) corresponding to the target IP address it is looking for. As we discussed in the previous section, prior to sending the Neighbor Solicitation, the node would have subscribed to the all-nodes multicast address and the solicited-node multicast address corresponding to the target address. The format of the Neighbor Solicitation message is shown in Figure 2.7.

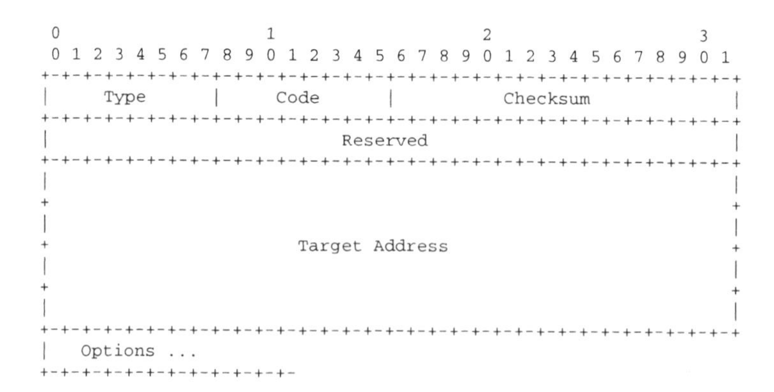

Fig. 2.7 Neighbor Solicitation Message Format

The node uses a valid IP address assigned to its interface as the source IP address. The node also typically includes its own link-layer address as one of the options.

The target node that receives the Neighbor Solicitation responds with a *Neighbor Advertisement* whose format is similar to that of the Neighbor Solicitation message but with some additional flags; see Figure 2.8. The 'R' bit specifies that the sender is a router, the 'S' bit indicates that the advertisement is sent as a response to a solicitation, and the 'O' bit indicates that the receiver should overwrite an existing neighbor cache entry using the link-layer option present in the advertisement. The 'S' and 'O' bits are particularly useful in determining the reachability of a neighbor. For instance, when a node receives an advertisement with the 'S' bit set, it knows that the neighbor (for which it sent the solicitation) is reachable and that the node itself is reachble to the neighbor, i.e., bidirectional reachability is confirmed. Hence, the node can create a neighbor cache entry in the "Reachable" state.

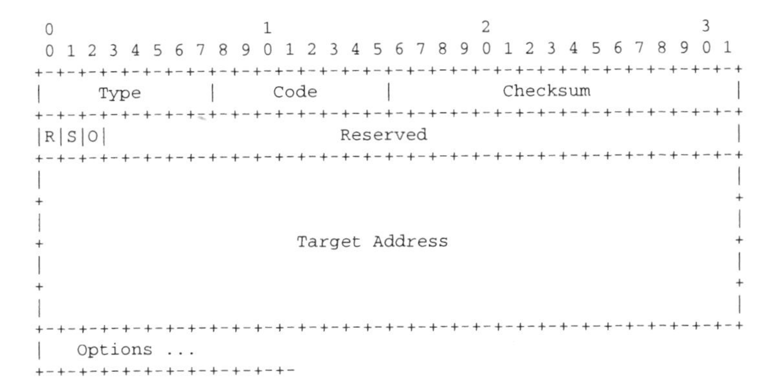

Fig. 2.8 Neighbor Advertisement Message Format

On the other hand, if only the 'O' bit is set, meaning that a neighbor sent the advertisement unsolicited, the node receiving the advertisement is unsure about bidirectional reachability; only the reachability from the sender of the advertisement to the receiver is certain. For this reason, the receiving node creates a neighbor cache entry in the "Stale" state, wherein the node probes the neighbor for reachability when it has packets to send but will simultaneously attempt to deliver the packets. These often subtle behaviors are specified in [6].

In summary, nodes resolve their neighbors using the Neighbor Solicitation and Neighbor Advertisement messages.

2.6.2 Duplicate Address Detection (DAD)

DAD is performed once an interface becomes available for IP communication. The purpose is to obtain and assign a unique IP address to the interface without requiring stateful or manual configuration. It is typically done only for the link-local addresses when the same Interface Identifier is used for configuring the global and site-local addresses.

DAD uses the same sequence of messages used for resolving the neighbor described in the previous section, but the semantics are different. The node performing

DAD is actually trying to resolve the uniqueness of its own address rather than resolve a neighbor's IP address to a link-layer address.

Before sending a DAD Neighbor Solicitation, often referred to as a *DAD probe*, for a tentative address, the node first joins the all-nodes link-local multicast address (represented by FF02::1) and the solicited-node multicast address corresponding to the target address it is attempting to assign to its interface. Subscription to the all-nodes link-local multicast address allows the node to receive Neighbor Advertisements from another node which may already be using the tentative address. The subscription to the solicited-node multicast address, which is formed by appending the low-order 24 bits of the address to the well-known prefix FF02::1:FF, allows the node to detect another node which may be trying to assign the same address.

Joining a multicast group means sending a *Multicast Listener Discovery* (MLD) Report [3, 11] for that group. Before sending such a report for the solicited-node multicast address, RFC 2462 recommends a random delay between 0 and 500 ms in order to avoid possible congestion and race conditions when multiple nodes start on the link at the same time.

The node performing DAD first creates a link-local address using the well-known prefix and its own IID, and makes the address tentative. The node sends a Neighbor Solicitation message (i.e., the DAD probe) to the solicited-node multicast address using the unspecified address as the source IP address and includes the tentative address in the Target Address field (see Figure 2.7). The node delays sending the DAD probe if that is the first message to be sent after an interface is initialized. The node may retransmit the probe multiple times, but each probe is separated by a specified timer. The default number of probes sent is, however, one.

After sending the DAD probe(s), the node waits for *RetransTimer*, whose value is one second [6] before it declares the DAD process over.

When a neighbor receives a DAD probe, processing varies, depending on the state of the tentative address. If the neighbor already owns the address specified in the Target Address field in the DAD probe, it multicasts a Neighbor Advertisement. The node performing DAD then realizes that the tentative address it was probing is not unique.

If the neighbor that receives the DAD probe is also trying to perform address configuration for the same address (i.e., neighbor's own interface address is tentative and is the same as in the Target Address in the DAD probe), then the neighbor knows that the tentative address is a duplicate. [5] The neighbor does not respond to the DAD probe in this case.

If any Neighbor Advertisements are received for the DAD probe, then the tentative address is already in use. The node stops performing address autoconfiguration. It may initiate a new stateless address autoconfiguration procedure with a different identifier.

[5]The neighbor may see its own DAD probe if the multicast packets are looped back. In such a case, the DAD probe does not indicate a duplicate.

In summary, a node needs to perform a few operations in order to become capable of IPv6 communication on a link. This includes the following: configuring link-local and globally-routable unicast addresses, router discovery and neighbor discovery. These operations are closely related. For instance, how does a node determine that it needs to configure a new globally-routable unicast address? It needs a Router Advertisement with prefixes which do not include its currently configured address. How does it test the uniqueness of its link-local address? Even though the link-local prefix is a constant, a node may need to test its assigned address if link conditions change, for instance, if the node detects that it has moved to another link (which may or may not imply movement to another subnet). In any case, the sequence of events involving address configuration and changing the default router can be summarized by the following:

- With a link-event, test if the event corresponds to new IP configuration. This can be done by sending a Router Solicitation. The node can use one of its current addresses as a source address only if it is certain that it is unique. When the node is testing uniqueness of its address, it has to use the unspecified address as the source address. As a result, the routers can only multicast a Router Advertisement, which is not sent until the next scheduled event in order to avoid congesting the link. Typically, the multicast advertisements are sent at an interval of 3.5 seconds. So, a node can receive an advertisement on average within 1.75 seconds. The received advertisement can help the node to initiate a new configuration but ND requires the node to perform the *Neighbor Unreachability Detection*(NUD) [6] before it switches the default router. This process adds additional delay.

  Another way to receive a Router Advertisement is by configuring a new link-local address as soon as a link change is determined and sending a Router Solicitation with the unique link-local address. In order to do this, the node has to first send an MLD Report staggered by a delay of up to 500 ms, send a DAD probe, wait for a second, declare that the address is unique, and subsequently send the Router Solicitation. The router can respond to this with a unicast solicited advertisement after a delay of 250 ms to avoid synchronization of routers.

  Router Advertisement provides, as we saw earlier, enough information for a node to configure a new IP address.

- The node may perform Neighbor Discovery for communication with on-link nodes. With a unique link-local address assigned to its interface, this is feasible. As we discussed in Section 2.6.1, the procedure involves sending a Neighbor Solicitation and anticipating a Neighbor Advertisement. There is another way a node can communicate with its neighbor without first performing Neighbor Solicitation: It can send all its traffic directly to the router, which then forwards the traffic to the destination. In such cases, the routers typically send a *Redirect* message to the sender providing details of the destination so that the peers can communicate with each other directly.

- Configure globally-routable unicast address(es) based on the prefix advertised in the Router Advertisement. This allows the node to communicate with all nodes, but especially those not on the same link. It is necessary to configure a global address for continuing the existing communication as well. The global address configuration is also subject to DAD [10].

- Switch the default router. This is done if Neighbor Unreachability Detection indicates that its current default router is no longer reachable. In order to do this, the node sends multiple, usually three, Neighbor Solicitation messages spaced a second apart. If it hears no response, then it declares the current router unreachable and switches to the new router.

As is evident from above, the entire procedure involves significant delay (for good reasons). This delay is especially undesirable for supporting real-time applications such as VoIP, video conferencing and many other applications yet undeployed. We investigate a set of protocols which are designed to specifically address performance.

Stateless address auto-configuration is considered one of the novel features of IPv6 address management. Indeed, it provides a simple and effective mechanism for nodes to become IP endpoints. Subsequently, the nodes can resolve their neighbor's IP addresses to MAC addresses for on-link communication and use the router for communication with nodes on the Internet. These operations are specified in the Neighbor Discovery protocol. So, we can see that with IPv6 address definitions, stateless address configuration and the ND protocol, nodes can communicate using IPv6.

2.7 SUMMARY

In this chapter, we reviewed some important parts of the IPv6 protocol suite, including the base IPv6 protocol definition, header and extension header formats, the ND protocol, Stateless Address Autoconfiguration, and DAD. IPv6 provides basic elements for the emerging era of mobile communication where mobility and security will be taken for granted in addition to the basic addressing support. It can readily solve the basic requirement of easily supporting IP addressing of millions of Bobs and Alices. Mobility depends on basic addressing support as well as neighbor and router discovery, as we will see in the forthcoming chapters.

REFERENCES

1. S. Blake, et al. "An Architecture for Differentiated Services," RFC 2475, Internet Engineering Task Force, December 1998.

2. S. Deering and R. Hinden. "Internet Protocol, Version 6 (IPv6) Specification," RFC 2460, Internet Engineering Task Force, December 1998.

3. S. Deering, W. Fenner, and B. Haberman. "Multicast Listener Discovery (MLD) for IPv6,", RFC 2710, Internet Engineering Task Force, October 1999.

4. R. Droms, J. Bound, B. Volz, T. Lemon, Perkins, C., and M. Carney. "Dynamic Host Configuration Protocol for IPv6 (DHCPv6)," RFC 3315, Internet Engineering Task Force, July 2003.

5. R. Hinden and S. Deering. "IP Version 6 Addressing Architecture," RFC 2373, Internet Engineering Task Force, July 1998.

6. T. Narten, E. Nordmark, and W. Simpson. "Neighbor Discovery for IP Version 6 (IPv6)," RFC 2461, Internet Engineering Task Force, December 1998.

7. D. Plummer. "An Ethernet Address Resolution Protocol Converting Network Protocol Addresses to 48.bit Ethernet Address for Transmission on Ethernet Hardware," RFC 826, Internet Engineering Task Force, November 1982.

8. J. Postel. "Internet Control Message Protocol DARPA Internet Program Protocol Specification," RFC 792, Internet Engineering Task Force, September 1981.

9. K. Ramakrishnan, S. Floyd, and D. Black. "The Addition of Explicit Congestion Notification (ECN) to IP," RFC 3168, Internet Engineering Task Force, September 2001.

10. S. Thomson and T. Narten. "IPv6 Stateless Address Autoconfiguration," RFC 2462, Internet Engineering Task Force, December 1998.

11. R. Vida (Editor) and L. Costa (Editor). ""Multicast Listener Discovery (MLD) for IPv6,", RFC 3810, Internet Engineering Task Force, June 2004.

3

IP Security

A small number of rules or laws can generate systems of surprising complexity... The rules or laws generate the complexity and the ever changing flux of patterns that follows leads to perpetual novelty and emergence. –John Holland

3.1 INTRODUCTION

The problem of providing private communication over a public communication medium has always been considered important, since privacy of communication can be considered as a basic necessity of every user. With the advent of the Internet and the World Wide Web, it has become even more important to provide private communication. The essential *security* problem could be summarized as consisting of the following: How do peers such as Bob and Alice authenticate each other over the Internet, how do mutually authenticated users establish means for pursuing private communication, and how is communication between mutually trusting users protected against tampering and eavesdropping? As one could imagine, each of these is a vast problem on its own, addressing such crucial topics as trust, authorization, and confidentiality. In this chapter, we will briefly review the mechanisms used for supporting security for IP communication.

Providing security for communication over the Internet is also crucial for enabling such important applications as e-commerce, including banking, payments, purchasing, buying and selling in e-market places (e.g., eBay), virtual private networks (VPNs) for enterprises, and interconnecting distributed sites, just to name a few. The term "security" refers to a collection of functions used to protect the privacy of communication over public networks such as the Internet. At a minimum, it involves peer authentication, which is sometimes followed by authorization to use certain resources, secret key derivation and exchange, and private communication by using agreed-upon cipher algorithms. In today's Internet, security is provided either at the transport layer or at the IP layer. For instance, the "secure http" commonly visible as the scheme "https:" on web pages is based on the *Secure Socket Layer* (SSL), which is now more generally referred to as *Transport Layer Security* (TLS). [1] The VPN applications familiar to roaming enterprise users are based either on SSL or on IP Security. Both TLS and IP security are self-sufficient in providing the minimum security functions necessary; indeed, both offer much more sophisticated and often an overwhelming set of features for a naive user. In this chapter, we focus on providing an overview of IP Security (IPsec) as it is commonly referred to. This is because IP mobility is primarily concerned with IP layer security. This chapter is not intended as a primer for Internet security or for cryptography itself. Readers should consult the literature for a comprehensive treatment of security. A brief, yet good introduction to cryptography presented in [7] may be useful for readers.

3.2 WHAT IS IPSEC?

IPsec broadly refers to key management, cryptographic algorithms for authentication and encryption, and security protocols for carrying IPsec-protected traffic and managing *security associations* between endpoints. Let us briefly consider each of these. Key management involves establishing secret keys between peers for private communication, once the peers are authenticated. It can be done by manual provisioning or it can be automated using a protocol. The *Internet Key Exchange* (IKE) [2] is an example of the keying protocol used to exchange keys dynamically. We will discuss IKE in detail in a later section.

Let us consider authentication for a moment. It is a process by which a peer proves its claimed identity. For example, Bob may have an identity provided by a well-known authority who is also known to Mable, who is interested in authenticating Bob. Bob has to prove that the identity belongs to him, and the identity of the authority itself must be verifiable. A common analogy often presented is the driver's license. A user's identity (name and address) and related attributes (date of birth, height) are vouched for by a government agency that everyone within a certain domain trusts. The *digital certificates* follow this model to provide user authentication.

The authentication of an IP endpoint is slightly different. There are no "certificates" for IP addresses in general, although the addresses can be used as attributes in user certificates. For an arbitrary IP device, the challenge is to prove that it is the same device which a peer has been communicating with even if the rightful ownership of the

[1] SSL was developed by Netscape Communications in 1994, and IETF used SSL version 3.0 as the basis for TSL.

IP address used cannot be verified. This is important especially when an IP address changes due to mobility. So, in general, device authentication at the IP layer is about proving continuity of liveness at the IP address used to begin the communication, whereas user authentication is more to do with proving the identity.

Authentication and authorization go together, and hence the distinction is often overlooked. Successful authentication proves identity, whereas authorization determines what the peer is allowed to do subsequently. Sometimes, authorization is implicit, e.g., once a roaming user is authenticated, the user is authorized to access the home network. No specific authorization step is necessary. In protocols, when a message can be authenticated, it can be authorized to perform the only operations within the boundary of message definition.

Returning back to the scope of IPsec, there are defined protocols for carrying traffic securely. The Authentication Header (AH) protocol [5] provides protection against tampering of data in transit, which is called *integrity protection* and authentication of the sender, which is called *data origin authentication*. The Encapsulating Security Payload (ESP) [3], in addition to the functions offered by AH, also provides confidentiality. Because of this, the IETF recommends ESP for supporting IPsec. These protocols, which are called "traffic service protocols," come into play once authentication and key exchange have taken place. The cryptographic algorithms used by these protocols are negotiated during the initial key exchange, typically performed by IKE. IPsec defines default algorithms to use but allows different algorithms to be negotiated.

Both the key management protocol and the traffic service protocols use the concept of a *security association* (SA), which is central to IPsec. Just as TCP defines a connection endpoint for reliable data transfer, IPsec defines a security association as a connection abstraction to render security services according to the key management and traffic service protocols. The SA is a simplex construct, meaning that a pair is required for bidirectional communication between the same two endpoints. IKE typically establishes the SA in pairs to facilitate bidirectional communication. An SA can be identified by a *Security Parameter Index* (SPI) alone, which is an unsigned integer. However, an SPI may be used together with the destination IP address, or with destination and source IP address fields, or with the traffic service protocol (AH or ESP) type. We will discuss them in the following sections.

3.3 SECURITY ASSOCIATIONS

3.3.1 SA Types

There are two types of Security Associations defined: transport mode SA and tunnel mode SA. See Figure 3.1. The transport mode SA is typically used for end-to-end

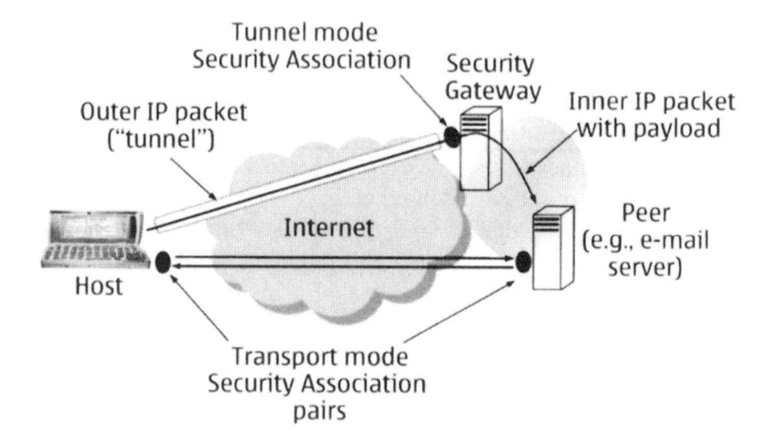

Fig. 3.1 IPsec Security Associations

security [2] between two hosts. It may also be used between a host and a security gateway. In transport mode, AH or ESP processing is done on a fully formed IP packet which is ready to be transmitted. In IPv4, the AH and ESP headers appear after the IP header and any options, but right before the next layer protocols such as TCP or ICMP. In IPv6, the headers appear after the IPv6 header and any extension headers with the exception of the destination option which may be protected by IPsec. The ESP protection applies only to the next layer protocols but not to the headers preceding the ESP header since an encrypted IP header cannot be forwarded using IP routing! With AH however, at least some of the header fields can be protected.

In tunnel mode SA, IPsec is applied to an entire IP packet which is part of a tunnel. Recall that a tunnel is simply one packet inside another, and is used to first visit an IP endpoint before the original packet is eventually delivered to the intended recipient. We can see this in Figure 3.1, where the packet first visits the Security Gateway before reaching the e-mail server in the tunnel mode. In a tunnel, the *outer* packet is meant for a node such as the Security Gateway, and the *inner* packet is meant for the eventual destination such as the e-mail server.

In tunnel mode, the SA applies to the inner IP packet of a tunnel. Since all traffic must go through the gateway which needs to be able to inspect the traffic before delivering it to the final destination, transport mode is inappropriate. The security

[2] Security could mean simply authentication (data origin authentication and integrity) or authentication and confidentiality.

protocol header appears after the outer IP tunnel header and before the inner IP tunneled header. Again, ESP does not protect the outer IP header.

3.3.2 Selectors

The selectors define the traffic set which is subject to IPsec processing. Each packet, once selected, is either protected, bypassed or discarded by the IPsec module based on the entries in a *Security Policy Database* (SPD). If it is afforded protection, then the corresponding entry in the SPD indexes the SA, which is maintained in a *Security Association Database* (SAD). The security association itself is established either manually or, typically, by using the IKE protocol.

The fields which are used as selectors are as follows:

- Local (or source) IP address: This can be a single IP address, including unicast, multicast, anycast, or broadcast (for IPv4 only), or a range of IP addresses.
- Remote (or destination) IP address: This can also be a single IP address, including unicast, multicast, anycast, or broadcast (for IPv4 only), or a range of IP addresses. This field is always the inner IP packet's destination or remote IP address when the tunnel mode SA is used.
- Next Layer Protocol: This is the Protocol field in IPv4, and the Next Header field in IPv6, once all the extension headers are skipped. A particular Next Layer Protocol of interest for us is the Mobility Header, which is used in Mobile IPv6. The Mobility Header Type identifies a particular type of a Mobile IPv6 message. The other fields used as selectors are the transport protocol source and destination port numbers.
- Name: It can be used as a symbolic name to IP addresses in the SPD entries. Four different Name formats are supported: a fully qualified user name (e.g., e-mail), a fully qualified domain name (e.g., www.google.com), a byte string and an X.500 [1] distinguished name. All these names can be carried as the *Identifiers* (IDs) in IKE during SA establishment.

3.3.3 The Databases

There are two databases that IPsec uses most often. [3] The Security Policy Database defines the policies that affect all the inbound and outbound traffic from the IPsec module. The Security Association Database contains all the entries as a result of successfully establishing a security association.

The SPD contains an ordered list of selectors, an indication of what treatment to afford, and, if IPsec processing is required, how to relate to security associations. Using the selector values in the packet or the SPD entries, new entries for the SAD

[3] There is a third database, the Peer Authorization Database, which is appearing in more recent versions of the IPsec specification and serves as a link between the SPD and IKE.

Selector	SPD Entry	PFP	Field in Packet	Value in SAD
MH Type	1 - 4	0	MH Type 1	1 - 4 (RR SA)
MH Type	1 - 4	0	not available	discard packet
MH Type	5	1	MH Type 5	5 (BU SA)

Table 3.1 Relationship Between SA Constructs

can be derived. A Populate From Packet (PFP) flag associated with each selector field specifies whether the value present in the packet should be used to create an entry in SAD.

The Security Association Database contains the values negotiated during SA establishment for each of the selector values. One of these values is the *SPI*, which is used to index into the correct SA for incoming traffic so that IPsec processing can take place. It is also used to construct the IPsec (AH or ESP) header for outgoing traffic. The SPI is selected by the receiver and needs to be unique. [4] Another entry in SAD indicates the type of SA (transport or tunnel mode), and the IP addresses of the inner packet if the SA is of tunnel mode. There is also the Lifetime field, expressed either as a time or bytecount, which specifies the time after which the SA needs to be replaced with a new one or is simply terminated. The SAD contains numerous other fields including an Anti-reply window, AH-specific fields, ESP-specific fields, the Sequence Number Counter for the security protocol headers etc.

The selector, the SPD entry, the PFP flag and the SAD entry can be illustrated using Table 3.1. We use the Mobility Header (MH) as the selector type example. Other selectors are described in [8].

The first row specifies that for the MH Type as the selector, the SPD entry is a range of Type values from 1 to 4, which refer to the Mobile IPv6 Return Routability (RR) signaling messages discussed in Chapter 8. For this range, the SA entry is not populated from the packet (PFP = 0), and the SAD entries correspond to the range of MH Types for which a single (RR) SA exists. The Field in Packet is ignored. The third row indicates that for MH Type 5, the Mobility Header field in the IPv6 packet must be looked (since PFP = 1) to derive the SA corresponding to MH Type 5, which is the Binding Update message (which is discussed in Chapter 7).

[4]The SPI uses 32 bits, which offers sufficient space for uniqueness.

Header Fields	Outer Header	Inner Header
version	6	6 (no change)
Flow label	copied from inner packet	as in application packet
Next Header	AH or ESP or routing header	as in application packet
Src Address	constructed independently	as in application packet
Dst Address	constructed independently	as in application packet

Table 3.2 Relationship Between Outer and Inner Headers

3.4 TRAFFIC PROCESSING

In an IPsec implementation, all traffic is subject to IPsec processing, since each packet must be verified if it needs IPsec protection. Those which do not need protection bypass any further processing. For incoming packets, each packet is verified for the presence of either AH or ESP headers and whether the packet is addressed to the host. In the tunnel mode, this verification is done using the destination IP address in the outer header. However, further IPsec processing (i.e., authentication, decryption) is done using the packet's inner header addresses as selectors for the SA. The SA lookup is done using the SAD, and the SPI is used to index into the correct SA.

For outbound packets, SPI is of no use. The selectors in the packet are matched against the policies in the SPD. If there is a match, the packet is afforded one of the bypass, discard or protect treatments. If IPsec processing is required, the SPD entry has a link to the SAD entry, which ultimately determines the security protocol (AH, ESP) and tunnel or transport mode SA treatments. For tunnel mode SAs, the outer packet fields are mostly constructed independently, except for the DSCP and ECN fields (see Chapter 2). The outer tunnel header addresses determine the tunnel endpoints, and the inner header addresses determine original sender and receiver. The inner header is untouched by IPsec, except to decrement the Hop Count or the TTL fields. [5] Table 3.2 shows how some of the header fields in the outer and inner packets are constructed. Tunnel mode header creation and processing are especially important for Mobile IPv6.

The IPsec module does not copy any extension headers to the outer header. This is important because Mobile IP uses an IPv6 extension header called Mobility Header, which contains signaling messages as well as the Mobile IP Home Address itself. Pro-

[5] When the encapsulator is not the source, it decrements the field when it forwards the packet, and the decapsulator decrements it if it forwards the packet.

tecting these fields may be considered important for certain types of traffic, including some of the Mobile IPv6 signaling itself.

3.5 INTERNET KEY EXCHANGE (IKE) PROTOCOL

As we have mentioned a few times above, IKE is a protocol for dynamically establishing an IPsec SA between two peers. In fact, IKE can establish an SA for any "service" or "child," provided an appropriate Domain of Interpretation (DOI) is used. It is an arguably complex protocol with various modes and phases. Hence, it is not feasible to describe it succinctly. Fortunately, mobility itself does not add to its complexity but uses certain specific parameters. Nevertheless, we need to understand some basic operations, so we describe them below.

In order to establish SAs for a service such as IPsec, IKE first needs to establish a secure channel which can then be used for establishing the child SAs, such as those for IPsec AH or ESP. Correspondingly, there are two phases in IKE. The purpose of phase 1 is to establish an IKE, or more precisely an Internet Security Association and Key Management Protocol (ISAKMP) [6] SA which forms the basis for creating IPsec AH or ESP SA. The first phase consists of SA negotiation, *Diffie-Hellman key exchange* and authentication of Diffie-Hellman key exchange. These distinct steps are performed in either the *Main* mode, which takes six messages, or the *Aggressive* mode, which takes three messages. Another difference between these two modes is that the number of SAs which can be negotiated in Aggressive mode is limited. In phase 2, an IPsec-specific SA is negotiated and new key material is derived. The phase 2 negotiation uses the shared secret established in phase 1 to secure the communication.

Logically, an initiator presents an "offer" of SAs it wishes to use. The responder replies back with an SA it can support. This is step 1. In step 2, the initiator presents the Diffie-Hellman public values. In order to establish that the initiator is not a bogus node, a cookie is also included in the offer. The responder presents its own Diffie-Hellman public value and a cookie. Both the initiator and the responder derive multiple shared keys [6] using the Diffie-Hellman public values and the cookies. In the third step, the initiator computes and transmits a *hash* over its identity, the cookies exchanged, the Diffie-Hellman public values, SA parameters and the key generated in step 2. A hash is a one-way function generated on known input parameters using a secret. A sender generates the hash and a receiver can verify it if it shares the secret. The responder also computes and transmit-ts a similar hash. The hash values authenticate the exchange and establish the IKE SA.

The Aggressive mode combines SA offer and Diffie-Hellman exchanges to reduce the number of messages exchanged.

Phase 2 is called *Quick* mode. This mode actually establishes an IPsec SA. The whole exchange, but not the header, is encrypted using a key generated in phase 1.

[6] Each peer derives keys for authentication and encryption of future IKE messages as well as a key which is later used for deriving IPsec keys.

The exchange minimally consists of a single SA and a nonce from each peer. The new key material is derived using the SA parameters, such as an SPI and a protocol (e.g., Mobility Header), the nonces, and the key established during phase 1. This new key material is then used when applying IPsec AH or ESP transforms. An optional parameter of interest for mobility is the identifier. The default identifier is the IP address used during phase 1. In Mobile IP, this IP address often tends to be the IP address from a visited network. Since a mobile node roams from a network to another, using the IP address from phase 1 as the identifier means performing both the phases upon every handover. In order to avoid this, a fixed identifier is used during phase 2 for the IPsec SA. Such a fixed identifier is the Mobile IP Home Address, which we will visit at numerous places in this book.

So, the end result of the entire IKE exchange is the establishment of two SAs. The first one is the IKE SA itself, which can be long-lived. The second one is the IPsec SA, which can be rekeyed multiple times as long as the IKE SA remains valid. That is, multiple phase 2 operations can be performed once a single phase 1 operation has been completed. Once an SA is established, the SPD and SAD databases are populated so that traffic processing with AH or ESP can begin using the selectors and other parameters agreed upon during the IKE exchange.

If the foregoing description is somewhat dense, it is both understandable and justifiable. As we mentioned, IKE is a complex protocol and really needs a rigorous treatment than we could afford here. Fortunately, we do not need to delve into its mesmerizing choices in this book. Nevertheless, IKE is widely used in Virtual Private Network (VPN) applications that use IPsec. It can also be used to establish a security association in Mobile IP between a mobile node and its Home Agent which provides forwarding support for the node when it is roaming.

3.6 SUMMARY

In this chapter, we have provided an introduction to IPsec, and reviewed various components. We described what constitutes an SA, how it can be set up and different databases used by IPsec. We also studied how selectors are used for either bypassing, discarding or imparting IPsec for user traffic processing. Finally, we reviewed the IKE protocol which is used to set up an SA for IPsec AH and ESP protocols. In a subsequent chapter, we will go into the details of the interaction between IPsec and Mobile IP.

REFERENCES

1. CCITT X.500 (1988/1993)/ISO Directory

2. D. Harkins and D. Carrel. "The Internet Key Exchange (IKE)," RFC 2409, Internet Engineering Task Force, November 1998.

3. S. Kent and R. Atkinson. "IP Encapsulating Security Payload," RFC 2406, Internet Engineering Task Force, November 1998.

4. S. Kent and R. Atkinson. "Security Architecture for the Internet Protocol," RFC 2401, Internet Engineering Task Force, November 1998.

5. S. Kent and R. Atkinson. "IP Authentication Header," RFC 2402, Internet Engineering Task Force, November 1998.

6. D. Maughan, M. Schertler, M. Schneider, and J. Turner. "Internet Security Association and Key Management Protocol," RFC 2408, Internet Engineering Task Force, November 1998.

7. P. Zimmerman. "Introduction to Cryptography," Pretty Good Privacy (PGP) 6.5.1 documentation.

8. S. Kent and K. Seo. "Security Architecture for the Internet Protocol," RFC 4301, Internet Engineering Task Force, December 2005

Part II

IP Mobility

First, there was voice that connected people all over the world for decades. It still does. Then almost simultaneously came the Internet and mobile voice technology. It is inevitable that personal mobile voice or *personal mobility* itself, without undue exaggeration, and the Internet are going to meet in quite unpredictable ways. It is only a matter of time, and indeed we have begun noticing the beginning of this trend. So, why would mobile voice, which works fine on its own, need the Internet Protocol? Perhaps the answer lies in recognizing that mobile voice networks were not designed to provide all the different applications that the IP itself was not intended to provide but is now doing anyway! The only feasible way, it might be argued, of supporting applications, imagined or otherwise, is by embracing IP. And those applications are increasingly mobile and often are only mobile in some parts of the world. Perhaps the day when mobile Internet applications will be taken for granted is not too far..

This part is dedicated to the discussion of concepts and principles behind supporting IP mobility. We begin with concepts and principles underlying mobility support on the Internet and review various approaches to supporting mobility in Chapter 4. We provide an overview of Mobile IP and related terminology in Chapter 5. In Chapter 6, we start discussing the details of Mobile IP, identifying the various parts and salient features of each. Chapters 7 through 12 describe in detail each part identified in Chapter 6, including the context and background of various design considerations. Chapter 13 is about network mobility, where the entity that undergoes mobility is itself a network rather than an end host. We hope that these chapters bring about the insights into concepts and principles buried under the mobility protocol design.

We cannot end this introduction without saying something about IETF (Internet Engineering Task Force), the organization that standardizes protocols that matter to the Internet. It is fair to say that without IETF, there would be no Internet as we know it. IETF has standardized IP, TCP, UDP, and http to name only a few. It has played a critical role in the development and deployment of the Internet. Yet, it is quite unlike many other standardization bodies around the globe. It is represented by individuals (as opposed to companies or governments) and is sometimes characterized by terms such as "informal," "irreverent," and "chaotic." Nevertheless, there is great technical depth and review, and "running code" and "rough consensus" have come to epitomize IETF over the years (although recent criticism of IETF has included switching the first words in these phrases!). In addition, there is, without a doubt, a great deal of political dynamics that influences protocol design in IETF. In some sense, this reflects the reality of the Internet as an open forum for myriad representations. It has also come to reflect the commercialization of the Internet. In any case, there is little question that it is an interesting place to witness the modern technology tussling with policy and commercialization!

As you read these chapters, you will come across some terms that have special connotations in the context of IETF. Three such terms commonly seen are "MUST," "SHOULD" and "MAY." Whenever a protocol operation is deemed absolutely necessary, it is specified using a "MUST." This means implementations are required to support the particular operation, assuming that they support the protocol itself. A SHOULD clause typically means a recommendation. It suggests that there are benefits to be gained by implementing the specific feature, but are not strictly required for inter-operability. However, quite often the usage is blurry; many recommended features become de facto MUSTS, and folks settle for SHOULD as a compromise in standardization (which quite often results in good progress). A MAY indicates that an operation or a feature is left optional for implementations. In this part and elsewhere in this book, we will use this terminology when we "must"!

4

Mobility Concepts and Principles

There are 10 types of people: those who understand binary and those who do not understand it. –Anonymous

4.1 INTRODUCTION

In this chapter, we will discuss in detail the basic concepts that define mobility in the Internet. Our intent is to expose the reader to concepts and principles that underlie mobility. We will look at the roaming and handover problems, and approaches being considered to solve them. We also look at the core Internet principles and how the different approaches compare against them.

4.2 ROAMING AND HANDOVER TOGETHER CONSTITUTE THE MOBILITY PROBLEM

When a mobile node moves to a new location, it must satisfy a number of requirements before it can begin to receive packets. Some of the requirements can, depending on circumstances and the roaming strategy used, be satisfied without additional operations on the mobile device. These requirements include:

- Establishing a link

- Fulfilling conditions for access control to the link
- Acquiring a topologically correct IP address
- Maintaining resolvability of its Domain Name Space name into a reachable IP address
- Enable traversal of the local firewall

The amount of work required to satisfy these requirement varies, depending on the roaming strategy selected. For the case of Mobile IP, for instance, the mobile device easily maintains resolvability of its DNS name since the point of Mobile IP is to enable continued reachability of the mobile device at its Home Address, which does not change. In contrast, if another solution is used which does not maintain reachability at any previous IP address, the node's communication partners cannot rely on any DNS services for making contact.

Arguably, there are other approaches to maintaining reachability in the event of mobile node movements. For instance, some approaches may never explicitly publish the IP address of a device to the external world but rely on some other identifiers such as *Uniform Resource Identifiers* (URIs) [2] to be publicly available. The particular mechanism would have to map the URI to a well-known agent of some sort that would in turn keep track of mobile node movements. Even such mechanisms stand to gain from the use of a fixed address mapped to a mobile node, since a fixed address also enables persistence of transport sessions which follow once a mobile node is reached by its peer. We will look at this in a later chapter.

4.2.1 Roaming Problem: How Packets Reach the Current Location of the Mobile Node

Given the hierarchical nature of Internet addressing, the mobile device has to get an IP address which is topologically correct for its new point of attachment within the hierarchy. In general, this new IP address will require a different routing path than its previous IP address, but usually this is not an issue since existing IP routing will forward the packets to the new IP address.

All the difficulty typically resides in the operation of getting a new IP address. If the local network supports IPv6 address autoconfiguration (see Section 2.6), then there is no direct difficulty in obtaining the IP address but there may still be an obstacle placed by the access network to prevent that IP address from being considered reachable to the Internet. For instance, the user may be presented with a web page that demands payment by way of a credit card before the local access router will be enabled to forward packets from the newly configured IP address.

Alternatively, the mobile node may have to use Dynamic Host Configuration Protocol for IPv6 (DHCPv6) [1] to acquire an IPv6 address. This may also be insufficient for full access until user supplies payment credentials or some other identification or authorization.

In fact, one can say that especially for IPv6, the whole difficulty really lies in satisfying any local constraints imposed for the purpose of access control. The existing mechanisms have not been integrated with Mobile IP, since the problem of access control predates the time when the Mobile IP standard was published. Furthermore, the existing access control mechanisms in the Internet have been designed for use in scenarios with either very low performance requirements or very low user expectations.

A better solution is to combine Mobile IPv6 operation with network access and authorization. For instance, when a mobile node attaches to a new access network, providing credentials so that the access network can contact the mobile node's home network for authentication and authorization purposes will enable easier deployment. The entities involved in providing actual mobility service (e.g., the home agent), authentication service (e.g., a "AAA" server), and access itself (e.g., the access router) share trust, depending on the actual deployment model.

Many other possibilities arise. The alternatives available in today's slow-roaming networks often rely on interactive procedures, and such methods are clumsy, aggravating, error-prone, and tedious. Furthermore, they are not at all appropriate for authorizing Internet access by computing resources embedded within other, larger machines such as cars or trains. For these newer application platforms, Mobile IPv6 offers a much improved way to provide the necessary authentication and thus identification for mobile devices.

4.2.2 Robustness Problem: Connection Must Withstand Change of IP Address

When a mobile device is running some Internet applications and it gets a new IP address, certain actions have to occur, depending on how it uses the new IP address. If the applications can continue to use the same IP address, then naturally no actions are required for those applications. However, as noted earlier, that same IP address then must be made reachable by way of the new IP address, for instance by way of Mobile IP.

If the applications cannot continue to use the same IP address, then there are only two other possibilities:

1. The new IP address has to be supplied to the applications. There are several ways to do this.

2. The application has to exit.

Case 2 is far more prevalent today. This leads to a disruption in service to the user, and the period of disruption can vary from a few seconds to dozens of seconds or even minutes, depending upon the time required to satisfy the constraints imposed at the new point of attachment.

Case 1 may become more common, especially for applications that are enabled to run Stream Control Transmission Protocol (SCTP) [3]. SCTP has a mechanism by which the transport interface to the application can be populated with multiple IP addresses, perhaps even dynamically. However, there is no clear indication that existing applications will be enhanced with SCTP functionality.

There are numerous proposed solutions by which applications are signaled directly with information about changes in the network environment. For instance, the Session Initiation Protocol (SIP) provides a mechanism to *Re-Invite* a peer when any parameter, including the IP address, changes [4]. Change of IP address is one clear example, but there are many more. Any such application interface would also include triggers related to change of network bandwidth, access rights, latency or jitter, and so on almost ad infinitum. At this writing, there is again no clear candidate for such a programming interface that would be available to applications. There is little current hope that existing web browsers and other Internet utility functions will transition to use such features even if a clear candidate were to emerge.

This effectively means that we are back to case 2, or perhaps figuratively back to square one, if applications cannot continue to use the same IP address. It's interesting to consider the further implications of this case, as discussed in Section 4.3.

4.2.3 Beyond Robustness: Supporting Real-time Mobility

We can define handover as a cooperative activity between two access routers for improving the perceived behavior when a mobile node moves to a new link. By this definition, Mobile IP is not a handover protocol, because there are no stated requirements imposed on the operation of the access routers. In fact, Mobile IP can be well understood as a highly effective roaming protocol. The protocols for enhancing access routers to become mobility-aware are well underway in IETF and other venues. With these protocols, one could imagine applications such as VoIP and video being easily supported on mobile devices undergoing handover, just as people are used to cellular voice technology.

It is easy to see that problems related to real-time mobility management are primarily delay and packet loss. Clearly, it is critical to make sure that applications do not perceive unacceptable latency and packet loss due to operations necessary to regain IP connectivity. As we observed earlier, the handover process involves establishing new (often wireless) link connectivity, sometimes performing access authentication, obtaining IP configuration and eventually notifying the communication endpoints about any change of IP address. All these operations are delay prone; meanwhile, the data arriving for the existing IP address will be lost. Unless these basic problems are addressed, real-time mobility management will not be effective.

Beyond making the handover *fast*, there is also the problem of making the handover *smooth*. For a number of reasons, the access routers often have to maintain state or "context" information about each mobile node and its ongoing packet streams. Examples of such contexts are access control and QoS. When handover takes place, the new access router has to reestablish these contexts in order to provide the desired forwarding treatment for the mobile node's packet streams. Indeed, in some instances, packet forwarding may not even take place without appropriate access control lists in place. Hence, techniques that facilitate smooth operation of mobile nodes in the presence of such context information have been investigated. We have gained many insights in this area, but more are definitely needed.

4.3 INTERNET PRINCIPLE: CORE NETWORK TRANSPARENCY

One of the tenets of the Internet design is an intelligent endhost, simple core network. According to this, the core network (consisting of routers and switches that understand IP) itself provides a minimum of functionality, such as forwarding packets from one place to another; it does not even promise reliable or in-order delivery. The task of building reliability is burdened on the transport protocols such as TCP. The network does not provide any intelligence, such as *Call Forwarding* in traditional voice networks. The intelligence is assumed to be built into the endhosts. For instance, the Call Forwarding feature is expected to be supported, perhaps with assistance from a cooperating application-layer proxy or server, by the endhost itself. This separation of functionality has allowed the Internet to scale very well while simultaneously supporting myriad applications.

According to this principle, who has the task of supporting mobility? Does mobility fall in the same category as reliability or in the category of routing? This is an interesting question. Mobility does involve routing change, but it's the change of the endhost. Internet routing as such can easily handle changes in routing information corresponding to a new network, but not that of the endhost itself. In addition, even if the routing substrate were to be used to propagate changes, it would involve potentially a large number of *host* routes that correspond to topologically inconsistent addresses. To understand why this would be the case, recall that IP routing is hierarchical and is based on prefixes which map to specific network topologies. When a host moves from one network to another and uses the address from the original network on the visited network (e.g., in order to be continuously reachable), the network has to propagate routing information just for that host across the routing topology so that packets for the host can be forwarded to its current point of attachment. This is already challenging, and will impose a heavy burden on the network as a function of the number of mobile hosts. It would be extremely difficult to imagine using such an approach to easily scale and address mobility.

So, we see that endhost mobility is perhaps best supported by *not* involving the core network itself. In other words, the routers and switches should see no difference when a host moves from one network to another. This is what we call *core network transparency*. This is one of the original design goals of the Internet. It does not mean that fully functioning mobile networks cannot be built without it. Indeed, cellular networks built using the *General Packet Radio System* (GPRS) [6] [1] use a network-based approach to mobility. These networks do not require the mobile node to perform IP routing updates, and hence follow endhost transparency (in a limited sense, since the mobile node actively performs link-layer specific operations). These networks are operational, and although they have not yet reached their full serving capacity, they are expected to serve millions of mobile nodes. So, we will briefly review the principles behind them in the next section.

[1]The 2.5G and 3G packet networks in the GSM-based cellular systems use the GPRS core network architecture.

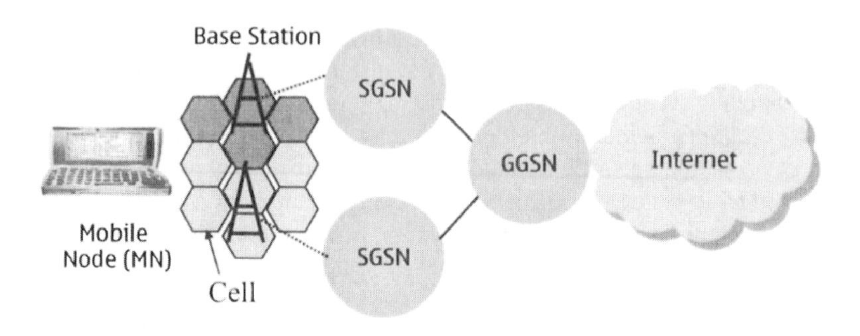

Fig. 4.1 Simplified 3G Network Architecture

4.4 NETWORK-CONTROLLED MOBILITY

In a network-controlled routing update model, the endhost does not participate in routing changes but only triggers a change. For instance, when an endhost moves from one radio base station to another, the base station has to inform an upstream gateway of a new mobile's attachment. The gateway then switches its forwarding to the new base station. If the mobile happens to attach to a base station which is connected to a different gateway, then the procedure is more complex; the new gateway has to determine the old gateway based on the information provided by the mobile, establish communication with the old gateway to obtain access control information, and finally perform routing update with a fixed gateway which never changes as far as the mobile is concerned. This is illustrated in Figure 4.1. The *Serving Gateway Support Node* (SGSN) and the *GPRS Gateway Support Node* (GGSN) cooperatively provide network-controlled mobility together with link-layer assisted protocols that operate between the Base Station and the SGSN. [2]. This is feasible where the access network infrastructure elements, including the base stations and gateways, as well as the mobile node itself cooperate in a "link-specific" manner to support mobility. Yet, the mobile node itself does not perform any routing update in order to send and receive packets.

An interesting corollary to the network-controlled routing update model is that the IP address of the mobile node does not have to change at all as a function of mobility.

[2]For clarity, we have omitted the Base Station Controller (which is called Radio Network Controller in 3G systems) from the figure. We have also left out the details of 3G mobility protocols, which are beyond the scope of this book.

As long as the mobile node is within the realm of the same network that provides network-controlled mobility, the IP address can be made to remain unchanged even when the mobile node traverses different subnets. A single IP address is sufficient since the network will adjust the routing to forward packets. Readers may wonder how this might scale given our previous discussion on the ability of Internet core routing being able to support changes in IP addresses. Indeed, the network-based mobility models do not involve changes to the core network at all. This is usually done by the edge nodes and an anchor node that resides in the mobile's home network. Such an *overlay* is a powerful concept, as we will see throughput this book. It is not that network-based models were the first to use it; the host-based models deploy it as well. Overlay in IP is achieved through *tunneling* of a packet within another, and this was first defined by the IETF. In any any event, tunneling has served the purposes of both host-based and network-based mobility models, as well as applications such as VPNs.

In summary, there are approaches which still adhere to core network transparency but do not involve the endhost at all for effecting routing changes. Such approaches also avoid the problem of change of IP address. To the extent that the network architecture and the elements support it, it is a feasible way to support mobility. However, on the global Internet, where a mobile node may move from one form of network (such as the one that supports network-based mobility) to another, uniformity of a single access network architecture and network-based support cannot be assumed. For instance, a mobile node may roam from a 3G network to a WLAN. So, we still require a mechanism that provides mobility on the global Internet.

4.5 APPLICATION LAYER AND SESSION LAYER MOBILITY

Many people claim that it's acceptable simply to require the applications to handle changes in the network environment. For some kinds of changes, this is probably correct. For instance, if a streaming application notices a change in latency or jitter, it might be worthwhile to allocate more space for the input buffer. In a very sophisticated system, one could even imagine that the application applies a smooth transform to gradually increase the playback delay until an optimal value is achieved according to current network conditions.

Routing problems, on the other hand, are traditionally considered outside the jurisdiction of most applications (except, of course, for applications which are responsible for obeying routing protocols and managing route tables). This, too, may change in the future. To see why, suppose that networks grow more sophisticated and that a network environment is able to be characterized in terms of its security features, speed, latency, QoS guarantees, cost, service connectivity, availability of multicast, or other features. Today, such features do not exist because there has not been enough demand for any but the most basic Internet connectivity services.

If an application is intelligent enough to be able to characterize its requirements on the network environment, then one can easily imagine the following general iterative scheme for managing change within the network:

Begin network connections according to user demands, default values, and initial network conditions.

while (still operating) do
{
1. Monitor network conditions *and* check whether user interactions require changes in network connectivity.
2. If network conditions have changed, then inform the user and make the smoothest possible adjustment.
3. If user requirements have changed, then invoke system routines to obtain updated network connectivity.

}

With this scheme, applications can offer a very responsive network experience for the user. In fact, with minimal change, this programming model can be adapted for embedded computers with a stimulus/response requirement based on their particular solution approach. With socket-based programming, implementing this kind of programming model is natural but may require the creation of many new socket options to control the various specific needs for security, cost, and so on.

That's one model in which each application handles mobility in its own way. With a rich enough abstraction set *between* link, IP, transport and session layers, it is feasible to implement mobility at an application layer. Imagine the challenges that arise when this is the *only* model to support mobility. Every application developer who wishes to develop mobile applications would have to be equipped to explicitly incorporate mobility into the design cycle. Depending on the parameters, such as the transport protocol and the operating system being used, the application response would vary. So, in addition to keeping the focus on end user experience based on the application behavior alone, the developer has to be concerned about the effects due to mobility. Finally, how would a mobile Internet work with hundreds (or thousands) of applications, each handling mobility in its own way? We do not have any evidence.

The other model is to devise a new session layer that exposes mobility to upper layers while providing mobility functions. For instance, the work in *Migrate* [5] makes a number of good observations regarding the deficiencies in supporting mobility on the Internet, including the dependence of applications on the lower-layer identifiers for basic communication (which effectively binds the application state to an IP address and a transport protocol port number) and the absence of a uniform model to handle network disconnectivity. Migrate proposes a session layer approach to address these deficiencies. In this model, the session layer treats any change to IP addresses as a network disconnectivity event and subsequently allows an application to withstand the effect. The actions performed by the session layer include informing each application about the change of IP address, buffering of data, blocking sockets, and selectively discarding unreliable packets. Furthermore, the session layer can provide transport protocol-specific actions such as disabling TCP keepalives. The session layer may also be chartered to reestablish communication and handle long periods of disconnectivity by managing the state per application. When connectivity is regained, the session layer informs its peer about the new IP address and resumes communication. For most transport protocols, this means restarting the connections while "preserving" the existing one by making use of a session layer identifier.

Approaches such as Migrate are certainly feasible. The end result for an application may still be the same, i.e., transparency of mobility from the stability of the connection endpoint. In addition, the *Pause and Resume* usage model is quite powerful, at least something that an IP layer cannot easily provide. However, such approaches require complex state management on endhosts to be effective, and often require knowledge of and modifications to transport protocols. Although deficiencies exist in supporting mobility, it is not clear why a session layer is best suited to support changes in IP routing. It seems logical that a session layer is a natural place to offer VCR-like functionality for applications during mobility events.

What we really need is a clear separation of concerns. Application developers and users should have a rich set of programming interfaces to respond better in a mobile Internet environment. For instance, applications may adjust their transmission based on the nature of the new network they are attached to. Applications may wish to have a Pause and Resume support. However, they do not have to restart or build their own *kludge routing logic* to handle changes in routing. Performance is another crucial problem which applications and transport protocols are not well-equipped to handle themselves, but they could readily make use of information about changes to performance. For instance, handovers with minimal delay and zero packet losses are almost impossible for an application or a transport protocol to design on its own, but each application (e.g., IP HDTV) would gladly benefit from such handovers. Just as IP has performed an invaluable service by providing minimal functionality to *all* layers above it, mobility extensions to IP should serve a similar purpose as well. Such minimal routing extensions and performance enablers, coupled with a rich set of APIs, can provide a foundation for the mobile Internet.

4.6 SUPPORTING MOBILITY USING IP

In the previous sections, we have discussed the basic problems presented by mobility on the Internet and how it could be addressed by different layers of the protocol stack. There are advantages and disadvantages present at each layer. For instance, mobility support at the IP layer is not best suited to provide application adaptation. It is perhaps bestsuited to handle routing changes and provide performance. This is what we will focus on in the rest of this book. At the same time, we recognize the need for providing better interaction between mobility events across different layers of the protocol stack. For example, the link layer could notify the IP layer about new radio establishment so that appropriate actions to regain IP configuration could be performed. Subsequently, the IP mobility software could inform transport protocols about the change of network connectivity so that appropriate actions, such as congestion control, could be performed. [3] Finally, applications may benefit from indications from the transport protocols so that they can conduct their own adaptation. This is a good vision, and we recognize that work is needed to implement it. Although we focus on IP mobility as supported by Mobile IP and real-time mobility management protocols, we also dedicate Part V of this book to research topics, where we share some insights and expose some of the problems that need to be addressed to eventually transform the vision into reality.

A few things are needed for IP mobility to work. These include reliably detecting movement to a new subnet, configuring a new IP address, informing peers about a new IP address, and forwarding traffic to and from the new location without breaking existing connections. In order to address these issues, an implicit network architecture is assumed. It includes a *mobile node*, which roams freely across networks on the Internet, its *correspondent node* which communicates with the mobile node, and a *Home Agent*, which is the mobile node's trusted partner in forwarding new and ongoing traffic. There is a *home network* where the mobile mode normally resides when it is not roaming across *visited networks*. These and many other elements and concepts will become clearer as we go through this book.

4.7 SUMMARY

In this chapter, we discussed the basic problems introduced by mobility on the Internet from a networking perspective. We considered how reachability of a device (and hence that of a user such as Alice) and persistence of ongoing connections and sessions form the basic requirements for supporting mobility on the Internet. Beyond these basic problems, mobility based on IP must be able to support performance in terms of making the handovers fast as well as smooth; the VoIP call between Bob and Alice must not only persist but also must meet the performance requirements during handovers. We also considered Internet principles of keeping the network, especially the core network, simple and expecting the endhosts to be intelligent. And we contrasted applying this principle to IP mobility with network-controlled mobility. Finally, we discussed supporting mobility at the application layer and the session layer. With this background, we can discuss supporting IP mobility with Mobile IP in more detail in the following chapters.

Exercises

4.1 Compare the advantages and disadvantages of host-based mobility and network-based mobility. What kinds of deployments favor network-based mobility? Is it sufficient on its own?

[3]To be clear, there is little evidence of what actions should be taken as a function of mobility.

4.2 How are reachability and session persistence related to each other? Can a solution for one work without addressing the other?

4.3 Extend the TCP connection establishment sequence to handle change of IP address. What security concerns must be addressed?

REFERENCES

1. R. Droms, J. Bound, B. Volz, T. Lemon, Perkins, C., and M. Carney. "Dynamic Host Configuration Protocol for IPv6 (DHCPv6)," RFC 3315, Internet Engineering Task Force, July 2003.

2. T. Berners-Lee, R. Fielding, and L. Masinter. "Uniform Resource Identifiers (URI): Generic Syntax," RFC 2396, Internet Engineering Task Force, August 1998.

3. R. Stewart et al. "Stream Control Transmission Protocol," RFC 2960, Internet Engineering Task Force, October 2000.

4. J. Rosenberg et al. "SIP: Session Initiation Protocol," RFC 3261, Internet Engineering Task Force, June 2002.

5. A. Snoeren, H. Balakrishnan, and F. Kaashoek. "Reconsidering Internet Mobility," in the Proceedings of the 8th Workshop on Hot Topics in Operating Systems, May 2001.

6. 3rd Generation Partnership Project. "General Packet Radio Service (GPRS)," 3GPP TS 23.060, Service description, Stage 2, Release 6 http://www.3gpp.org

5

Mobility Support Using Mobile IP

An elephant is a mouse with an operating system. –Donald Knuth

5.1 INTRODUCTION

In this chapter, we first discuss the mobility events and actions that take place in the seemingly simple process of a mobile node moving from one network to another. We then discuss how Mobile IP addresses these events, emphasizing the concepts. Many of these events and actions are then presented in great detail in the subsequent chapters.

As always, the IP address used by the mobile node on a visited network is called the "care-of address"

5.2 MOBILITY EVENTS AND ACTIONS

We can identify several relevant events that are important for mobility management at the IP layer:

- Link detections
- Link establishment
- Subnet detection

- Care-of Address allocation
- Access authorization
- Enabling packet forwarding
- Informing correspondent nodes about the new location
- Upon returning home, taking appropriate actions

Some of these do not require network-layer support in every instance. For instance, the method used for link detection is typically specialized for the kind of physical link involved. However, when a link is detected, it may trigger certain network-layer activities.

Some mobility events are not really within the design space of Mobile IP. In particular, due to process considerations and in order to focus the Mobile IP effort within the IETF, Mobile IP does not offer the best possible support for handover events. However, additional protocol mechanisms have been designed to work with Mobile IP in order to provide better handover support, as described in the next chapter.

5.2.1 Detecting Movement to a New Subnet

When a mobile device visits a new network, it has to establish the typical subnet parameters needed by all IP-addressable devices. In particular, the mobile node needs to determine the routing prefix or subnet prefix at the new point of attachment, along with the size (or number of bits) of the routing prefix. The smaller the prefix, the bigger the address space allocated to that routing prefix. Equipped with the routing prefix, the mobile node is then able to carry out operations needed for autoconfiguring or requesting a new IP address. Without this new IP address, the node cannot perform certain operations expected of all IP-addressable devices, such as responding appropriately to various multicast (or, in the case of IPv4, broadcast) messages. Since the new IP address is topologically correct for the mobile node's new point of attachment, it is useful for *locating* the mobile node. It is a *locator* address. Since the new IP address will enable access at the new network, it has also been called an *IP access* address. We will use that terminology in this book from time to time.

5.2.2 Regaining IP Connectivity

After a new link has been established, the mobile device has managed to determine subnet parameters as just described. However, in many circumstances, this determination cannot proceed unless the the mobile node is able to provide some credentials to prove that it is authorized to make use of the physical resources of the visited access network. Many kinds of such credentials have been designed, and there are just as many or more ways to check their validity. A full discussion of this subject might require another entire book. Contributing to the diversity of such mechanisms is the fact that access authentication is typically performed in an access-dependent way. Even with generic AAA architecture [4], the link access is often tied to it. For instance, the

IEEE 802.11 system, popularly known as WLAN, defines a framework in which the link layer messages for discovering security parameters and agreeing with particular suite of security parameters is combined with the Extensible Authentication Protocol (EAP) [7]. Another example is the use of the Point to Point Protocol (PPP) [6] in obtaining connectivity in CDMA systems. We do not discuss access authentication in detail here but defer it to later chapters.

Once access is regained, the mobile node can acquire an IP address. In IPv6, this can be done using stateless address autoconfiguration, which we discussed in a previous chapter. In the absence of stateless autoconfiguration, the DHCP is typically used. Address configuration follows router discovery, which is used for detecting movement to a new subnet. Once an address is configured, the mobile node can send and receive IP packets.

5.2.3 Packet Forwarding Subsequent to Movement

When a mobile device has overcome all the hurdles mentioned in the previous section, it can begin to receive packets at its new locator address, called the *care-of address* in Mobile IP, which is topologically valid for the network link at which the device has made its attachment. However, this locator address is not used to identify the mobile node. The mobile device maintains ownership of a long-lived IP address by which it can be identified to its communications partners and which is typically associated with a well-known name resolvable by way of DNS (or by other means) into this identifying IP address, which is called the *home address*. The home address is topologically valid for a network link known as the *home network*, and when the mobile node is not present on its home network it then needs the assistance of its home agent, which permanently resides on the home network.

The challenge of continuous packet forwarding is being able to send and receive packets at the new locator address without losing the ongoing sessions. This can be done with the help of a home agent which can intercept packets arriving for the home address and forward them to the care of address. However, the operation of changing care-of addresses cannot always be quick enough to avoid misdelivery of packets to the previous care-of address, which can be invalid if the mobile node is no longer reachable at that address. Unless additional steps are taken, these misdelivered packets can be lost and unrecoverable.

With additional support within the access networks, such packet losses can be avoided. We discuss them in Part III of this book.

5.2.4 Route-optimized Communication Between a Mobile Node and Its Correspondents

As described, packets are typically sent to the mobile node's home address. Then the home agent tunnels the incoming packets to the mobile node's locator address. This way of delivering data to the mobile device is understood to happen in two parts, along two independent segments of the overall routing path. See Figure 5.1. In most configurations, these two routing paths require a longer time (sometimes much longer) to traverse than a direct routing path between the mobile node and its communication partners. In the extreme case, the communicating partner might be on the same link as a mobile node but might still be sending packets to the mobile node's home network. The time taken to deliver each packet will be many times greater than that required for a one-hop routing operation.

It would be a lot better for the mobile node to supply its locator address to its communication partners and let them tunnel the packets directly to the mobile node at that locator address. Notice that the correspondent node is still delivering packets to the mobile node's home address, but it is doing so by causing them to first be delivered to the appropriate care-of address. Since the mobile node owns both the home address and its care-of address, this does not introduce any additional segments to the routing path. The mobile node gets the packets, at its home address, in the quickest possible way. This is illustrated in Figure 5.1.

5.3 HOW MOBILE IP SUPPORTS IP MOBILITY

The following sections introduce the basic operations of Mobile IPv6. Each of these sections is then discussed in a separate chapter.

Each of the operations in the previous sections is supported by protocol operations specified within Mobile IPv6. Some of the protocol elements are fully self-contained within that document. Other parts of Mobile IPv6 define some minimal upgrades to the base IPv6 specification, especially as related to features of the Router Solicitation and Neighbor Discovery. We have gained additional experience with wireless IPv6 devices and the ways in which their mobile operation changes the assumptions that had been made when only static devices needed IP service.

5.3.1 Mobile IP Terminology

The basic entities involved in the operation of Mobile IPv6 are the mobile node, home agent, and correspondent node are illustrated in Figure 5.1. Here are some definitions for those terms and other useful terminology. Some of them have been adapted from the Mobile IPv4 specification, and others are pertinent only to Mobile IPv6.

- *Binding*: An association between a mobile node's home address and its claimed care-of address.
- *Binding Cache*: A set of cache entries maintained by a home agent or a correspondent node, each of which makes an association between a mobile node's home address and its current care-of address.
- *Binding Management Key* (K_{bm}): The secret key needed to verify security information that must accompany the message which indicates a change to an entry in the Binding Cache.

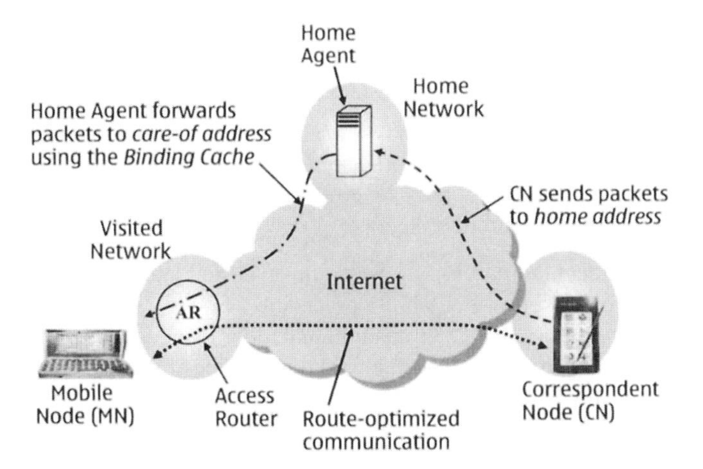

Fig. 5.1 The Mobile IP Components

- *Care-of address*: The termination point of a tunnel toward a mobile node for datagrams forwarded to the mobile node while it is away from home. In Mobile IPv6, a mobile node has its own care-of address, as opposed to Mobile IPv4 where a care-of address could either belong to the foreign agent or to the mobile node.
- *Correspondent Node*: A peer with which a mobile node is communicating. A correspondent node may be either mobile or stationary.
- *Visited network*: Any network other than the mobile node's home network. Sometimes, it may also be called a *foreign network*.
- *Home address*: An IP address that is assigned for an extended period of time to a mobile node. It remains unchanged regardless of where the node is attached to the Internet.
- *Home network*: A network, possibly virtual, having a network prefix matching that of a mobile node's home address. Note that standard IP routing mechanisms will deliver datagrams destined to a mobile node's Home Address to the mobile node's Home Network.
- *Mobility Security Association (SA)*: A collection of security contexts between a pair of nodes, which may be applied to Mobile IP protocol messages exchanged between them.
- *Return Routability*: A procedure by which a correspondent node uses known properties of Internet routing and addressability to carry out minimal verification that a mobile node owns an address (home address) and is reachable at another (care-of address).

- *Tunnel*: The path followed by a datagram while it is encapsulated. The model is that, while it is encapsulated, a datagram is routed to a knowledgeable agent, which decapsulates the datagram and then correctly delivers it to its ultimate destination.

- *Virtual network*: A network with no physical instantiation beyond a router (with a physical network interface on another network). The router (e.g., a home agent) generally advertises reachability to the virtual network using conventional routing protocols.

Additional definitions will be given as needed later in the book, especially those related to the Return Routability procedure.

In Mobile IPv4, there is an additional mobility agent called a "foreign agent," but this functionality is not needed given the plenitude of IPv6 addresses and the enhanced functionality of IPv6 mobile devices.

5.3.2 Subnet Movement Detection in Mobile IP

Surprisingly, subnet movement detection is not easy. There are many factors that come into play, and deployment variations add to the problem. First, every change of link connection does not imply change in subnet. For instance, a mobile node may move from a WLAN Access Point to another, both connected to the same subnet. However, the mobile node typically has no knowledge of this. Hence, it is required to ascertain if the subnet has indeed changed every time there is a clear indication of change of a link connection from the lower layers. Second, a router may advertise different prefixes on different links. So, even if the mobile node detects a change in the prefix, it may not necessarily imply that it has to change its default router; the current router may still be reachable on the new link. Third, there may be multiple routers on the same link. Hence, hearing a new router does not imply subnet change. Finally, even having multiple routers all advertising different prefixes does not imply subnet change; they may all be part of the same subnet. Movement detection in Mobile IPv6 is performed using a combination of link-specific events, such as an indication by the link layer that a new link has been established, and IPv6 router discovery and neighbor discovery operations (see Section 2.5). The general logic is to test for change of the supported prefix and then subsequently verify the reachability of the current default router. This is done by looking at the Router Advertisements that contain the prefix information and then performing the Neighbor Unreachability Detection (NUD) operation specified in [5]. The overall operation is not fast enough, especially for real-time applications such as VoIP. Hence, mechanisms are being developed to provide such fast movement detection. For instance, Fast Handovers [3] makes use of the neighborhood subnet information to quickly detect movement to a new subnet. The Detecting Network Attachment (DNA) protocol is being worked on by the IETF working group [8]. We go through the details of the movement detection process in Chapter 11.

Movement to a new subnet leads to configuring a new IP address. As mentioned previously, the IP address of the mobile node on a visited network is called the care-of address. This address has to be conveyed to the mobile node's home agent. We consider this in the next section.

5.3.3 Location Update to the Home Agent, and Correspondents

Since all the packets arrive for the home address, the home agent has to know the current location of the mobile node in order to forward those arriving packets. The mechanism used for this is a special kind of route management, in which the home agent binds the home address to the care of address. Until Mobile IP was designed, all route management was done by the routers and switches. Mobile IP introduces a host-based route update mechanism.

Mobile IP supplies the signals by which the mobile node should inform its home agent when its care-of address changes. These are described in Chapter 6. The home agent stores the care-of addresses in a conceptual data structure known as the *Binding Cache*. Following this model, the message by which a new care-of address is sent to the home agent is called a *Binding Update*, and maintaining the organization of the cache is called *Binding Cache management*. The process of creating the Binding Cache has to be necessarily secure; otherwise, any host could bind a known home address to any arbitrary IP address. This would lead to stolen traffic or traffic that could be redirected to unsuspecting nodes. Mobile IP specifies how the security between a mobile node and its home agent is established and maintained. Since the home agent and the mobile node are expected to belong to the same domain of trust, the security credentials and keys used for securing the Binding Update can be configured either manually by the network administrators or by means of the Internet Key Exchange (IKE) protocol [1] and Internet Protocol security (IPsec) [2], both of which we discussed in Chapter 3.

When a home agent has a valid Binding Cache entry, all arriving packets to the home address are forwarded to the mobile node, and all the packets from the mobile node are forwarded to the Internet. In this mode of communication, the correspondent nodes do not realize the mobility of the mobile node. All operations are handled by the mobile node and its home agent. This mode of home agent assisted communication is called *bidirectional tunneling*, referring to the technique used between the mobile node and the home agent for forwarding packets.

Mobile IPv6 also allows mobile nodes and correspondents to communicate directly with each other without the home agent being involved in forwarding packets. For this, a mobile node has to establish bindings with its correspondent nodes directly. The same Binding Update message is used to create the binding, but the security model is significantly more complex. Since a mobile node may communicate with any arbitrary correspondent node, establishing a binding between the two addresses needs to address two distinct problems. First, the mobile node has to be able to prove to an arbitrary correspondent that it actually *owns* the home address. Second, the mobile node has to demonstrate that it is *present* at the care-of address. These actions are necessary in order not to create any more vulnerabilities than those already possible on the current Internet. This is an important, yet often subtle, design principle that has particularly influenced the Mobile IPv6 protocol significantly. According to this *do no harm* principle, the Internet routing infrastructure is considered to be robust in forwarding packets to their rightful destinations. So, if packets can be sent to and received from a specific IP address, that would attribute the ownership of that IP address to the host in question. This loose sense of address ownership based on the current Internet infrastructure is used as the basis for establishing the binding between home address and care-of address at a correspondent node.

5.3.4 Packet Forwarding

Once a home agent has the Binding Cache entry, it is equipped to forward packets to the care-of address. However, it needs to ensure that it is first able to obtain the packets arriving for the home address. This is not a problem with packets arriving from links other than the home network itself, especially when the home agent is a router. All packets have to reach the home agent which is a router, and the home agent uses the Binding Cache entry to forward packets. The traffic originating from the home network itself does not have to traverse the router; two nodes on a link can communicate directly using the Address Resolution Protocol (ARP) or Neighbor Discovery (ND). Hence, the home agent has to announce to everyone on the home link that nodes should use the home agent's link layer address when sending packets to the mobile node's home address. The home agent acts as a "proxy" for ARP or ND. This allows it to intercept packets arriving for the home address and forward them to the care-of address.

Since a mobile node is not on the home network, it cannot be directly reached by means of on-link forwarding using ARP or ND. Any packets forwarded using the normal IP routing will bring the packet right back to the home agent. So, the packet must be forwarded to the care-of address, but the original packet must be preserved. In order to do this, the home agent performs *IP-in-IP encapsulation*, placing the original packet meant for the home address in another IP packet whose destination is the care-of address (and the source is the home agent's IP address). In a similar fashion, the mobile node also encapsulates the packet meant for the correspondent in another IP packet whose destination is the home agent and whose source is the care-of address. This kind of routing forms a bidirectional tunnel between the mobile node and the home agent for packet forwarding.

When communicating directly with a mobile node at a visited network, a correspondent node uses a IPv6 Routing Header (see Section 2.3.2.1) in order to forward a packet to the home address. The destination address is the care-of address but the routing header processing rules force the mobile node to swap the address in the routing header (home address) with the destination address. In the reverse direction, the mobile node uses an IPv6 destination option (see Section 2.3.2.2) to place its home address and sends the packet to the correspondent's IP address. Processing rules for the destination option force the correspondent to swap the home address for the care-of address. The correspondent processes the routing header and the destination option only when it has valid Binding Cache entry for the home address.

We go more deeply into these protocol operations in the following chapters.

5.4 LIMITATIONS OF THE BASE MOBILE IP PROTOCOL

As we indicated earlier, the basic Mobile IPv6 protocol provides support for reachability and persistence in the presence of mobility on the Internet. Without these, very basic forms of mobility cannot be supported for *all* applications on the Internet. There are many limitations which are not addressed by the basic protocol itself, but instead by a few companion protocols. One of the crucial problems, especially for real-time applications, is handover latency, which is the time required to switch from one network to another as seen as by the application layer datagrams. Several factors come into play in handovers, including link acquisition and switching, movement detection, IP configuration and location update. Together, they constitute operations that incur latencies typically unacceptable for real-time applications such as VoIP. Tightly coupled to latency is packet loss. Since the mobile node is without IP connectivity during a handover, packets sent to the mobile node would be lost. So, the higher the latency, the higher the likelihood of packet loss. Handover latency and packet loss are two crucial metrics that determine the feasibility of real-time applications of future.

If handover latency and packet loss can be considered routing-related problems, there are problems related to the state maintained at the access router, which needs to be re-created at the new point of attachment. For example, consider the access control state or *context* maintained at an access router. Without an equivalent state, the new router may not forward any packets for the mobile node. Consider the context corresponding to QoS. Without an appropriate context, the new router has no means to provide the desired QoS. One way to establish these contexts is by re-initiating the signaling from scratch. However, other techniques that eliminate the need for mobility-aware signaling are worthwhile investigating. Such mechanisms should provide a smooth handover experience for end users.

Both fast handover and smooth handover can be considered to be part of the overall IP mobility protocol suite necessary for providing comprehensive support for Internet mobility. We go more deeply into these topics in Part III.

5.5 SUMMARY

In this chapter, we discussed the mobility events and the associated networking operations that come into play in IP mobility. We identified different problems and provided an overview of what the Mobile IP protocol does in order to solve these problems. Subsequent chapters discuss individual operations in detail, providing detailed protocol descriptions, packet formats and so on.

Exercises

5.1 Tunneling is an important concept in IP. In a sense, it introduces an exception to the normal IP routing because the tunnel originator itself *routes* the original packet by constructing the tunnel. It's the outer packet that is subject to IP routing until the tunnel terminator is encountered. Draw a picture of the IP-in-IP tunnel, identifying all the fields. Enumerate the steps that need to be taken by the originator and the terminator. Can the destination address of the inner and outer packets be the same? How should the other IP header fields such as DSCP, ECN and Flow Label handled on the outer packet?

5.2 We briefly discussed what it takes to bind an address to another at an arbitrary correspondent. See Section 5.3.3. We identified the security problem where a correspondent needs to verify the ownership of home address and reachability at the care-of address. We also studied IPsec in a previous chapter. If Alice and Bob can establish an IPsec security association (for instance because they both work at the same company), what part of the security problem in binding the home address to care-of address is resolved? What are the similarities and differences between the trust relationship between a mobile node - home agent, and mobile node - correspondent node pairs?

REFERENCES

1. D. Harkins and D. Carrel. "The Internet Key Exchange (IKE)," RFC 2409, Internet Engineering Task Force, November 1998.

2. S. Kent, and K. Seo. "Security Architecture for the Internet Protocol," RFC 4301, Internet Engineering Task Force, December 2005

3. R. Koodli (Editor). "Fast Handovers for Mobile IPv6," RFC 4068, Internet Engineering Task Force, July 2005

4. C. de Laat et al. "Generic AAA Architecture," RFC 2903, Internet Engineering Task Force, August 2000.

5. T. Narten, E. Nordmark, and W. Simpson. "Neighbor Discovery for IP Version 6 (IPv6)," RFC 2461, Internet Engineering Task Force, December 1998.

6. W. Simpson (Editor). "The Point-to-Point Protocol (PPP)," RFC 1661, Internet Engineering Task Force, July 1994

7. "Medium Access Control Security Enhancements," IEEE Standard 802.11i, http://standards.ieee.org/getieee802/download/802.11i-2004.pdf

8. Detecting Network Attachment Working Group, Internet Engineering Task Force, http://ietf.org/html.charters/dna-charter.html

6

Mobile IPv6 Protocol

The secret of getting ahead is getting started. The secret of getting started is breaking your complex overwhelming tasks into small manageable tasks, and then starting on the first one. –Mark Twain

—

In this chapter, we begin to examine the details of the Mobile IPv6 protocol, which can naturally be broken down into several related parts:

- Binding Cache management
- Return Routability procedure for route optimization
- Security management and RFC 3776
- Delivering packets to the care-of address
- Movement detection and link establishment
- Home agent discovery

These will provide a good recipe for organizing the protocol details. Our intention is not to duplicate the protocol specification in [9] but instead to explain each part, along with the design rationale, in a way that does not typically belong in the specification itself.

Certain details about IPv6 will also need explanation along the way. In addition to locating these details separately in a preceding chapter, we will explain the relevant IPv6 features along the way as they are needed.

6.1 BINDING CACHE MANAGEMENT

Mobile IPv6 can be understood as a way of managing route table entries in nodes that need to send packets to a mobile node. These route table entries have the mobile node's home address as the destination and provide information about how to deliver packets to that address. When the mobile node is located at a care-of address, the relevant route table entry should contain information about the care-of address. Using this model, we can see that the route table entry should cause packets targeted at the mobile home's address to be redirected towards the mobile node's current care-of address.

A great deal of effort has been invested to make the Binding Cache operations secure. As mentioned before, an incoming binding cache message cannot be allowed to modify any of a node's internal Binding Cache entries unless the node can verify with a high degree of certainty that the information presented is accurate and reflects the authoritative information provided by the the implied originator of the message data in the signaling packet. In particular, a home agent or correspondent node must never update the care-of address for a mobile node unless it is positively verified that the mobile node has authorized the information.

In Chapter 7, we will explore the Binding Cache messages in detail. The messages are as follows:

- Binding Update (BU) (Section 7.3)
- Binding Acknowledgement (BAck) (Section 7.4)
- Binding Error (BERR) (Section 7.6)
- Binding Refresh Request (BRR) (Section 7.5)

The Binding Update is really the crucial protocol element which directs the whole operation of Mobile IPv6. Almost all other protocol elements and operations within Mobile IPv6 should be understood specifically in terms of their relationship to the Binding Update. That is, usually what matters is how each message is related to the operation of updating a mobile node's care-of address.

The Binding Cache management messages (and others described later) are structured as message types carried in a new IPv6 header known as the *Mobility Header*.

There are some important extensions to these binding management messages. They are as follows:

- Alternate Care-of Address (Section 7.7.3)
- Binding Authorization Data (Section 7.7.1)

- Binding Refresh Advice (Section 7.7.4)

If a mobility header message type has an extension, the extension header and data for that extension are appended after the message data for that particular type of message within the Mobility Header.

Finally, there is one new IPv6 Destination Option which has been defined for use with the Binding Cache management messages, known as the *Home Address destination option*. This option, also described in Chapter 7, carries the mobile node's home address, which is important since typically the IPv6 header itself will only show the mobile node's care-of address.

6.2 RETURN ROUTABILITY DEVELOPMENT

Providing the necessary security for a Binding Cache management message requires that the sender and receiver of the message share a security association. This security association (SA) can include many details but can be expected to at least include:

- identification for the communications partner (e.g., IP address)
- expiration time for the security association, and
- a key (or key generation material) to be used for a cryptographic algorithm

If the operation is specified so that a new key is generated from data stored as part of the SA, then we may as well include the key generation steps as part of the cryptographic algorithm. With that proviso, we may then consider that the SA typically includes the key itself.

Many different cryptographic algorithms are usable, with varying degrees of safety and processing load. We can say that an SA offers more safety if it enables a higher degree of certainty that an unauthorized packet of data can be detected. As a general rule, in order to increase the safety of an SA, the algorithm in use must spend more processor cycles encoding and decoding the cryptographic data to be supplied with the protocol elements. This is only a very general rule, however, and researchers are on a continual quest to provide new algorithms that make equal or better security available using the same or fewer processing resources.

There are two popular kinds of keys that are often used: public key and symmetric key. Public key cryptography, while not specified for use with Mobile IPv6, is nevertheless very attractive in many situations. Briefly, this method works by associating a public key and a private key. The receiving node keeps its private key secret, and makes its public key generally available for use by any potential communications partner. This is very nice, but there is no automatic connection between the public key supplied and the actual identity of the node supplying that key. So, for instance, I may claim to be Alfred E. Neumann and provide you my public key. Then you can be assured of secure communications, but you do not have any assurance that you are communicating with Alfred E. Neumann. In the wide-area Internet, such assurances are only available by making use of security services known as *Certificate*

Authorities (CAs) who attest to the identity of the key holder. Then, for example, all the subscribers or clients of a particular CA can more confidently make use of public key cryptography. The general security model provided by public-key cryptography has enabled great advances in computer security.

However, there have been difficulties in scaling the use of Certificate Authorities to hundreds of millions, or even billions, of users, who are envisioned to be the target numbers for the IPv6 Internet. For this reason, Mobile IPv6 is geared to the use of symmetric keys. Furthermore, as a general rule, public-key cryptography is quite expensive computationally, so that it is often disfavored for use between small, mobile, and typically power-constrained wireless devices.

But there are still quite substantial difficulties in exchanging symmetric keys between two communications partners on the Internet who are not previously acquainted. This isn't welcome news, since we fully expect to have beneficial communications with many other people and services within the Internet regardless of their proximity or any personal involvement. In addition, symmetric keys are even less likely to be useful for the purpose of verifying the identity of the endpoints, especially when created dynamically.

In the case of home agents, it is easier to imagine how to set up the needed SA, since quite often the home agent and the mobile node have some common point of reference or belong under the same administrative jurisdiction. For arbitrary correspondent nodes, however, this is not true. Since Mobile IPv6 specifies route optimization (see Chapter 8) for arbitrary correspondent nodes (i.e., arbitrary IPv6 communications partners), it cannot be assumed that there is any common administration or even shared online toolset.

About the time when the IETF working group was substantially finished with the base protocol design, a new requirement was presented: the working group had to solve this key exchange problem for arbitrary IPv6 nodes. In other words, for a mobile node running Mobile IPv6 and an arbitrary correspondent node, a method had to be designed so that secret data could be exchanged between the two nodes. Then a mobility SA could be constructed using this secret data.

This new requirement caused great consternation within the working group, because no such wide-scale key exchange mechanism had ever been worked out before for use on the general Internet. Previous schemes for widespread key distribution were oriented towards military command and control, and participation was enforced in typical military fashion. These techniques are not at all appropriate for the Internet, and it is not easy to find an authority agreeable to all Internet users. Fortunately, the security-minded members of the working group were able to identify a very useful simplification of the problem, and in fact a useful "authority" that could be relied on for one very specific purpose.

The simplification turns out to involve exactly the same question of identification discussed before for Alfred E. Neumann. The first insight is that, for the purposes of controlling mobility, the identity of the endpoints is not at issue. Rather, the only point to be verifiably maintained is that the identity of each endpoint does not change. This means that symmetric keys are sufficient even though they wouldn't offer any verification of identity. Then, in order to securely exchange the secret data to be used for generating the keys, what is needed is a way to allow the two endpoints of a key exchange protocol to acquire information that is not readily available to other nodes in the Internet.

In order to accomplish this ambitious goal, it is important to determine the degree of precision which has to be reached. In other words, we have to determine how completely the key generation data has to be kept unavailable to other nodes within the Internet. As an example, suppose we had an algorithm that allowed only 10% of the nodes of the Internet to inspect the secret contents of the key exchange protocol. Then this algorithm would be completely unacceptable because that would still allow millions of potential attackers to gain an avenue of attack against the mobile node and its communications partners.

We do, in fact, have a fairly precise guideline that defines the required degree of secrecy to be maintained during the exchange of the key generation secrets. It is the time-honored (perhaps even time-worn) dictum, that new protocols in the IETF should not "harm the Internet." In other words, new protocols should not open up new vulnerabilities, or present new mechanisms for congestive failure, and so on. In particular, for us this means that any key exchange mechanism that may be devised has to avoid enabling attacks by nodes that otherwise could not mount attacks against the mobile node and its correspondent nodes. Let's refer to this as the "Harm Not" directive.

To understand the implications of this guideline, we have to understand something about the ways that attacks can be mounted against static nodes in today's Internet. For this purpose, imagine a client C in communication with a distant server S within the Internet. Suppose that these two nodes do not have an SA, so that all packets are exchanged without privacy encapsulation, authentication data, or other such security measures.

Between C and S, there will be several nodes that can overhear and inspect the payloads of packets that are being exchanged. For example, all the routers between C and S can see the data. Since these are typically backbone routers and routers belonging to a service provider, we implicitly trust that these routers are not malicious. Otherwise, we shouldn't send money to the service provider! We also have to take the Internet backbone on faith, but acquaintance with the engineers running the backbone does serve to justify that faith. Nevertheless, governments can and do get involved, so even the backbone is not completely immune to inspection. Then you have to decide whether you trust your government to disrupt or subvert your typical Internet communications. Let's just agree that these issues are not serious enough to be a source of concern in today's Internet, even though there are in fact jurisdictions where one should truly be concerned.

The nodes on the local subnet by which the correspondent node and mobile node are addressed may be much more of a problem. To fully understand the impact of this on the security design of Mobile IPv6, we need to take a close look at how nodes are addressed on subnets within the IPv4 Internet.

When a node on an IPv4 subnet wants to establish communications with another node on the same subnet, it first has to obtain that node's IP address and its layer-2 or MAC address. The IP address is usually obtained by access to DNS, or alternatively (and nowadays much less often) by some sort of static configuration. Of course, the DNS names are often themselves discovered by static configuration (e.g., following a link on a web page), but that additional level of indirection has proved to be enough for handling a wide range of communication needs.

Once the IP address is known, the node has to find an appropriate layer 2 address in order to be able to frame the packets for delivery. If the IP address belongs to another node on the same IPv4 subnet, then the node uses the Address Resolution Protocol (ARP) [15] to obtain the desired layer-2 address. ARP basically works by broadcasting the desired IP address and waiting for someone to respond that it owns the IP address and is reachable locally at some corresponding layer-2 address.

For our discussion, the important point is that this communication is completely insecure. On a shared Ethernet subnet, any of the other nodes on the same Ethernet can falsely claim to own the IP address. Fortunately, this is less of a problem on switched Ethernets which are very commonly in use today.

In IPv6, the protocols which most closely correspond to ARP are known as *Neighbor Discovery* (ND) protocols, and have already been discussed in Chapter 2. Although deployment is not yet as widespread as that of ARP, ND is usually considered not any more secure than ARP. Perhaps on a switched Ethernet for IPv6 there will also be no real vulnerability.

In order to determine whether or not Mobile IPv6 has met the security requirement imposed by the "Harm Not" directive, the solution in the form of *Return Routability* procedure has to be carefully analyzed in the context of messaging on a local subnet, switched or shared. We will refer to this again, and it is a good idea to keep it in mind during all of the discussion about Return Routability. [1]

More important still, one needs to understand the basic model of routing in the Internet today. Briefly, an application has almost no control over the actual route taken by the packets of its communication streams. The node hosting the application has a little more control, because it can select a default router that will relay the application's packets to the next stop along the way across the Internet to the destination (i.e., the node hosting the application's desired communication partner). Neither the application nor the node hosting the application is likely to have any control over what happens after the local default router has completed its job. Any malicious nodes that happen to have inserted themselves into the routing fabric of the Internet will be able to carry out arbitrary mischief on the application payloads or on any protocol headers within the packet.

Now, consider two routing paths. One routing path (call it route A) goes from the mobile node to its home agent, and then from the home agent to the correspondent node. Suppose route B, on the other hand, goes directly to the correspondent node without any constraint to pass through the home agent. Route optimization would change the communications path to use route B instead of route A.

In today's Internet, any node along the routing path can harm the communications, but other nodes cannot. For communications using IPv6 route A just described, any

[1] It is also useful to note that we discuss the Return Routability in detail in Chapter 8.

malicious node along the way from the home agent to the mobile node, or between the home agent and the correspondent node, will be able to damage the communications. Packets flowing along route B would be vulnerable to a typically different set of malicious nodes. In the usual case, there would be few malicious nodes with the potential to attack both of these two routes.

We want to determine whether Return Routability does anything to violate the "Harm Not" directive. It would do so if the messaging involved in enabling the route optimization has the effect of increasing the number of existing attacks or introduce new attacks in communications where at least one endpoint is mobile. As it turns out, because the route optimization uses the so-called Return Routability tests, which traverse both routing paths (not just one of them), we have not increased the vulnerability to such attacks along the routing paths. We defer the description of the procedure to Chapter 8, but for now the readers should place their faith in the procedure.

These Return Routability tests are designed to make sure that the mobile node is reachable (as it claims) at both its home address and its care-of address. As described in detail in Chapter 8, the mechanism works because a secret key is generated that can only be produced by nodes that have two separate data items which are transported along route A and route B, respectively.

The messages that implement the Return Routability protocol are as follows:

- Home Test Init (HoTI)
- Care-of Test Init (CoTI)
- Home Test (HoT)
- Care-of Test (CoT)

Also used is the Nonce Indices option, which is appended to the Binding Update message which is properly considered to be a protocol element of the Return Routability protocol and thus is also described in Chapter 8.

6.3 SECURITY MANAGEMENT

In order for network nodes to gain protective benefits from their security association, they have to make sure of at least two things:

- The key has to be kept secret
- The cryptographic algorithm has to be strong

Clearly, if the key is disclosed, the nodes lose their ability to exchange unforgeable and secret data. Even assuming that the key remains private, however, does not assure success. If a weak cryptographic algorithm is chosen, an adversary can more easily recover the plaintext associated with encrypted text by various brute force or clever mathematical methods. If a cryptographic algorithm is used that takes too much computational power, the battery on the mobile device will be depleted rapidly or the protocol signaling will take too long.

Over time, the requirements for the strength of cryptographic algorithms have increased dramatically, mostly as a result of the *Moore's Law*. Since available computational power is improving so rapidly, methods that were formerly considered secure are now falling victim to new technology. When Mobile IPv4 was designed, MD5 [16] was considered to be a very reasonable basis for the mobility security association between the home agent and the mobile node. Now MD5 is deprecated, and SHA-1 [18] is mandated instead. Although SHA-1 has recently been weakened somewhat, it is still considered to be strong enough to protect low-volume control signal messaging like that of Mobile IPv6 Binding Cache management. In the future, it is possible that SHA-256 [19] or AES [3] will be mandated, at some cost in computational complexity and power consumption. As a rule of thumb, longer key lengths provide exponentially more security but only require linearly more computational expenditure by the participants who wish to make use of the key for their security association.

However, designing a protocol so that two nodes can share key material is a remarkably tricky feat of engineering and depends crucially on what information can be taken as given. The design of the key exchange is typically completely separate from the design of the actual use of the key. The way that the key is used depends on protocol operations that are typically not closely related to the way that the key is generated. Unfortunately, in the case of Mobile IPv6, the final design of the key generation operation was integrated with the design of the key utilization procedure in a way that does not conform to this design guideline.

The reasons for the more integrated approach have mostly to do with the perceived danger arising from the way that the key generation material is exchanged. Even though the key generation tokens are delivered by way of two routing paths, a single malicious node can still obtain both pieces by:

- having visibility of both routing paths, or
- being able to inspect all packets sent on the network occupied by the correspondent node, or
- being able to inspect all packets received on the network occupied by the mobile node

While these three cases are often discussed separately, they are really three subcases of the same overall phenomenon whereby a malicious node is able to inspect all packets between two end nodes. In other words, it is the same problem that exists in today's Internet and should be covered by the same "Harm Not" dictum. Unfortunately, the "spectre" of mobile nodes (and, in particular, mobile malicious nodes) has made many engineers quite paranoid about any such exposures, with the result that the Mobile IPv6 key exchange protocol is specified to provide keys with very short lifetimes. For some relief from this injunction against reasonably long-lived dynamic keys, we at least have the possibility of installing preconfigured key information according to the design given in [14], which we discuss in Section 8.8.

When Mobile IPv6 was initially specified, the simplest possible standard method was provided for the verification of control messages between the mobile node and other Internet nodes, in particular its home agent. This simple method involves the use of the Authentication Header (AH) for IPv6, as specified in RFC 2402 [8]. The other security header specified for IPv6 is the Encapsulating Security Payload (ESP) header, specified in RFC 2406 [7]. ESP was specified for keeping data private by means of encryption.

The AH protocol is relatively small compared to ESP, and it includes the IPv6 header information as part of the calculation for the data used for authentication. Of course, it is also possible to secure Mobile IPv6 control messages by encrypting the message contents according to a secret shared by the mobile node and the home agent. For this reason, it was always thought to be allowable that, if desired by the home agent and the mobile node, ESP could be used for encrypting the Binding Cache management messages and thus providing the necessary security feature.

Over time, two schools of thought seemed to develop within the IPsec working group. One camp held that AH was a perfectly fine and more compact protocol, and that AH deserved the continued support of the IETF as a protocol standard in good standing. The other camp held that, since ESP could function in place of AH, and since it had the additional advantage of keeping the data private, there was little need to continue to support both AH and ESP. Perhaps it would be simpler for the IPv6 security implementation teams to settle on a single standard mechanism. The above-mentioned advantages ascribed to AH did not seem to provide sufficient motivation for the maintenance of two standards instead of just one.

By the time these two schools of thought were evolving, Mobile IPv6 was already well along the way toward standardization. But then the new scalability requirements for route optimization were levied against the protocol specification as it existed at that time, and new members joined the working group who seemed mostly interested in security. These new members provided impetus not only for the design of Return Routability, but also for greater acceptance of ESP as a standard mechanism for securing the Mobile IPv6 control messages between the home agent and the mobile node.

Consequently, the version of Mobile IPv6 which was finally standardized has language which favors the use of ESP. It is still possible to use AH, but the mobile node and the home agent are required to support ESP as the means for securing Binding Updates and so on.

As it turns out, using IPsec has been interpreted to mean conformance with existing deployments and implementations of IPsec, and these existing implementations can in some ways be very restrictive. The information about the security association has to be used only in certain ways and at certain times during the marshaling of protocol fields and, conversely, during the parsing of incoming packets. These ways are not compatible with the original design of Mobile IPv6, and many changes had to be made. In particular, the original design of Mobile IPv6 provided for inserting Binding Cache management signaling as part of an ongoing flow of traffic between nodes. Somehow, the implementations of IPsec were seen to make this particular feature very difficult to carry out. As a result, all Mobile IPv6 protocol operations have to be performed out of band, and presumably data transfer will come to a halt or at least be at risk of loss during the time required to establish a new care-of address at the home agent and the correspondent nodes.

Since IPv6 itself is not yet widespread compared to IPv4, we do not have the daily experience to know whether the more restrictive designs will degrade the perceived user experience of mobile applications under IPv6. However, we can make use of long engineering experience with protocol design to predict certain outcomes. In any event, the interaction between IPsec and Mobile IPv6 processing has been detailed enough that there is a companion specification [1] that describes how to use the two together. We go into the details in Chapter 9.

6.4 DELIVERING PACKETS TO THE CARE-OF ADDRESS

Mobile IPv6 provides a way to deliver packets to a mobile node at its care-of address instead of its home address without changing the application payload (or even the transport protocol data) in any way. There are many ways of doing this, even at the network layer, but they can all be understood as variations on the theme of encapsulation. By encapsulation, we can change the destination of a packet (and its payload) without disturbing the packet in any way. There are some measurable effects, but they mainly have to do with the relative spacing between two successive packets, various round trip times, and other timing matters. There is also a potential global effect, since the path of the encapsulated packet will characteristically be longer than the path of a more simply delivered packet. When a network uses various forms of network-layer encapsulation for most of the traffic being delivered, it will reduce the total capacity of the network. If the encapsulation causes packet fragmentation, then this reduction in network capacity will become much more pronounced.

One specific effect that has been studied for Mobile IP has to do with *triangular routing*. This means that a packet making a round trip from the mobile node to a correspondent node will travel along three identifiable legs instead of the usual two-way travel. This is because packets from the correspondent node to the mobile node go through the home agent before they are encapsulated for further delivery to the mobile node. Packets from the mobile node to the correspondent node could be encapsulated (in any of several possible ways) and delivered directly without any need for the services of the home agent. This depends, of course, on the ability of the correspondent node to decapsulate the packet (thereby reversing the temporary effect of the mobile node's encapsulation). The net effect of this triangular routing is usually some small disruption of the round-trip timing used by TCP and possibly other streaming protocols.

While very interesting, these detailed timing and capacity effects are beyond the scope of this book. Furthermore, triangular routing as such is no longer specified in Mobile IPv6 due to the possibility of using an unverified home address without authorization. Instead, we will discuss the various kinds of encapsulation that are used, and their relationship to each other.

There are several basic kinds of network-layer encapsulation:

- IP-within-IP encapsulation [2, 11]
- source routing
- minimal encapsulation schemes [12]
- special-purpose options
- flexible minimal encapsulation schemes

IP-within-IP encapsulation is pretty self-explanatory. The only protocol detail to keep in mind is that the encapsulating header needs to have a way to identify that the "upper-level protocol" header following the encapsulating header is actually the encapsulated IP header, not really a upper-level protocol. Some network protocol stack implementations do not readily enable this kind of encapsulation; at the network layer, the program control flow has to be modified so that IP can be called, with a new route-table lookup, before completing the incoming or outgoing protocol processing. These implementation details, while important, do not have any impact on the conceptual model that we use for IP-within-IP encapsulation in this book.

Source routing can be shown to be equivalent to a special kind of encapsulation, for which the source IP address stays the same in the encapsulating header but the destination address changes. This is, of course, perfect for Mobile IP packet delivery to a mobile node. We want the correspondent node's address to stay the same, but we want the destination address to first be the care-of address and subsequently (after delivery to the care-of address) the home address. IPv4 has two kinds of source routing, both encoded as IP options within the IPv4 header. The two kinds are "strict source routing" (SSR) and "loose source routing" (LSR). SSR requires that the source route include *every* intermediate routing point; LSR is much more flexible and useful, because it only requires that the source route identify *some* of the intermediate routing points, perhaps even just one other intermediate point. This is perfect for Mobile IP, because with LSR we can identify the care-of address as an intermediate routing point for the mobile node's home address, and IP will then automagically cause the packet to be delivered to the care-of address on its way to the mobile node. There was an attempt to provide route optimization [13] for Mobile IPv4 by using IPv4 LSR option in exactly this way.

IPv6 has replaced the IPv4 source route options with a Routing Header extension to the IPv6 packet header. The operation of the IPv6 Routing Header was intended to be used much as described above. Using the design experience from Mobile IPv4 route optimization, the packet delivery feature for Mobile IPv6 route optimization was designed using IPv6's Routing Header (type 0). Later in the evolution of Mobile IPv6, there was some thought that using a newly specified Type 2 Routing Header would provide operational advantages. These details and many more are explained in depth in Chapter 10.

Just as route optimization can use source routing to provide a kind of encapsulation that maintains the same destination, it is possible to design a type of encapsulation that maintains the same destination but changes the source address. This is a very good way to understand Mobile IPv6's Home Address destination option 7.1. In this case, we want the mobile node to send packets directly to a correspondent node, but the local network may effectively require that the IP source address field be different than the mobile node's home address. The desired kind of encapsulation in this case would enable the source address to be constrained by the visited network until the outgoing packet departs the visited network. Then, when the packet is received, it should enable the receiver to change the source IP address to match the mobile node's home address, which is the intended IP address for communication with the application. That is exactly what the home address destination option is supposed to do. Since it is a destination option, only the correspondent node is expected to perform the action on the option. The option carries the mobile node's home address, and the receiver is expected to replace the source address with the contents of the option data (i.e., the home address). This is perfect for delivery of data from the mobile node's application to the correspondent node's application.

Mobile IPv4 offers two kinds of care-of address to a mobile node: *foreign agent* care-of address and *co-located* care-of address. The former means that the mobile node uses a care-of address provided by the foreign agent; typically, this care-of address can be the same for every mobile node visiting the same foreign agent. A co-located care-of address, on the other hand, is owned by the mobile node, not the foreign agent. Two mobile nodes will never share the same co-located care-of address at the same time.

When a Mobile IPv4 mobile node is using a co-located care-of address in a domain that enforces *ingress filtering*, it has to use the local care-of address as the source IP address in the IPv4 packet header that is visible to the routers in the visited network. However, when the packet departs the visited network and is eventually routed to the receiver, the source IP address in the visible IPv4 header has to be the mobile node's home address, just as in the case described above. For IPv4, no IP option was ever defined for this case, so for mobile nodes visiting ingress-filtering domains, a new IPv4 header using the care-of address is required to encapsulate the IPv4 header containing the mobile node's home address.

Other schemes have been devised to reduce the overhead of encapsulation, notably *minimal encapsulation* [12] and, for IPv6, a more flexible encapsulation scheme due to Steve Deering and Brian Zill [4]. These have features that allow savings in the overhead of encapsulation, usually by some very precise semantics about how to construct or reconstruct a desired IP header as needed when the modified or extended IPv4 header arrives at the appropriate point in its routing path.

It is worthwhile to understand some of the perceived risks attending the use of encapsulation and its variants. For all variations except IPv6 Routing Header type 2 (which is described in detail in Section 10.3), the perceived risks are similar. The typical encapsulation scheme enables packets to be sent to some destination which is not disclosed in the usual field of the IP packet header; of course this information could be present in a field which is plainly detectable, very nearby, and according to a globally accepted standard. Of course this should not be any problem whatsoever, except when one starts to consider the characteristics of network administration and firewall product features. If the firewall product in use at the visited enterprise has the wrong features, then there can be serious damage or complete interruption of protocol operations.

Since packet encapsulation typically causes the destination to change, firewall policies have often been put in place which drop encapsulated packets. This is because insufficient effort is often invested in the firewall rules to admit packets based on the secondary destination. Since the network administrator does not have the right tool to make sure that decapsulation does not expose vulnerable internal network nodes, often the administrator makes it impossible for internal nodes to receive decapsulated packets. Unrestrained packet redirection might be seen to open up a security hole which could be exploited to infect or degrade service to otherwise safe internal systems which were not prepared to be exposed to the many dangers present in the Internet today.

Until the time comes when such dangers no longer exist, we will probably continue to confront fairly restrictive firewall policies. If firewall vendors made sufficient tools available to properly control the flow of packets after decapsulation, we would probably be better able to count on encapsulation as a method for flexible packet delivery. As it is, these tools have to be used very carefully during protocol design, and it is important for firewall product designers and network administrators to become aware of what is needed to properly manage perimeter security in a way that enables the convenience and performance of Mobile IPv6.

6.5 HOME AGENT DISCOVERY

It is possible that, during its travels, the mobile node does not have a valid address for its home agent. For instance, the home agent could have been renumbered with a different IP address during the absence of the mobile node. While we can hope that these misfortunes do not happen very often, during the development of Mobile IPv6 it was felt that a safety net should be created. Otherwise, it will be the user of the mobile device who suffers and is faced with the requirement to place a service call so that the necessary information can be reconfigured.

The details of the currently specified mechanism can be found in Chapter 12. It is our belief that this is an area which will undergo significant future revision. In fact, since the publication of RFC 3775, there has been a large amount of discussion within the IETF working group related to bootstrapping for the mobile node and home agent. The outcomes of these related discussions will almost surely impact the ways in which home agent address discovery as well as configuration of home network parameters are finally carried out.

6.6 MOVEMENT DETECTION AND LINK ESTABLISHMENT

IPv6 defines some basic methods for determining the network layer parameters needed when a node has to establish a link to a new network. These basic methods are the ones that have been mentioned previously, collectively known as Stateless Address

Autoconfiguration specified in RFC 2462 [17]. There is also a stateful approach using DHCPv6 [6], but it may be too heavyweight to offer the kind of speedy response time that Mobile IPv6 offers. Thus, the stateful address configuration approach would hide many of the engineering advantages of Mobile IPv6.

We can consider movement detection as the process of determining when a mobile node is required to reinitiate its link establishment and address configuration procedures. As a mobile node moves away from its previous point of attachment, its previous link parameters and address become irrelevant. Further communication will require the mobile node to detect the disengagement from its previous point of attachment and the consequent need to repeat the appropriate configuration steps according to, say, RFC 2462, as mentioned above.

Unfortunately, the basic IPv6 methods were designed mainly with the goal of satisfying the needs of statically located devices, although there was no shortage of advice from working group members about the future importance of mobile devices. Since the IETF is typically dominated by concerns arising from current network deployments, these future concerns about mobility had only a secondary effect on the evolution of the IPv6 protocols. At least we were able to engineer the autoconfiguration protocols so that some future enhancement could possibly serve the needs of IPv6 mobile devices.

The stateless address autoconfiguration procedures specified in RFC 2462 work by using information provided in Router Advertisements as defined within the ND protocol (RFC 2461 [10]). Router Advertisements are used by the mobile node for the following purposes:

- whether or not stateless address autoconfiguration is made available on the local network
- detection of a new link
- detection that a previous link is no longer reachable
- layer-2 and layer-3 address details about the advertising router
- address range from which a new address has to be selected
- Maximum Transfer Unit (MTU) for the link, which is necessary information for configuring packet size limitations at layer 3

The Router Advertisement message is the crucial element that enables IPv6 devices to establish any new connection to the Internet.

Only a few changes were needed to the IPv6 Router Advertisement to provide appropriate details to mobile nodes. First, the advertisement definition was expanded to include a new flag indicating whether or not the router was a home agent for the subnet. There are likely to be many routers that would not provide home agent services to mobile nodes. It would be nice if the mobile node could have some assurance about the availability of the home agent service before trying to send Mobile IPv6 signaling messages to the router. Along with this basic functionality, a new extension was added to the basic advertisement message, which allows a home agent to inform the mobile node about its global IPv6 address. This was added because the Router Advertisement message definition constrains the local router to use its link-local address for the Source IP field in the IPv6 header, but the mobile node has to be able to send packets to its home agent by using its global IPv6 address instead.

One more important functional improvement had to be made for reasonable performance. When a mobile node moves to a new link, its communications will often be disrupted by the amount of time required to establish the new link, new network-layer addressing and so on. If the Router Advertisement is used as a primary indicator for the presence at a new link, this discovery process will be dominated by the frequency of transmission for the advertisements. The original IPv6 document specified that the advertisement interval was, at minimum, some number of seconds (*MinRtrAdvInterval* in [10]), and that is enough so that users of mobile devices would be seriously irritated whenever establishing a new point of attachment to the Internet. For this reason, Mobile IPv6 specifies that the advertisements can be sent out as often as once every 30 ms, which is a much better solution.

To further enable a reasonable response for the mobile nodes, a new *Prefix Interval option* has been defined so that the mobile node will be able to tell how often it should expect advertisements. The most important effect is that the mobile node can then tell when in the future it will lose connectivity to the local router. The information may also be used to save battery power by adjusting the duty cycle of operation at a mobile node when it is quiescent.

It is worthwhile to mention here the on-going work in the *Detecting Network Attachment* (DNA)" [5] working group in the IETF. The group was founded to address the deficiencies of the current protocols in reliably and quickly performing IP configuration when nodes unplug and plug themselves back into the Internet. The working group has produced multiple solutions which are being harmonized as of this writing. One of the key problems being addressed is how to reliably identify links and subnets. Often, these have been used interchangeably, but increasingly people are realizing the need for a separation. For instance, how does a node detect that it has actually traversed subnet boundaries when it has changed a link? Alternatively, there may be multiple subnets on a single link, and a node may be using one of them. How does the node detect and act when to effect a new IP configuration (and when not to do so) in the presence of this link and subnet ambiguity? This question, along with fast router discovery and reducing the delay in IP configuration, are being addressed by the working group.

6.7 SUMMARY

In this chapter, we introduced the readers to the sketch of the Mobile IPv6 protocol. We identified various functional parts of the protocol and provided a basic introduction and context. The rest of the chapters in this part describe in detail each of the sections in this chapter.

Exercises

6.1 In route-optimized communication, a mobile node sends packets to a correspondent using the home address present in the destination option. Why does the design use a routing header in the reverse direction? Why not use destination option in both ways?

6.2 Can IP addresses ever provide proof of identity of a user? What are the benefits? What are the drawbacks and computational requirements?

6.3 Even though we have not discussed the theory of Return Routability yet, it is instructive to ask how it could satisfy the "Do No Harm" directive, which forms a strong basis to judge protocols against perceived vulnerabilities. [2] Assuming that IP routing works on the Internet today and we send two different secrets along two different routing paths, why a proof that combines the two secrets sufficient for binding one address to another?

REFERENCES

1. J. Arkko, V. Devarapalli, and F. Dupont. "Using IPsec to Protect Mobile IPv6 Signaling Between Mobile Nodes and Home Agents," RFC 3776, Internet Engineering Task Force, June 2004.
2. A. Conta and S. Deering. "Generic Packet Tunneling in IPv6 Specification," RFC 2473, Internet Engineering Task Force, December 1998.
3. J. Daemen and V. Rijmen. "AES Proposal: Rijndael," http://csrc.nist.gov/CryptoToolkit/aes/rijndael/Rijndael-ammended.pdf, 03/09/1999.
4. S. Deering and B. Zill. "Redundant Address Deletion when Encapsulating IPv6 in IPv6," draft-deering-ipv6-encap-addr-deletion-00.txt (work in progress), November 2001.
5. "Detecting Network Attachment (DNA)" Working Group, Internet Engineering Task Force, http://ietf.org/html.charters/dna-charter.html
6. R. Droms, J. Bound, B. Volz, T. Lemon, C. Perkins, and M. Carney. "Dynamic Host Configuration Protocol for IPv6 (DHCPv6)," RFC 3315, July 2003.
7. S. Kent and R. Atkinson. "IP Encapsulating Security Payload," RFC 2406, Internet Engineering Task Force, November 1998.
8. S. Kent and R. Atkinson. "IP Authentication Header," RFC 2402, Internet Engineering Task Force, November 1998.

[2]In fact, it is more appropriate to call it "Do No More Harm", since some weaknesses are already assumed to exist.

9. D. Johnson, C. Perkins, and J. Arkko, "Mobility Support in IPv6," RFC 3775, Internet Engineering Task Force, 2004.

10. T. Narten, E. Nordmark, and W. Simpson. "Neighbor Discovery for IP Version 6 (IPv6)," RFC 2461, Internet Engineering Task Force, December 1998.

11. C. Perkins. "IP Encapsulation within IP," RFC 2003, Internet Engineering Task Force, October 1996.

12. C. Perkins. "Minimal Encapsulation within IP," RFC 2004, Internet Engineering Task Force, October 1996.

13. C. Perkins and D. B. Johnson. "Route Optimization in Mobile IP," draft-ietf-mobileip-optim-11.text (work in progress), September 2001.

14. C. E. Perkins. "Securing Mobile IPv6 Route Optimization Using a Static Shared Key," RFC 4449, Internet Engineering Task Force, June 2006.

15. D. Plummer. "An Ethernet Address Resolution Protocol Converting Network Protocol Addresses to 48.bit Ethernet Address for Transmission on Ethernet Hardware," RFC 826, Internet Engineering Task Force, November 1982.

16. R. Rivest. "The MD5 Message-Digest Algorithm,", RFC 1321, Internet Engineering Task Force, April 1992.

17. S. Thomson and T. Narten. "IPv6 Stateless Address Autoconfiguration," RFC 2462, Internet Engineering Task Force, December 1998.

18. National Institute of Standards and Technology, "Secure Hash Standard," FIPS PUB 180-1, April 1995, http://www.itl.nist.gov/fipspubs/fip180-1.htm

19. National Institute of Standards and Technology, "Secure Hash Standard," FIPS PUB 180-2, August 26, 2002, http://csrc.nist.gov/encryption/tkhash.html.

7

Binding Cache Management

The primary purpose of the 'Data' statement is to give names to constants; instead of referring to Pi as 3.141592653589793 at every appearance, the variable Pi can be given that value with a Data statement and used instead of the longer form of the constant. This also simplifies modifying the program, should the value of Pi change. – Fortran manual for Xerox Computers

—

In this chapter, the parts of Mobile IPv6 that handle Binding Cache management and related messages are described in detail. This includes all aspects of processing by the home agent, the correspondent node, and the mobile node.

There are three Binding Cache management protocol elements:

1. Home Address destination option

2. IPv6 Mobility Header messages

3. Options

First, the Home Address destination option will be described; IPv6 destination options have the characteristic feature that they are only to be interpreted by the destination. This destination option is used by mobile nodes in several cases, including transmission of normal data packets to correspondent nodes. The new destination option is also used when sending mobility-related messages that require information about the mobile node's home address if that address isn't available in the IPv6 packet header.

After the description of the new destination option, the new IPv6 mobility messages are described. They are identified as types of a general IPv6 Mobility Header. These

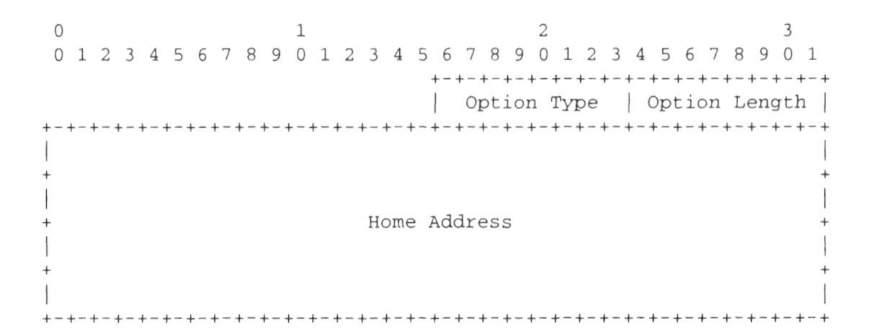

Fig. 7.1 Home Address Destination Option Format

mobility messages can have options. The options are described in detail after all of the mobility messages have been completely described. This approach is preferable, since several of the options can be applied to multiple mobility messages.

Binding Cache management is dependent on the presence of security association at the home agent and the correspondent node. Mobile IPv6 specifies how a security association can be set up between a mobile node and its home agent using IPsec; we discuss this in Chapter 9. Return Routability is the default mechanism for establishing such a security association between a mobile node and an arbitrary correspondent node; we discuss Return Routability in detail in Chapter 8. However, we are obliged to refer to Return Routability, since Binding Cache messages are closely linked to it.

7.1 HOME ADDRESS DESTINATION OPTION

A mobile node uses the Home Address destination option to provide its home address to its communication partners in any packet which does not already contain that information in the Source IP address of the IPv6 packet header. In many cases, when the mobile node is located within a network that is subject to *ingress filtering* [4], it is not free to transmit packets with its home address used as the Source IP address. Ingress filtering requires devices using the local network to have IP addresses which are within the range of addresses that are assigned to the local network. Using the home address would cause the mobile node to claim connectivity from a nonlocal network (i.e., its home network); usually, this goes against the policy enforced by ingress filtering.

The format of the Home Address destination option is illustrated in Figure 7.1. The Home Address option has the simplest possible structure with only a single field, which contains the mobile node's home address. The option type is 201 (0xC9 in hex), and the option length is 16 (i.e., the length of the IPv6 address which constitutes the option data).

The Home Address option is frequently used with Binding Update messages and also in packets sent by the mobile node to the correspondent node. A correspondent node can require that a security association be established between itself and a mobile node before accepting a packet with the Home Address option, in order to reduce the possibility that a malicious node in the Internet might try to impersonate a mobile node. This is not foolproof, but it is pretty effective overall. On the other hand, there is a significant protocol overhead in establishing the appropriate security association. It remains to be seen whether the required operations will be carried out in the usual case, or if communications will suffer decreased performance, or if correspondent nodes do not enforce the requirement for additional security associations. The address used in the Home Address option has to be a routable unicast address – multicast addresses are explicitly disallowed.

For any IPv6 destination option, the top 3 bits of the type field have special significance. For unicast addresses, the first two bits have the following meanings for any destination that does not recognize the destination option:

00 skip over this option and continue processing the header.

01 discard the packet and take no further action.

10,11 discard the packet and send an ICMP *Parameter Problem* (with Code 2) message [1] to the packet's Source Address, pointing to the unrecognized Option Type.

The type of the Home Address destination option is 0xC0, which has the first two bits equal to '11'. So, if a destination node does not recognize the Home Address destination option, it has to send the ICMP Parameter Problem message as indicated.

The third bit of the type field for the Destination Options header specifies whether the option data is mutable in transit. Some IPv6 headers could have data that is changed by intermediate routers, and in those cases the Authentication Header has to treat the mutable data as all zeroes. However, the option data for the Home Address destination option is the mobile node's home address, and that is not designed to change en route. This is shown by the fact that the third bit of the type field is '0', indicating nonmutability.

7.2 MOBILITY HEADER

The Binding Cache management messages are allocated as message types of the new IPv6 extension called the *Mobility Header*. The general form of Mobility Header messages is illustrated in Figure 7.2. All type fields, length fields, and other numeric values are unsigned integers unless otherwise specified.

In the figure, the message data is arbitrary, and the number of bytes in the message data has to be such that the total length of the Mobility Header is a multiple of 8 bytes. The Header Len field is calculated as one less than one-eighth of the length of

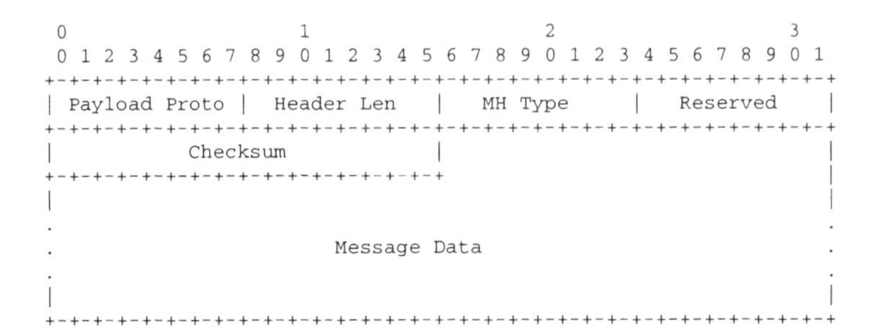

Fig. 7.2 Mobility Header General Message Format

the entire Mobility Header message. Since the total length of any Mobility Header message has to be a multiple of 8 bytes, we are safe in using 8 bytes as the basic unit of length for the Header Len field. Since the minimum size of any Mobility Header message has to be at least 8, no additional information is imparted by accounting for the first 8 bytes. By implicitly acknowledging that 8 bytes will always be present, we can maximize the amount of information available for use in the field of the message header. Thus, the total length of any Mobility Header message (including options) is 2048 bytes.

The Reserved field is ignored on reception and sent as zero by Mobile IPv6 implementations conforming to the current protocol specification. Future protocol documents may use this field if necessary.

Mobility messages that are specified for use with Mobile IPv6 use the following values in the MH Type field:

0 Binding Refresh Request (BRR) message (Section 7.5)

1 Home Test Init (HoTI) message (Section 8.2)

2 Care-of Test Init (CoTI) message (Section 8.3)

3 Home Test (HoT) message (Section 8.4)

4 Care-of Test (CoT) message (Section 8.5)

5 Binding Update message (Section 7.3)

6 Binding Acknowledgement message (section 7.4)

7 Binding Error Message (section 7.6)

Other documents may specify other message types for the Mobility Header. For instance, new message types have been allocated for use with Fast Handover (see Chapter 14).

The checksum is calculated by first obtaining the one's complement[1] sum of the octets of the Mobility Header (preceded by the IPv6 *pseudo-header* (see the IPv6 protocol specification RFC 2460 [3]). The one's complement of this summation is then used as the value of the Checksum header field.

The Payload Proto is the next protocol number, which is expected to be IP-PROTO_NONE until further specification enables the use of the Mobility Header as a "non-final" IPv6 extension.

When the Header Len indicates that the length of Message Data is longer than required for the value of the message type (i.e., the MH Type field), the remainder of the data is treated as one or more Mobility Options. These are described later in Section 7.7.

There are certain processing steps that must be performed when any node receives an IPv6 packet containing a Mobility Header. These include the following:

1. The checksum field must be verified as correct.
2. The MH type field has to correspond to a message type known to the receiver
3. The length of Message Data has to correspond to the minimum length possible for the message type as indicated in the MH Type field

These steps are in addition to the usual checks performed for all IPv6 packets. Moreover, any IPsec header processing is supposed to be done before processing the Mobility Header.

Next, we will provide details about the Binding Management message types that use the Mobility Header. The other messages types will be treated in Chapter 8, which describes the protocol for verifying Return Routability.

7.3 BINDING UPDATE

The Binding Update message is the most essential part of the Mobile IPv6 protocol specification. Using this message, the mobile node supplies its care-of address to the desired recipient, which can be either the home agent or any correspondent node – that is, any IPv6 node. It uses MH type 5 in the Mobility Header. The format of the fields of the Binding Update message is illustrated below in Figure 7.3. These fields follow the fields of the Mobility Header, as shown in Figure 7.2.

The *Lifetime* field is 16 bits and determines how long the care-of address is to be associated with the mobile node's home address. This lifetime is called the *Binding*

[1]One's complement arithmetic treats zero the same as all ones. When adding two numbers generates an overflow, the generated bit is carried over into the least significant bit of the result.

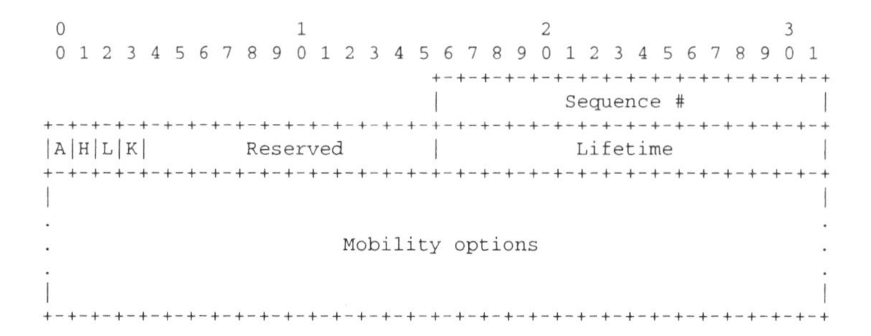

Fig. 7.3 Binding Update Message Format

Lifetime. Each unit is 4 seconds, so the maximum value for the Binding Lifetime is 262,140 seconds, or about 3 days. Requiring a renewal Binding Update every 3 days, even when the mobile node does not change to a new care-of address, is not considered to be a significant overhead.

The *Sequence #* field enables a recipient to put incoming Binding Update messages in proper chronological order. This is very important, because such messages are not otherwise guaranteed to be delivered in order. The "best-effort" nature of IP does not prohibit rerouting, reordering or even duplicating IP packets at any intermediate router, although the latter two events are not very likely. Nevertheless since two successive Binding Update messages can travel along different routes, they can arrive out of order. If they are applied out of order, the recipient would typically lose contact with the mobile node, so this has to be prevented. The sequence number does this for all foreseeable circumstances. The probability of 65,535 legitimate Binding Updates being active in the Internet from a single mobile node has to be taken as practically zero.

However, not all packets are legitimate. A malicious node somewhere in the Internet could try to keep track of messages transmitted by the mobile node and at some future time replay those messages for processing by a home agent or correspondent node. The message authentication data which is required to be present in Binding Update messages will protect against tampering with the data but not against replay attacks, because the data is just the same as originally generated by the mobile node.

The sequence number is useful to protect against replay attacks, since a mobile node cannot reasonably change its care-of address 65,535 times within the lifetime of the security association used to generate the authentication data.

There are four single-bit flags which may be set in the Binding Update message.

'A': The 'A' bit is set when the mobile node wants to get an acknowledgement that the Binding Update has been received. Acknowledgements from correspondent nodes are not required, because the mobile node can tell whether or not the correspondent has a current binding. If not, then packets from that correspondent node will be encapsulated and delivered by the home agent. When that happens, the mobile node can arrange to send another Binding Update.

'H': The 'H' bit is set when the mobile node considers the recipient to be its home agent. The home agent has special duties for delivering packets to the mobile node, and the recipient of the Binding Update with the 'H' bit set cannot approve it unless it is willing to carry out these duties (see, for example, Section 7.3.1). When the Binding Update does not have the 'A' bit set. it is an error for the Binding Update to have the 'H' bit set.

Setting the 'H' bit also protects against the rare but not impossible situation in which a mobile node already has a home agent, but wishes to establish communications with another node on the same subnet that just happens to be able to perform the duties of a home agent. Since the mobile node already has a home agent, and since it wants to establish communications with the new node in some role purely as a correspondent node, the recipient has to be able to determine what action is expected from it. This is possible even when the home-agent-capable correspondent node has a mobility security association with the mobile node.

'L': When the 'L' bit is set in the Binding Update, the mobile node provides additional information to the home agent about its addressability on its home subnet. Namely, setting the 'L' bit informs the home agent that the mobile node's home address shares the same interface identifier as its link-local address; this latter address is valid only on the home subnet.

The home agent should use this information to protect the mobile node's link-local address against use by other nodes on the home subnet. The mobile node cannot protect its link-local address during times when it is not connected to the home subnet.

'K': When the 'K' bit is set, the mobile node can maintain a security association with the recipient across changes of care-of address. Unless there is some likelihood of a security breach by doing so, this bit should be set in most typical circumstances for home registrations. RFC 3775 specifies that security associations with correspondent nodes should be quite short-lived, but this was mainly due to uncertainty about the safety of the Return Routability procedure. Perhaps these concerns may turn out to be unrealistic after all, and improvements could be made to reduce the performance overhead imposed by the Return Routability procedure. Furthermore, newer specifications [9] allow static configurations of security associations even between a mobile node and a correspondent node. This is described in Section 8.8. The 'K' bit should also be set in such circumstances.

Note that the mobile node's care-of address is not part of the message data for the Binding Update. Instead, the recipient must look in the IP Source Address field of the IPv6 header to find this information. Using the care-of address in that field of the IPv6 header enables proper handling by ingress-filtering routers, because such routers demand a topologically correct address in the Source IP address field. The care-of address is topologically correct by definition; the home address, by contrast, is rarely or never topologically correct for any visited network.

Therefore, for Binding Updates, the mobile node's home address can no longer be located in the Source IP address field of the IP header, where it might traditionally be expected to reside. Instead, the mobile node supplies its home address within the Home Address destination option. Since the care-of address is a routing address and the home address is used for identifying the mobile node, these design provisions do make sense. The Home Address destination option is described in detail in Section 7.1.

Processing of the Binding Update follows fairly straightforward rules, as detailed in the next few sections. Whenever any node receives a Binding Update, two general rules apply.

First, the sequence number of the Binding Update has to be greater than the sequence number for the previously received Binding Update. Actually, it is a good idea to make sure that it is greater than the value of any recently received Binding Update, for some reasonable value of "recently." So, for instance, suppose five Binding Updates were received in the last 10 minutes, with sequence number values 190, 181, 182, 184, 188. Then, by the less restrictive rule, the last three Binding Updates might be considered acceptable, but using any reasonable value of "recently" would lead to invalidating each of the last four Binding Update messages.

Second, the home address of the mobile node has to be a unicast routable IPv6 address. This means that it is not a multicast address, a loopback address, the null address, or any link-local address. The specification technically allows for the use of a site-local home address, but later developments have strongly suggested a prohibition on the use of site-local addresses as well.

7.3.1 Home Agent Actions for Receiving Binding Updates

When a home agent receives a Binding Update, it takes the actions necessary to register the incoming care-of address provided by the mobile node – in other words, to associate the care-of address with the mobile node's home address. In this way, future packets addressed to the mobile node's home address can be then tunneled to the current care-of address as required.

The main things that the home agent has to do as a result of receiving a valid Binding Update are to:

- Verify the authenticity of the message to avoid the possibility that another node could steal or divert traffic that is supposed to go to the mobile node.
- Make sure that the new care-of address is the tunnel endpoint for packets that will be tunneled to the mobile node (or not tunneled at all in the case of a deregistration). In many implementations, this will involve updating some fields in the home list or route table.
- Make sure that it can intercept packets that are addressed to the mobile node on the home network (or no longer intercepted in the case of a deregistration).
- In the case of a new or updated registration, defend the mobile node's home address (and, potentially, its link-local address) from being claimed by any other node on the home network.

- Alternatively, in the case of a deregistration, cease defending the mobile node's home address and link-local address.

- Send a Binding Acknowledgement to the mobile node, as described in Section 7.4.

First, the verification steps will be described. They are very straightforward but tedious.

7.3.1.1 Verification In order to process the message, the home agent will have to identify the home address of the mobile node that sent the message. If the Home Address destination option is present, then the home agent uses the information contained within that option for the home address. Otherwise, the home address is as shown within the IP source address of the IPv6 packet header. When there is no home address option, since the IP source address is also the care-of address, it must be the case that the care-of address is the same as the home address, and therefore that the mobile node has returned to its home network. In this case, the mobile node will have also set the lifetime of the binding to equal zero, so that the Binding Update message represents a deregistration operation for the given care-of address. After deregistration, the mobile node behaves just as any other IPv6 node and the home agent performs no special duties for the mobile node. Note that the mobile node may also provide a home address equal to its care-of address within the Home Address option.

The home agent first makes a number of initial verifications:

- It verifies that there is an appropriate SA with the mobile node identified by its home address, typically in the Home Address option.

- It verifies the authentication data supplied with the Binding Update, using the SA just determined.

- It makes the general verifications for Mobility Headers, as indicated in Section 7.2.

- It makes the two general verifications for Binding Updates, as described in section 7.3.

- It verifies that the home address belongs to a mobile node which has been approved for the service, unless by default every mobile node on the home network is eligible to receive service.

In order to verify the authentication data supplied, the home agent needs to have an SA, which, as we saw in Chapter 3, specifies how to identify and process a packet for security processing. Mobile IP specifies how to establish such an SA between a mobile node and its home agent using the IKE protocol. The details of this specification deserve a separate chapter, and hence we discuss it in detail in Chapter 9.

7.3.1.2 Binding Cache Entry Management If the Binding Update is valid, then the home agent proceeds to use the information provided to update its binding for the mobile node. The relevant information includes the following:

- the new care-of address
- the lifetime of the binding
- whether or not the mobile node's link-local address also needs to be defended during its absence
- potentially, disposition of existing security information between the mobile node and the home agent.

The new care-of address is first supplied to the mobile node's entry in the Binding Cache. If the care-of address is the same as the home address, and/or the lifetime of the binding is shown as zero, then the Binding Update is a deregistration, and the home agent has to stop carrying out all of its former duties for the mobile node. In this case, then the home agent deletes the Binding Cache entry for the mobile node and takes further actions to discontinue intercepting packets destined to the mobile node. It also stops responding to any further address autoconfiguration messages intended for the mobile node, so that the mobile node once again becomes responsible for defending its own addresses.

Otherwise, for the case of registration packets, if there is no existing entry for the mobile node in the Binding Cache, the home agent creates one and initializes it according to the values given within the Binding Update. The home agent is allowed to limit the lifetime of the binding to be less than that requested by the mobile node, according to administrative policy. This restriction should rarely be required, since the configured parameters on the mobile node itself should be compatible with the values configured for the mobile node at the home agent.

Some Binding Update messages contain the Alternate Care-of Address option. For such messages, the home agent no longer uses the Source IP address as the care-of address and instead uses the address supplied in the option. For such packets, the home address of the mobile node is still supplied in the Home Address destination option.

7.3.1.3 Intercepting Packets Addressed to the Mobile Node The result of arranging the proper care-of address in the Binding Cache entry for the mobile node is to enable tunneling packets to the mobile node, using IPv6-within-IPv6 encapsulation. In order to get those packets in the first place, the home agent has to intercept them on the home network. Typically, the home agent does that by causing its own MAC address to be associated with the home address of the mobile node in the Neighbor Cache entries of all nodes on the home network that maintain such entries for the mobile node.

In IPv6, the process of resolving an IP address to a local MAC address is done by the process of Neighbor Discovery [8], which we studied in Chapter 2. The two relevant messages here are Neighbor Solicitation and Neighbor Advertisement.

Typically, a node locally broadcasts a Neighbor Solicitation in order to acquire the layer-2 address for some destination of interest on the local network. The target node then responds with a Neighbor Advertisement message.

However, the home agent needs to generate the correct association between its own layer-2 address and the mobile node's IPv6 home address, even though it does not receive a Neighbor Solicitation message corresponding to the home address of the mobile node. For this purpose, the home agent sends a Neighbor Advertisement message to every node on the home network, by using the link-local all-nodes multicast address which is FF02::1 (see Section 2.4).

The Neighbor Advertisement multicast is constructed as follows:

- The Source IP address of the IPv6 header is set to the IPv6 address of the home agent.
- The Destination IP address of the IPv6 header is set the all-nodes link-local address.
- The Target Address field of the Neighbor Advertisement is set to the IPv6 home address of the mobile node.
- The home agent's layer-2 address is located in the Target Link-Layer Address option of the Advertisement message so that it will be associated with the mobile node's IPv6 address
- The 'R' bit of the Advertisement is set to zero, since the home agent is not advertising that the mobile node is a router.
- The 'O' bit of the Advertisement is set, because the home agent is instructing every recipient of the multicast message to override any existing Neighbor Cache entry that may already be present for the mobile node.
- the 'S' bit is set to zero, since the multicast message was not in response to any Neighbor Solicitation message (in other words, the multicast message is unsolicited).

If the home agent receives a Neighbor Solicitation message for the mobile node, it responds with an appropriately constructed Neighbor Advertisement, with the following differences from the multicast message just described:

- The 'S' bit is set.
- The Destination IPv6 address in the IPv6 header is the address of the soliciting node.

As just described, the home agent utilizes Neighbor Discovery protocol messages to enable its interception of packets that are addressed to the mobile node. These same messages are also used for the process of address autoconfiguration in IPv6 (see RFC 2460 [3] for details). The basic idea for stateless address autoconfiguration in IPv6 is that an IPv6 node can construct a candidate address for its own use; before committing that address for use as its endpoint identifier, the node has to make sure that it is not already in use. The way to do this is to issue a Neighbor Solicitation for that address. If any other node is already using the candidate address, the node will then receive a Neighbor Advertisement from the node already claiming the address; then another candidate address formed and the process repeated.

We have seen earlier in Chapter 2 that this process of detecting whether a candidate address is already in use is called Duplicate Address Detection (DAD). It has received much attention in the context of use by mobile nodes, as we will discuss in several other parts of this book.

7.3.2 Mobile Node Actions for Sending Binding Updates

A mobile node needs to send a Binding Update to its home agent when it acquires a new care-of address. There are multiple ways in which a mobile node might acquire a new care-of address, including stateless address autoconfiguration (which we discussed in Chapter 2) or by means of DHCP stateful configuration.

In order to register a new care-of address, the mobile node sends a Binding Update to its home agent typically constructed as follows:

- The new care-of address is placed in the Source IP address field of the IPv6 header.
- The mobile node puts its home address in a Home Address destination option (see Section 7.1).
- The 'H' bit is set in the Binding Update to denote that the mobile node requires home agent services from the recipient.
- The 'A' bit is set to indicate that the recipient is expected to respond by sending a Binding Acknowledgement message back to the mobile node at its care-of address.
- The mobile node identifies a suitable value for the duration of its ability to use the new care-of address and places this value in the Lifetime field of the Binding Update. The Lifetime has to be less than the remaining valid lifetime of its home address and not greater than any lifetime associated with the new care-of address it has obtained.
- The mobile node increments its current sequence number value that is used for Binding Update messages and sends the new value to the home agent in the Sequence Number field.

In addition, the mobile node has to assign suitable values for the 'L' bit and the 'K' bit of the Binding Update.

For setting the 'L' bit, the mobile node compares the Interface Identifier bits of its link-local address on the home network, to its global home address. If the IID bits of both addresses are the same, then the mobile node sets the 'L' bit to one; otherwise, it sets that flag to zero. The bit value of the IID will often be the same whenever the mobile node uses stateless address autoconfiguration (as specified in RFC 2462 [10]), because the identifier value is derived from the mobile node's layer-2 address (e.g., its MAC address). However, if the mobile node is using a home address derived by some other means, then this correspondence is no longer assured or even likely. For instance, the mobile node may have configured a randomized address as its home address (usually to avoid traceability, often by using RFC 3041 [7]).

In order to determine the setting for the 'K' bit, the mobile node has to determine the nature of its security association with the home agent. If this security association survives movements, then the 'K' bit should be set. Otherwise, the bit is cleared. According to the specification, the bit should also be cleared if there is a manual IPsec key configuration, but in the opinion of at least one of the co-authors of this book, that part of the specification is in error, and long-lived static key configurations should be allowable. Such manual configuration is also suitable for long-term security associations designed for use with the Binding Authorization Data option (see section 7.7.1).

Each Binding Update message must be able to be authenticated by the recipient, in this case the home agent. This can be accomplished in the following ways:

- The mobile node can use an IPsec Authentication Header Extension for IPv6 [5] to supply the authentication data.
- The mobile node can encrypt the Binding Update by using the IPsec Encapsulating Security Payload extension for IPv6 [6].
- The mobile node can use its security association with the home agent to calculate the appropriate data for use with the Binding Authorization Data suboption (see Section 7.7.1).

If the mobile node wishes to send its care-of address to a correspondent node, it also uses the Binding Update message for this purpose. In order to do so, the mobile node must have a security association established with the correspondent node. Given the existence of this security association, the general method for sending the Binding Update to a correspondent node is the same as for sending one to the mobile node's home agent. However, there are some differences. Moreover, when the security association is established by way of Return Routability some additional information is required. Specifically, the way the shared secret key K_{bm} for binding management is derived is quite different from the typical IKE and IPsec protocols used between the mobile node and the home agent. In order to fully understand the details, we must understand the return routability operation, which we describe in the next chapter. For the time being, it suffices to understand that the binding management key K_{bm} is derived by exchanging nonces between the mobile node and the correspondent using two different routing paths that correspond to the home address and care-of address of the mobile node. With the shared secret, the mobile node computes a Message Authentication Code and supplies it as part of the Binding Authorization Data Mobility Option, which is described in Section 7.7.1.

Here are the other main differences when the recipient is a correspondent node.

- The 'H' bit is not set.
- The 'A' bit is set only optionally, when the mobile node needs immediate assurance that the correspondent node has accepted the Binding Update.

Of course, the Destination IP address is that of the correspondent node, not the home agent.

7.3.3 Correspondent Node Actions for Receiving Binding Updates

The actions taken by the correspondent node upon receiving a Binding Update are somewhat different than the actions taken by a home agent. The main differences are as follows:

- When the binding update is accepted, there is no need for the correspondent node to begin home agent services for the mobile node. So, the correspondent node does not carry out any proxy ND operations involving multicasting Neighbor Advertisement messages.
- The correspondent node uses the care-of address in a different way than the home agent. Instead of tunneling packets to the mobile node using IPv6-within-IPv6 encapsulation [2], a routing header (type 2) is used. See Chapter 10, Section 10.3, for details about the use of this new type of routing header.

As specified, the security association which is established using Return Routability procedures is typically handled a bit differently than security associations for use with IPsec or statically configured associations. The main differences are due to two aspects of the design philosophy imposed on the Return Routability solution:

- The SA should have a short lifetime.
- The creation of the SA should be carried out in a way that causes the least requirement for keeping track of the state at the correspondent node.

Because of the first design point, the Return Routability procedure has to be run quite often, as often as every few minutes. Because of the second design point, the Binding Update is often required to be accompanied by additional information that might otherwise be stored on the correspondent node.

When the SA is established using Return Routability, usually the Binding Update must also include information that the correspondent node uses to retrieve its own stored data that is needed to create the actual shared key. Two such bitstrings (called *nonces*) are needed in the usual case; they are among several (or perhaps many) distinct nonces that are considered by the correspondent node to be stored in a logical array. So that the correspondent node can retrieve the two specific nonces that the correspondent node has earmarked for the SA, the mobile node has to supply two *Nonce Indices* to the correspondent node. These nonce indices are to be located in the Nonce Indices option (see Section 7.7.2) that the mobile node has to send with the Binding Update.

After retrieving the nonces, the correspondent node performs the same algorithm as the mobile node to recreate the key to be used to authenticate the data supplied with the Binding Update, in the Binding Authorization Data (see Section 7.7.1) option. We defer the description of key computation method to Chapter 8.

7.4 BINDING ACKNOWLEDGEMENT (BACK)

The main function of the Binding Acknowledge (BAck) message is to inform the mobile node about the status of its recent request to update its care-of address, in other words, the results of its Binding Update message. A BAck message is sent in response to a Binding Update whenever the latter has the 'A' bit set to one. It uses MH type 6 in the Mobility Header. The format of the BAck message is illustrated in Figure 7.4.

```
 0                   1                   2                   3
 0 1 2 3 4 5 6 7 8 9 0 1 2 3 4 5 6 7 8 9 0 1 2 3 4 5 6 7 8 9 0 1
                                +-+-+-+-+-+-+-+-+-+-+-+-+-+-+-+-+
                                |    Status     |K|  Reserved   |
+-+-+-+-+-+-+-+-+-+-+-+-+-+-+-+-+-+-+-+-+-+-+-+-+-+-+-+-+-+-+-+-+
|         Sequence #            |           Lifetime            |
+-+-+-+-+-+-+-+-+-+-+-+-+-+-+-+-+-+-+-+-+-+-+-+-+-+-+-+-+-+-+-+-+
:                                                               :
:                      Mobility Options                         :
:                                                               :
+-+-+-+-+-+-+-+-+-+-+-+-+-+-+-+-+-+-+-+-+-+-+-+-+-+-+-+-+-+-+-+-+
```

Fig. 7.4 Binding Acknowledgement Message Format

The mobile node is required to perform the usual validation operations to authenticate the BAck message. See Section 7.2 for details.

The *Status* field specifies the result of processing the Binding Update. There are numerous values defined, but in the most typical case, the mobile node should receive a Status of zero indicating that the Binding Update was accepted.

If set to 0, the 'K' bit states that the protocol used for establishing the SA between the mobile node and the home agent would have to be rerun. It means that the home agent cannot support modifying the IKE SA based on the Binding Update processing.

The Sequence # field contains the value copied over from the Binding Update message. This helps the mobile node to match its currently outstanding Binding Update message.

The Lifetime field specifies the actual time granted in units of 4 seconds.

In addition, the Binding Acknowledgment can contain Mobility Options, including the Binding Authorization Data and Binding Refresh Advice, which are described in Section 7.7.

7.5 BINDING REFRESH REQUEST (BRR)

A correspondent nodes sends a Binding Refresh Request (BRR) message to a mobile node in order to ask for a new Binding Update. In response to the BRR, the mobile node is expected to construct a new Binding Update packet, typically so that the lifetime of the binding can be extended by the correspondent node. This message uses MH type 0 in the Mobility Header, perhaps in recognition of its relative importance. The format of the BRR message is shown in Figure 7.5.

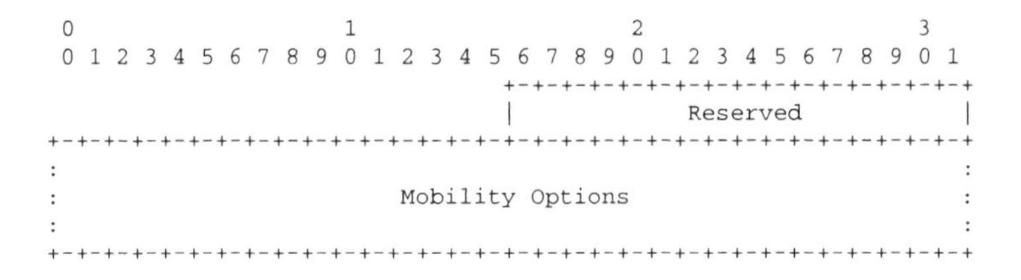

Fig. 7.5 Binding Refresh Request Message Format

As can be seen, the message does not have any functional data fields, especially since typically no mobility options are needed.

The BRR is not typically sent by the home agent, since the home agent is unlikely to be running applications that need to remain in communication with the mobile node. However, the lifetime of the binding at the correspondent node cannot exceed the lifetime of the binding with the same care-of address at the home agent. For this reason, the mobile node may be required to also send a Binding Update to its home agent if the remaining Lifetime at the home agent is too short to offer any meaningful extension to the correspondent node. There is no precise specification for this particular parameter, so it is effectively platform dependent. In other words, it is up to each implementation to decide how much remaining lifetime for the binding at the home agent can be considered sufficient.

7.6 BINDING ERROR (BERR)

The Binding Error (BERR) message is used to transmit error information about Binding Cache management operations in situations that are not already covered by the BAck message. This message uses MH type 7 in the Mobility Header. The format of the BERR message is shown in Figure 7.6.

Currently, there are only two identified uses for the BERR message:

- Status = 1: A BERR message with status = 1 is sent by a correspondent node to a mobile node. It handles the case when the correspondent node receives a packet with a Home Address destination option, but has no existing Binding Cache entry to authorize the use of that option. See the discussion on Re-

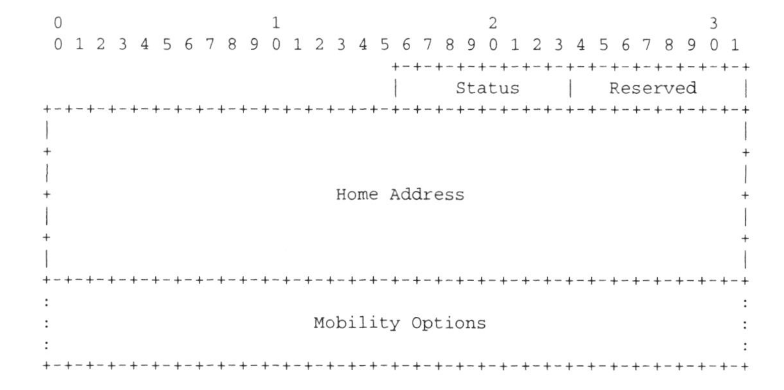

Fig. 7.6 Binding Error Message Format

turn Routability in the next chapter for more information on the need for this authorization.

- Status = 2: A BERR message with status = 2 is sent by any node if it receives a packet with an unknown Mobility Header type. In this case, the receiving node should make the best guess it can about the Home Address of the sender. If no information is available, the field should be set to 0:0:0:0:0:0:0:0 (the *unspecified* IPv6 address). This case should be very rare or almost nonexistent.

No Mobility Options are typically expected to be used with the BERR message in either of these cases.

7.7 MOBILITY MESSAGE OPTIONS

Mobility Options are data fields that are appended to the message data of a mobility message (i.e., a message identified as a type for the Mobility Header, like a Binding Update message). The length of all such appended options is determined by considering the overall length of the Mobility Header and the length of the fixed data fields that are defined for that specific message type. The extra length is then used as the length of all options together; each option has a type field and a length field. In this way, the options can be parsed and handled separately in turn.

Mobility Options are generally in type-length-value format as follows:

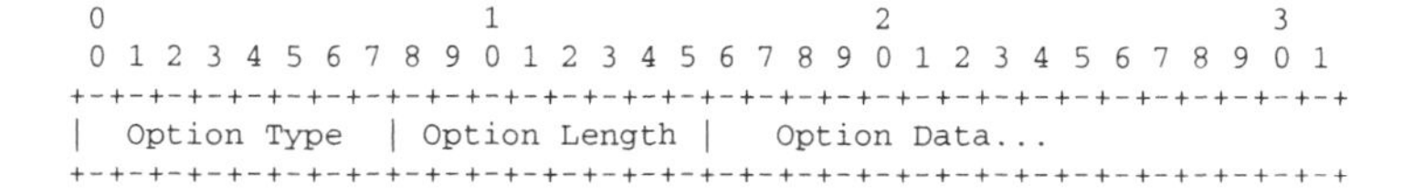

Fig. 7.7 General Mobility Option Format

The Option Type identifies the option, and the Option Length is used to indicate the length of the Option Data field. In contrast to the situation for IPv6 header options, unrecognized Mobility Options are quietly skipped. Subsequent options are processed as if no error had occurred.

The following Mobility Options are available:

0 Pad1 (Section 7.7.5)

1 PadN (Section 7.7.5)

2 Binding Refresh Advice (Section 7.7.4)

3 Alternate Care-of Address (Section 7.7.3)

4 Nonce Indices (Section 7.7.2)

5 Binding Authorization Data (Section 7.7.1)

Mobility Options may have an alignment requirement, which is specified so that Option Data fields that are longer than a single byte can be fetched into memory according to the most natural processor instructions. This usually offers improved performance. The alignment requirements are provided according to a notation such as "An + B", where A, n, and B, are non-negative integers. This means that the Option Type must appear at an integer multiple of A octets from the start of the extension header (in this case, a Mobility Header), plus B octets. Typically the value A is one of 2, 4, 8.

For some options, fulfillment of alignment requirements may require use of the special "pad" options, as explained in Section 7.7.5.

7.7.1 Binding Authorization Data option

The Binding Authorization Data option supplies data to be used when authenticating Binding Cache messages (especially Binding Update and Binding Acknowledgement messages). The term Binding Authorization Data is often shortened to BAuth, or sometimes just BAD, depending on the mood, even though the option is basically quite GOOD. The format of the Binding Authorization Data option is shown in Figure 7.8

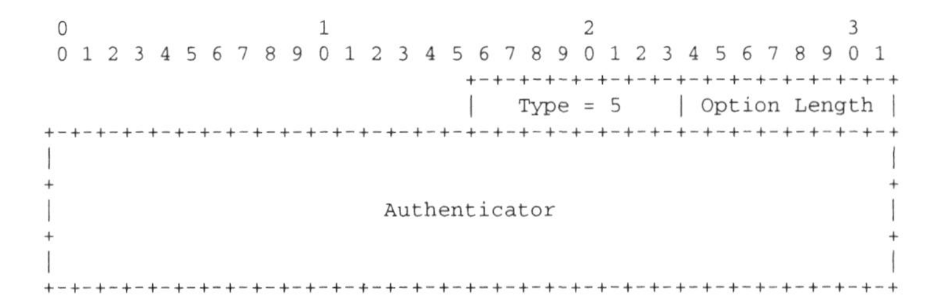

Fig. 7.8 Binding Authorization Data Option Format

The option data for this Mobility Option is always calculated over a byte string containing the data from the Mobility Header containing the option as well as the mobile node's care-of address and the IP address of the destination node to which the Mobility Header will be delivered. Currently, the Binding Authorization Data option can be used as an option within the Binding Update and the Binding Acknowledgement Mobility Headers. The rules for calculating the cryptographic value which will be used as the Authentication data depend on the security association in force between the mobile node and the recipient of the option.

All IPv6 devices have to support HMAC_SHA1 as a hash function useful for producing authentication data for this option. Other hash functions and cryptographic transforms can be used instead. The authentication data is 96 bits long, which is shorter than the output of HMAC_SHA1; the output is truncated and only the first 96 bits are used. According to RFC 3775, the following formula is used with HMAC_SHA1 to product the authentication data:

Mobility Data = care-of address '|' destination IP '|' MH Data

Authenticator = First (96, HMAC_SHA1 (K_{bm}, Mobility Data))

where K_{bm} is the current binding management key. This key can be established by multiple means, including IKE (especially with the home agent), Return Routability (with an arbitrary correspondent), and by means of preconfiguration (with a known correspondent). The symbol '|' denotes concatenation. MH Data includes the entire Mobility Header but not the Authenticator itself. The Checksum field in the Mobility Header includes the Authenticator, but the Checksum field is treated as zero in computing the Authenticator.

7.7.2 Nonce Indices

The Nonce Indices option supplies the information needed by the correspondent node to retrieve the nonces it had assigned for use when creating a new binding management key K_{bm}. The mobile node sends this option along with the Binding Update so that a correspondent node can retrieve these nonces, and the indices are given to the mobile node by the correspondent node in the Home Test and Care-of Test messages which were previously sent as part of the Return Routability procedure. The format of the Nonce Indices option is shown in Figure 7.9

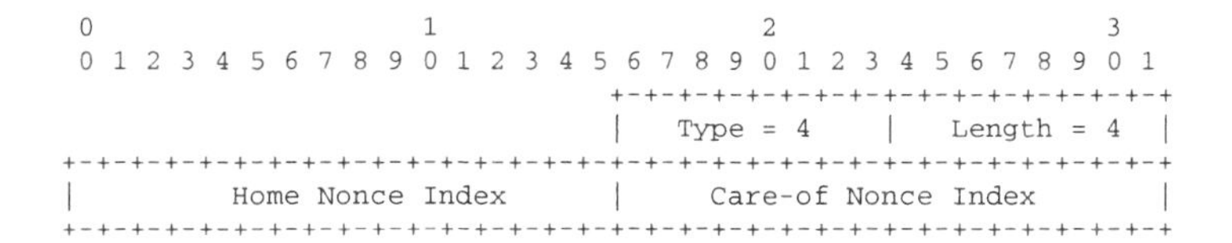
```
 0                   1                   2                   3
 0 1 2 3 4 5 6 7 8 9 0 1 2 3 4 5 6 7 8 9 0 1 2 3 4 5 6 7 8 9 0 1
                                +-+-+-+-+-+-+-+-+-+-+-+-+-+-+-+-+
                                |   Type = 4    |   Length = 4  |
+-+-+-+-+-+-+-+-+-+-+-+-+-+-+-+-+-+-+-+-+-+-+-+-+-+-+-+-+-+-+-+-+
|       Home Nonce Index        |     Care-of Nonce Index       |
+-+-+-+-+-+-+-+-+-+-+-+-+-+-+-+-+-+-+-+-+-+-+-+-+-+-+-+-+-+-+-+-+
```

Fig. 7.9 Nonce Indices Option Format

In this way, the correspondent node can store relatively few indices and still participate in the creation of binding management keys (K_{bm}) for potentially very many mobile nodes.

The Nonce Indices option is aligned on 2n byte boundaries, a very mild alignment requirement, so that the 16-bit index values are located on 16-bit word boundaries in memory.

7.7.3 Alternate Care-of Address

Usually, when a mobile node sends a Binding Update, the care-of address is found in the Source Address field of the IPv6 header. For times when that is not feasible or desirable, the mobile node can use the Alternate Care-of Address to supply its care-of address. The format of the Alternate Care-of Address option is shown in Figure 7.10

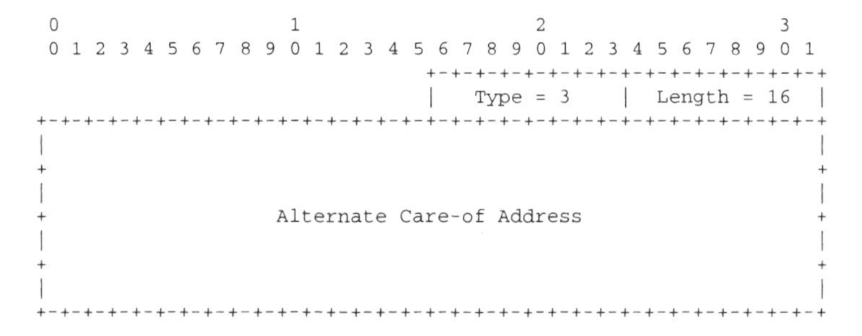

Fig. 7.10 Alternate Care-of Address Option Format

Most of the use scenarios for this option involve deregistering a previous care-of address. If the mobile node is no longer present on the visited network containing the care-of address, ingress filtering policies may preclude the use of its previously registered care-of address. This option is also useful when the information in the IPv6 header is vulnerable to undetectable tampering, say because the header fields are not included as part of the authentication calculation. This could happen when ESP is used instead of AH, since the IPv6 header fields are not included in ESP's encryption calculation. One could, on the other hand, use the Binding Authorization Data option along with ESP to solve this problem.

7.7.4 Binding Refresh Advice

The Binding Refresh Advice option is used to inform the recipient mobile node how often the sending node would like to receive Binding Updates. The format of the Binding Refresh Advice option is shown in Figure 7.11

The Binding Refresh Advice option should be aligned along 16-bit boundaries. Using this option enables the home agent to reduce the negative effects that would arise if the home agent lost the care-of address information it had recorded for the mobile node. In this way, the home agent could grant a very long binding lifetime for the mobile node's care-of address and still get more frequent periodic updates. Binding Refresh Advice option is currently specified for use only with Binding Acknowledgements coming from the home agent, but one could easily imagine that it

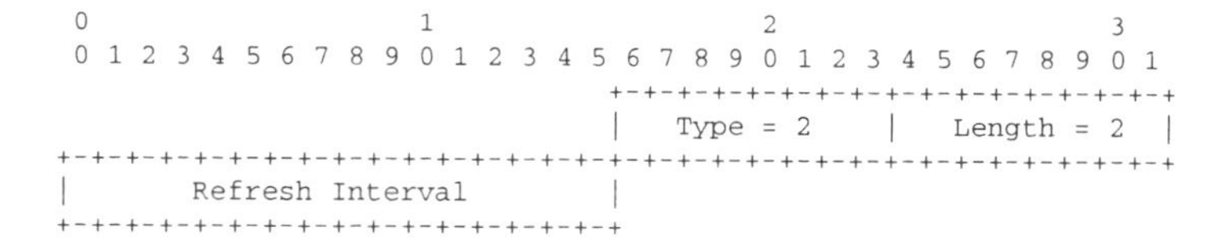

Fig. 7.11 Binding Refresh Advice Option Format

should also be used with a precomputable binding management key, as specified in Section 8.8. However, if a correspondent node loses its Binding Cache entry for the mobile node (i.e., care-of address information), it is not as serious a problem as if the home agent were to lose its Binding Cache entry. In the former case, the correspondent node would revert to sending packets to the mobile node by way of the home network, but in the latter case the mobile node would effectively lose its main connection to the rest of the Internet.

7.7.5 Pad Options

There are two pad options, which have to be treated in a slightly exceptional way compared to other Mobility Options. These options are used to fulfill alignment requirements and do not themselves impose any additional alignment requirement.

The first pad option is called *Pad1* because it enables the insertion of a single octet of padding. It has the following format:

```
 0 1 2 3 4 5 6 7
+-+-+-+-+-+-+-+-+
|   Type = 0    |
+-+-+-+-+-+-+-+-+
```

Fig. 7.12 Pad1 Mobility Option Format

The Pad1 option has no Option Length field, and no Option Data. The single byte Option Type field is itself the padding. When N copies of the Pad0 option are used, the effect is the same as simply zero fill for N bytes.

The other pad option is called *PadN* because it enables the insertion of any number up to 257 of octets of padding. It has the following format:

For the PadN option, the Option Length is shown as the length of the Option Data field, if present. Otherwise, it is shown as zero since, for the PadN option, the Option Type and Option Length fields themselves both provide the function of padding. The amount of padding, then, is two more bytes than the value of the Option Length field.

Padding can be supplied by using either the PadN option or (for $N > 1$), N occurrences of the Pad1 option. When the PadN option is used, the receiver MUST ignore

```
 0                   1                   2                   3
 0 1 2 3 4 5 6 7 8 9 0 1 2 3 4 5 6 7 8 9 0 1 2 3 4 5 6 7 8 9 0 1
+-+-+-+-+-+-+-+-+-+-+-+-+-+-+-+-+-+- - - - - - - - -
|   Type = 1    | Option Length | Option Data
+-+-+-+-+-+-+-+-+-+-+-+-+-+-+-+-+-+- - - - - - - - -
```

Fig. 7.13 PadN Mobility Option Format

the contents of the Option Data field, but no one has ever explained a way to enforce this rule.

7.8 SUMMARY

In this chapter, we have explored in great detail the operations and protocol messages corresponding to the Binding Cache management. Since Mobile IP introduces the concept of host-triggered route management, it is quite important to understand the protocol machinery and principles behind it. As we saw, the Binding Update and Binding Acknowledgment messages form the basis of such route management and contain the necessary elements to securely effect and manage the process. We did not go into the details of how the security association between the mobile node and the home agent, as well as between the mobile node and its correspondent, are established. We did, however, cover all aspects of the Mobility Header and the Mobility Options, and how the nodes have to process them. We also described the Home Address destination option used by the mobile node to provide its home address to the recipient.

REFERENCES

1. A. Conta and S. Deering. "Internet Control Message Protocol (ICMPv6) for the Internet Protocol Version 6 (IPv6) Specification," RFC 2463, Internet Engineering Task Force, December 1998.

2. A. Conta and S. Deering. "Generic Packet Tunneling in IPv6 Specification," RFC 2473, Internet Engineering Task Force, December 1998.

3. S. Deering and R. Hinden. "Internet Protocol, Version 6 (IPv6) Specification," RFC 2460, Internet Engineering Task Force, December 1998.

4. P. Ferguson and D. Senie. "Network Ingress Filtering: Defeating Denial of Service Attacks Which Employ IP Source Address Spoofing," RFC 2267, Internet Engineering Task Force, January 1998.

5. S. Kent and R. Atkinson. "IP Authentication Header," RFC 2402, Internet Engineering Task Force, November 1998.

6. S. Kent and R. Atkinson. "IP Encapsulating Security Payload," RFC 2406, Internet Engineering Task Force, November 1998.

7. T. Narten and R. Draves. "Privacy Extensions for Stateless Address Autoconfiguration in IPv6," RFC 3041, Internet Engineering Task Force, January 2001.

8. T. Narten, E. Nordmark, and W. Simpson. "Neighbor Discovery for IP Version 6 (IPv6)," RFC 2461, Internet Engineering Task Force, December 1998.

9. C. E. Perkins. "Securing Mobile IPv6 Route Optimization Using a Static Shared Key," RFC 4449, Internet Engineering Task Force, June 2006.

10. S. Thomson and T. Narten. "IPv6 Stateless Address Autoconfiguration," RFC 2462, Internet Engineering Task Force, December 1998.

8

Return Routability

We can't solve problems by using the same kind of thinking we used when we created them. – Albert Einstein

—

In this chapter, we describe the Return Routability protocol and the function and protocol operations of the Mobility Header messages used with it. Return Routability refers to the method for establishing that the mobile node is able to receive packets delivered to the IPv6 addresses that it claims to own. This is just about all that can be expected from a protocol like Mobile IPv6 that is modeled as running at the network layer; if Bob and Alice could converse using the Internet's (optimal) routing in spite of mobility, then the mobility protocol has served a very basic need. If packets are routed correctly, then the protocol is considered to be "working" at that level. If the mobile device can receive data at the IPv6 addresses it claims, then routing is working. For scenarios where Return Routability is considered to be too expensive, or where even higher confidence in the security of the Binding Update process is required, a static key configuration process has been devised, as described in Section 8.8.

There is no provision for verifying that the mobile node truly does "own" the IPv6 addresses. In fact, the Internet as such itself does not have an "address ownership" mechanism. This means, for instance, that a device could usurp the use of an IPv6 address for which it has no authorization. Mobile IPv6 Return Routability procedures do not protect against this, and it would be quite inappropriate for them to carry out such operations. Protection against this sort of abuse belongs elsewhere and is not related to the mobility of network devices. Exactly the same sort of abuse could occur if none of the devices involved were ever mobile.

For IP mobility purposes, the core problem may be summarized as follows: how to establish that the mobile node actually owns the home address and is currently at a care-of address using the *existing* Internet infrastructure. The solution to the problem has to be deployable on the current Internet and preferably should not introduce new elements. This is what is outlined in the Return Routability protocol.

The basic idea is to make use of Internet routing infrastructure to verify if an address actually belongs to a node. If a node originates a message using an address in the source IP field of the packet, then it must be able to receive packets at that address. If a node can demonstrate routability of two addresses, then Return Routability allows the node to create a shared secret that *binds* the two addresses in some sense. To do this, the mobile node sends initiation messages in order to establish a direct routing path between itself and the correspondent node. There is no specification that establishes the exact conditions under which a mobile node may determine that such a direct routing path is necessary. We can intuitively see that whenever an application initiates a high-volume or real-time session with some communications partner, improvements in the routing path can bring real benefits. For applications that expect only a few packets to be exchanged, on the other, the route optimization process is likely to be much less beneficial. It is expected that, as experience is gained with mobile devices running such high-volume or real-time applications, appropriate application controls will be engineered. The actual mechanisms are very straightforward in the operating system, but the policies determining the use of the mechanisms may be somewhat less straightforward to discover.

This chapter is organized as follows. First, the overall operation of Return Routability is described. Next, the messages are described that initiate the Return Routability procedure. These are the Test Initiate messages; Home Test Initiate (HoTI) and Care-of Test Initiate (CoTI). They are issued by the mobile node to elicit a response from a target correspondent node. Then the Test messages from the correspondent node are described, namely, the Home Address Test (HoT) and Care-of Address Test (CoT) messages. They are sent in response to the request by the mobile node to initiate the procedure. After this, the method by which the mobile node passes the tests posed by the correspondent node is described in detail. In order to prove its ability to fulfill the test conditions, the mobile node uses the Nonce Indices and the Binding Authorization Data extensions to the Binding Update.

The protocol messages described in the above-mentioned sections are the ones that make up the signaling procedures for Return Routability. All four of them are messages types of the Mobility Header, as described in Chapter 7. After the protocol details are presented, the chapter concludes with some discussion about the security models and threat analysis as perceived by the Mobile IPv6 working group during the design of the Return Routability protocol messages.

8.1 RETURN ROUTABILITY – THEORY OF OPERATION

The design of the Return Routability subprotocol for Mobile IPv6 has followed the following guidelines:

- "Do No Harm."
- Avoid vulnerability to denial-of-service attacks aimed at a network node rightfully owning the the care-of address claimed by the mobile node.
- Avoid vulnerability to stealing traffic meant for the home address, which may actually belong to any node.
- Avoid vulnerability to denial-of-service attacks aimed at causing a high processing burden at the correspondent node.
- In particular, strictly limit the requirement for maintaining an intermediate state at the correspondent node.

It is believed that the mobile node itself does not have a strong need to be protected against any denial-of-service attack, since the route optimization is requested by the mobile node, not the correspondent node. This may be viewed as an additional incentive for the mobile node to initiate the Return Routability protocol whenever it does not already have the appropriate Binding Management security association with the correspondent node. Since the correspondent node does not currently have any way to inform the mobile node that route optimization is preferred for the performance of the applications of interest, it must be up to the mobile node to take the appropriate steps. There is little downside involved if the mobile node usually decides in favor of enabling route optimization.

The overall design of the Return Routability is as follows:

- The mobile node asks the correspondent node to issue the two test messages (one for the home address, one for the care-of address). The request is made by sending the two associated *Test Init* messages. These two messages contain cookies that the mobile node expects the correspondent to return.
- Assuming all is well, the correspondent node responds by sending the Home Test (HoT) message and the Care-of Test (CoT)message, containing the data that will be used to construct a Binding Management key. This is the key that will be used to secure the Binding Update. The correspondent node also echoes the cookies it received from the mobile node.
- The correspondent node also sends some indexing information to the mobile node to allow the correspondent node to make the key calculation using the same input data that the mobile node uses.
- The mobile node has to be addressable at its home address to receive the key material from the HoT message. This key material is called the *Home Keygen Token*.

- Likewise, the mobile node has to be addressable at the care-of address to receive the rest of the key generation material, which is available in the *Care-of Keygen Token* field of the CoT message.

- Once the mobile node acquires the two fragments of the key generation material (i.e., the Home Keygen Token and the Care-of Keygen Token), it performs a hash function over the input data to calculate the Binding Management key. This hash function operates only on data that is available to the correspondent node, including the IPv6 addresses claimed by the mobile node. In this way, the key calculation is made to be dependent on the addresses involved in the protocol. In addition, the mobile node verifies that the correspondent returned the cookies it sent in the Test Init messages to match the corresponding Test messages, as well as confirming that it is actually communicating with the same correspondent which has sent the key fragments.

- Once the key is calculated, it can be used to secure future Binding Updates.

- The mobile node also has to remind the correspondent node how it has stored the key generation materials. This enables the correspondent node to minimize the amount of storage it maintains for Return Routability calculations, and thus reduces vulnerability to certain denial-of-service attacks.

Figure 8.1 shows the messages used in the Return Routability protocol. We also show the Binding Update (BU) and Binding Acknowledgment (BAck) sequence which usually follows the Return Routability exchange. Note that the HoTI and CoTI are likely to be issued by the mobile node at practically the same time and that the order of transmission is not important. Also, it is not important in which order the correspondent node transmits the matching HoT and CoT messages.

Once the keygen tokens are obtained, the following simple calculation is used to obtain the Binding Management key K_{bm} :

$$K_{bm} = \text{SHA1 (Home Keygen Token | Care-of Keygen Token)}$$

SHA1 is the well-known hash algorithm [6] that is used in many IETF specifications when irreversible message digests are needed. The Home Keygen Token and Care-of Keygen Token are computed using similarly simple formulas using the material present in HoT and CoT messages; see Section 8.4 and Section 8.5 for the keygen token calculations. K_{bm} is then used for the necessary authentication steps for the binding cache mechanisms. The messages in this chapter are all intended to enable the exchange of these tokens so that K_{bm} can be securely established.

8.2 HOME TEST INIT (HOTI) MESSAGE

A mobile node sends the HoTI to request that a correspondent node initiate its test to make sure that the mobile node is reachable at the home address. This message

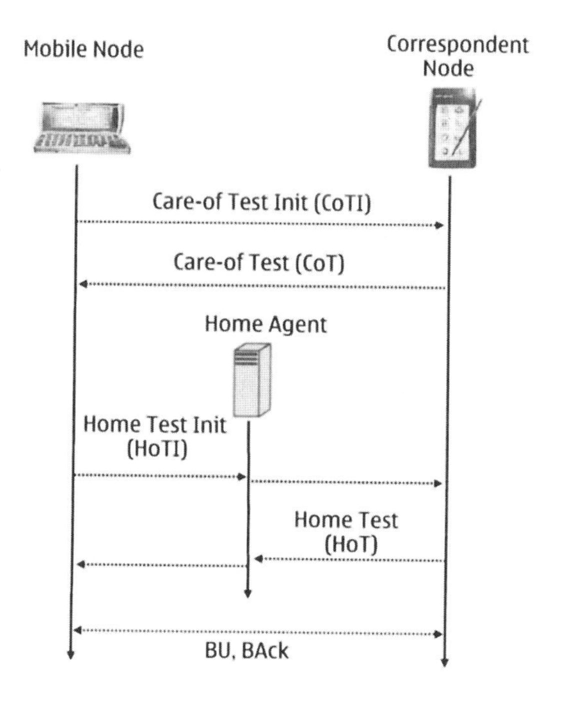

Fig. 8.1 Return Routability Protocol

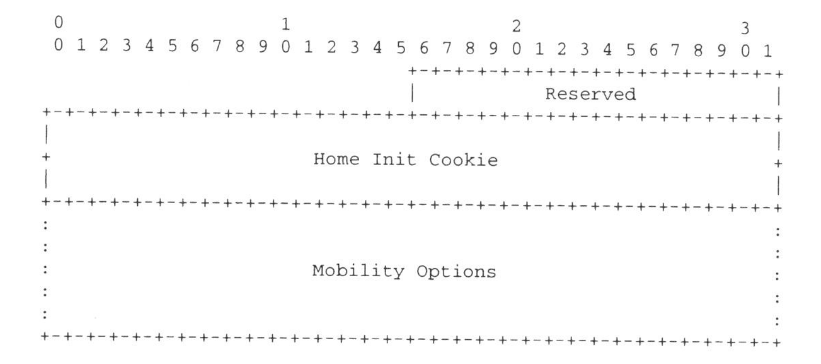

Fig. 8.2 Home Test Init Message Format

is identified by the value 1 in the Type field of the Mobility Header. The format of HoTI is illustrated in figure 8.2.

The *Home Init Cookie* is a random number 64 bits long. This is the only important field. The idea is that the mobile node is supplying something to the correspondent node that will be returned in the follow-up message (namely, the Home Test message, see Section 8.4). The mobile node has to reverse tunnel the Home Init message back through its home agent, because there may not be any current security association between the mobile node and the correspondent node which would enable the use of the Home Address destination option. This requirement makes it much more difficult to establish the security requirement with the correspondent node when for some reason the home network is dysfunctional or not reachable.

Although there may be some Mobility Options in the future, right now there are none defined for the HoTI message. As usual, the Reserved field (here, 16 bits long) has to be initialized as zero and ignored upon reception.

8.3 CARE-OF TEST INIT (COTI) MESSAGE

A mobile node sends the CoTI at about the same time that it sends the HoTI message. The CoTI message requests that a correspondent node initiate the similar test to make sure that the mobile node is reachable at the home address (i.e., as well as at its home address). This message is identified by the value 2 in the Type field of the Mobility Header. The format of the CoTI message is illustrated in Figure 8.3. It is identical in format to the HoTI message described in the previous section and serves an almost identical function.

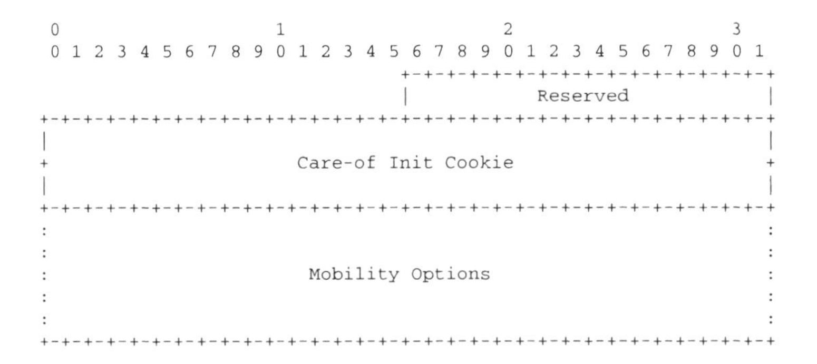

Fig. 8.3 Care-of Test Init Message Format

The *Care-of Init Cookie* is another random number 64 bits long. As in the case of the HoTI message, the cookie field is the only important field. This cookie is intended to be returned in the associated field of the Care-of Test message (see Section 8.5).

As before, there aren't any Mobility Options defined for the CoTI message, although there may be some in the future. Again, the Reserved field (16 bits long again) has to be initialized as zero and ignored upon reception. In contrast to the HoTI message, the CoTI message is required to be sent from the mobile node's care-of address and must not be reverse tunneled to the home agent.

Next, the corresponding Test messages will be described. They are a little more interesting.

8.4 HOME TEST (HOT) MESSAGE

The HoT message is sent from the correspondent node after it receives the HoTI from a mobile node (see Section 8.2). The HoT message is identified by the value 3 in the Type field of the Mobility Header. The format of the HoT message is illustrated in Figure 8.4.

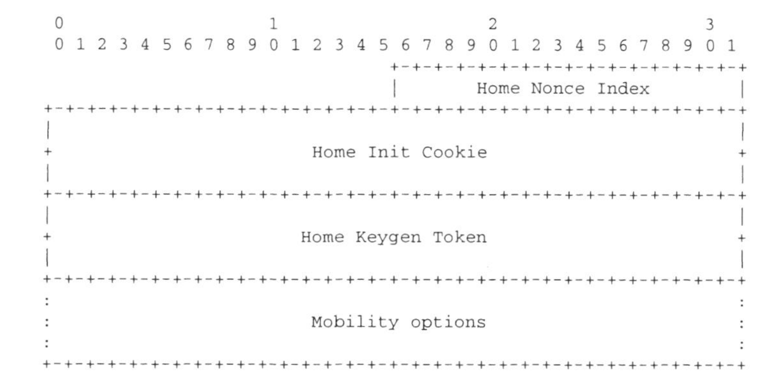

Fig. 8.4 Home Test Message Format

When the correspondent node sends the HoT message, it is offering the mobile node the chance to prove itself. The home address test requires the mobile node to verify that it has received a packet at its claimed home address. If the mobile node does own its home address, and if it receives the data in the HoT message, it will have half of the data it needs to calculate the cryptographic key it will need for sending Binding Update messages to that correspondent node. The other half of the data will be supplied by the correspondent node in another message, namely, the CoT message which is described next.

In order to understand the significance of the data fields for the HoT message, recall that the binding key K_{bm} is built using the Home Keygen Token and the Care-of Keygen Token. K_{bm} is never disclosed in protocol messages, and when used properly it will provide unforgeable proof that the mobile node has received the two tokens. The tokens are constructed in a way that depends on the values of the IP addresses and other parameters involved with the optimization transaction. That way, tokens that are meant to secure the binding operations have values that depend on the IP addresses involved, and will not be the same if other (or wrong) IP addresses are used by the correspondent node or the mobile node.

For the Home Keygen Token, the important values are the home address, a random nonce called a "node key" picked by the correspondent node for the computation, and the index that the correspondent node has to use to retrieve the value again. The node key is denoted K_{cn} (Kcn in RFC 3775). The Home Init Cookie value serves to help the mobile node match up an incoming HoT message with the previously transmitted HoTI message, and to thwart any spoofing by parties who have not seen the Init message.

Here is a short description of the fields of the HoT message:

- *Home Nonce Index*: The Home Nonce Index is included so that the mobile node can provide it to the correspondent node in Binding Update messages. The correspondent node needs it to retrieve the Home Keygen Token so that it can compute K_{bm}.
- *Home Init Cookie*: This 64-bit field should contain a value which is the same as the Home Init Cookie in the HoTI message.
- *Home Keygen Token*: This 64-bit field is the main point of the HoT message, and is used to compute K_{bm}. The calculation is described below.
- *Mobility Options*: As of this writing, Mobile IPv6 does not define any mobility options valid for the HoT message.

In order to compute the Home Keygen Token, the correspondent node needs the following values:

- K_{cn}: The node key.
- nonce: The correspondent node should generate new nonces at regular intervals and keep track of them in an indexed table.
- home address: The mobile node's home address

The Home Keygen Token is then taken from the first 64 bits of the following computed value:

HMAC_SHA1 (K_{cn}, (home address | nonce | 0)))

where HMAC_SHA1 is the cryptographic hash function based on SHA-1, as described in RFC 2104 [3]. The trailing 0 is an 8-bit value used to make sure that the result is different from the corresponding calculation for the care-of address.

In the next section, the analogous CoT message will be described.

8.5 CARE-OF TEST (COT) MESSAGE

The CoT message is sent from the correspondent node after it receives the CoTI from a mobile node (see Section 8.3). The CoT message is identified by the value 4 in the Type field of the Mobility Header. The format of the CoT message is illustrated in Figure 8.5.

The operational ideas behind the use of the CoT message are very much like those for the HoT message, challenging the mobile node to verify that it has received a

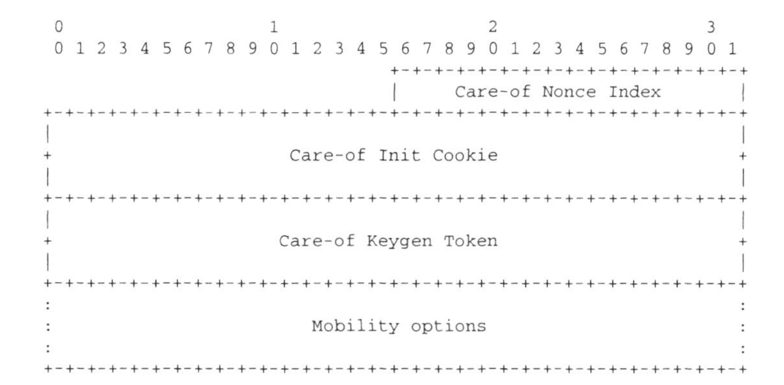

Fig. 8.5 Care-of Test Message Format

packet at the its claimed care-of address. If the mobile node does own its care-of address, and if it receives the data in the CoT message, it will have half of the data it needs to calculate the cryptographic key it will need for sending Binding Update messages to that correspondent node. The other half of the data will be supplied by the correspondent node in another message, namely, the HoT message which is described in Section 8.4.

For the Care-of Keygen Token, the important values are the care-of address, the random nonce (i.e., the Care-of Keygen Cookie) picked by the correspondent node for the computation, and the index that the correspondent node has to use to retrieve the value again. The Care-of Init Cookie value serves to help the mobile node match up an incoming CoT message with the previously transmitted CoTI message as well as helps thwart any spoofing.

Here is a short description of the fields of the CoT message:

- *Care-of Nonce Index*: This is included so that the mobile node can provide it to the correspondent node in Binding Update messages. The correspondent node needs it to retrieve the Care-of Keygen Token so that it can compute K_{bm}.

- *Care-of Init Cookie*: This 64-bit field should contain a value which is the same as the Care-of Init Cookie in the CoTI message.

- *Care-of Keygen Token*: This 64-bit field is the main point of the CoT message, and is used to compute K_{bm}.

- *Mobility Options*: As of this writing, Mobile IPv6 does not define any mobility options valid for the CoT message.

In order to compute the Care-of Keygen Token, the correspondent node needs K_{cn}, the care-of address, and, as before, a nonce that can be retrieved from an indexed table. The Care-of Keygen Token is then taken from the first 64 bits of the following computed value:

HMAC_SHA1 (K_{cn}, (care-of address | nonce | 1)))

The trailing (i.e., least significant) 1 is an 8-bit value (0x01) used to make sure that the result is distinguishable from the corresponding calculation for the mobile node's home address shown in the previous section.

8.6 USING THE BINDING MANAGEMENT KEY

Once the mobile node has collected the keygen tokens, it creates K_{bm} using the tokens according to the previously shown formula:

K_{bm} = SHA1 (Home Keygen Token | Care-of Keygen Token)

SHA1 returns a 160-bit result, and has up to the time of this writing withstood the rigors of cryptographic attacks well enough to provide the desired level of security.

Once K_{bm} is available, it can be used to secure subsequent Binding Update messages from the mobile node to the correspondent node. This is done by using the Binding Authentication Data suboption to the Binding Update message (see Section 7.3). The calculation for the authentication data (AuthData) is as follows:

HMAC_SHA1 (K_{bm}, (care-of address | correspondent | BU))

where BU is the contents of the Binding Update message up to but not including the Binding Authorization Data option header. The "correspondent" is the IP address of the correspondent node. AuthData is taken as the first 96 bits of the result of this calculation. The Binding Update to the correspondent node naturally includes the following data:

- sequence number
- home address: either in the IPv6 Source Address field or, if different, within the Home Address destination option (Section 7.1)
- home nonce index: provided in the Nonce Indices option (Section 7.7.2)
- care-of nonce index: provided in the Nonce Indices option

Since these values are part of the input data for the authenticator calculation (the "BU" part), they cannot be replaced by an active attacker without detection.

Unfortunately, RFC 3775 specifies terribly short values for the lifetime of the security association based on K_{bm}. Until this is improved, the procedure has to be reiterated every few minutes (controlled by MAX_TOKEN_LIFETIME, currently 240 seconds). An alternative is to use the Preconfigured Key mechanism described in Section 8.8.

When the correspondent node receives a Binding Update, it may send a Binding Acknowledgement to the mobile node in response (this is required if the 'A' bit is set in the Binding Update). K_{bm} is used for generating AuthData for the Binding Authorization Data option, in practically the same way as for the Binding Update. The calculation for the authentication data (AuthData) is as follows:

HMAC_SHA1 (K_{bm}, (care-of address | correspondent | BA))

where BA is the contents of the BAck message up to but not including the Binding Authorization Data option header.

8.7 THREAT MODELS TO WORRY ABOUT

During the evolution of the Mobile IPv6 specification, as noted previously, there was a demand for the creation and development of a security mechanism deployable to a future, immensely large-scale Internet. Along with this demand, there was a pervasive atmosphere of distrust in the ability of the Mobile IP working group (or, indeed, any other working group) to adequately grasp and overcome the threats of the future. During this time, many observers believed that if the Internet itself had been subjected to the same level of scrutiny during its early formative years, we would never have had an Internet. As it is, given the delays in the development of the Mobile IP specifications, we may find that true mobility will be long delayed, maybe even by many years, as far as most people are concerned. Of course, this book is written with the optimistic expectation that such a long delay will not come to pass.

In this section, we will summarize some of the worries and threat models that occupied the attention of the working group during the time when the route optimization feature was developed to enable use by all nodes within the future immensely mobile Internet.

One of the common (even overriding) concerns during the Mobile IP security discussions had to do with distributed attack scenarios. It is well known that in today's Internet, many computers are controlled by nefarious, shadowy characters. These computers can be programmed or directed to carry out actions remotely and sometimes synchronized to produce very serious effects. For instance, if a million computers were all programmed to send a gigabyte of data to your enterprise servers at the same time, over and over again, most likely your enterprise would suddenly find itself completely unable to communicate over the Internet for the duration of the attack. This so-called *distributed denial-of-service* (DDOS) attack has happened more than once and is a very serious problem in today's Internet. We were tasked to try very hard to avoid creating any new vulnerabilities to such attacks.

8.7.1 Hijacking Data

Most of the initial worries about the security of Mobile IP have centered on the potential for hijacking data which should be delivered to a mobile node when it is attached to an access network other than its home network. An obvious mode of attack would be for a malicious node to supply a spurious Binding Update message to the home agent which included an IP address of the malicious node (or some node controlled by the malicious node) as the care-of address.

This attack is rendered infeasible by the requirement for inclusion of unforgeable authentication along with the care-of address, such that the authentication data is strongly dependent on the contents of the Binding Update message (especially the care-of address). Unless the authentication data has this strong dependence, a malicious node could modify various parts of the Binding Update message without the knowledge of the transmitter or receiver. The receiver would not notice, because the authentication data would (by assumption) be the same regardless of what value were present in those parts of the Binding Update message. As it is, to be safe, the Mobile IPv6 specification requires the authentication data to have the stated strong dependence on *all* of the message data, including the IPv6 header, so that none of it can be changed while the packet is in flight.

Since the Binding Update message also includes a Sequence Number, the message data is also protected against replay attacks. Imagine for a moment that an attacker were to store some valid Binding Update messages, and then wait until the mobile node moves to a new location. Shortly after the mobile node moves away, the malicious node could transmit one of its stored Binding Updates in an attempt to divert data from the mobile node's new care-of address to a previous care-of address. If the malicious node were also able to receive the data at the old care-of address, then that node could quite conceivably continue to masquerade as the mobile node for quite a while until the care-of address timed out. It is even conceivable that the malicious node could continuously replay the old Binding Update so that traffic for the mobile node would be continuously diverted to the old care-of address.

All of these dangers are practically eliminated by the inclusion of the Sequence Number in the Binding Update. Since the Sequence Number always has to be different for each Binding Update, and since that will produce a different value for the authentication data, no previous Binding Update will be accepted again by the receiver.

The only conceivable dangers could come from either weakness in the hash function that produces the authentication data or rollover of the Sequence Number field of the Binding Update. In other words, if the mobile node sends 65,536 Binding Updates to a particular correspondent node, the next Binding Update will use the same Sequence Number as one of the previous Binding Updates.

In order for the rollover event to represent any kind of threat to the mobile node or correspondent node, the attacker would have to wait a long time, because it takes a long time for the mobile node to move that many places. This benefit is decreased somewhat by the requirement for the mobile node to reestablish its bindings at each correspondent node every 7 minutes (i.e., MAX_RR_BINDING_LIFETIME parameter in [1]). Even so, this means that rollover problems require a malicious attacker to store information for a long time, and by that time K_{bm} will be invalid anyway, so there is no real threat. For cases where K_{bm} has a longer lifetime (see Section 8.8), more care has to be taken, but in those cases the binding lifetime can be longer anyway, so there is still no real likelihood of danger.

8.7.2 Address Ownership

Many of the vulnerabilities investigated during the evolution of Mobile IPv6 have to do with address ownership. A control message containing an IP address of some node can usually be construed in some way that implies that the node sending the control message has the authority to effect some change of state at the node identified by the IP address. Whenever this is not true, the control message should be associated with some security feature to ensure that this authority is not maliciously usurped. For the Mobile IP care-of address, the security features are derived from cryptographic properties of hash functions, and often (for messages between the mobile node and the home agent) encryption functions. For other addresses, or for care-of address information in use before the establishment of a security association, abuses could still be possible.

For instance, suppose a malicious node attempts to mount an attack by masquerading as a mobile node at a new care-of address, and the correspondent node does not have a security association with the mobile node (or the malicious node). Further, suppose that the purpose of the malicious node is to bombard an unsuspecting victim with a lot of unnecessary, irrelevant traffic (e.g., a DDOS, as described above). It could proceed as follows.

Each attacking node could supply a correspondent node with a CoTI and a HoTI message, with the care-of address set to be that of the intended victim. If a huge number of Care-of Test Init messages were sent to various correspondent nodes, each correspondent node would add to the pain of the victim by sending to the victim an unwanted CoT message.

This attack is really not very much different from having the individual attack nodes send unwanted data to the victim, except that in this case the victim will initially consider that the correspondent nodes are the attackers instead of the true perpetrators of the mischief. On the other hand, the control message to the correspondent node could have been an ICMP "ping" message with the victim's IP address as the putative Source IP address, achieving much the same effect.

Other scenarios center around inserting false data into the Home Address destination option. This could cause the correspondent node to make incorrect assumptions about the source of a packet, so that the destination option itself would be specified to be invalid for use until a binding has been established at the correspondent node. This, of course, often requires the establishment of K_{bm} and the overhead of the Return Routability checks. Avoiding these overheads and enabling the use of the Home Address destination option will likely constitute an important motivation for exploitation of the preconfigured keys alternative [5].

8.7.3 Use of the Home Address Option

Another vulnerability often mentioned is the so-called *reflection attack*. Essentially it involves a malicious node using the home address in route-optimized communication without proper authorization. For instance, such a node might use an unsuspecting node's address as the home address in the destination option and elicit one or more correspondents to send unwanted traffic to the victim. If the correspondent has no binding for the address in the destination option, the response would go directly to the unsuspecting node's IP address. Ingress filtering does not catch this since the source

IP address may be topologically consistent. As can be seen, this attack is similar to what we have seen above.

In order to address the reflection attacks, the correspondent is required to process the Home Address destination option only if it has a Binding Cache entry for the address used in the destination option. This Binding Cache is typically created by means of Return Routability or by means of preconfiguration. In either case, the correspondent has some proof that the home address actually belongs to the node that is using the destination option.

Previous versions of the Mobile IPv6 specification allowed the use of triangular routing, in which a mobile node could send traffic directly to the correspondent using the Home Address option, but the correspondent had the option of simply replying back to the home address. This allowed a compromise between reverse-tunneling and route optimization without having to perform any additional operations. However, the current specification effectively removes this option due to the perceived vulnerability to reflection attacks.

8.7.4 Use of Routing Header Type 2

The IPv6 routing header has vulnerabilities of its own. For instance, the node that processes the routing header could be used as a reflector to send unwanted packets to addresses in the routing header. This is generally not a problem when the processing nodes are routers. It becomes a problem with endnodes even with firewalls which typically enforce rules based on IP addresses in the source and destination fields of the IP header. Since route optimization involves the use of routing header with endnodes, there is a possibility of reflection attacks.

The routing header in Mobile IP uses a far more restrictive type than the IPv6 routing header. Basically, the only address used in the routing header has to be the address used by the receiving node. That is, the home address in the routing header must belong to the mobile node currently at some care-of address. In this way, no other node can effectively receive this packet.

8.8 SECURING ROUTE OPTIMIZATION USING A STATIC SHARED KEY

As we discussed at length in the foregoing sections, periodic Return Routability test is used to verify both the right of the mobile node to a use a specific home address and the validity of the claimed care-of address. That mechanism requires no configuration and no trusted entities beyond the mobile node's home agent.

For years before the final protocol document was approved by the IETF, the Mobile IPv6 protocol also enabled the use of statically configured keys for Binding Updates. Such keys are very naturally used with the protocol and can be configured in about the same way that has been in use for quite a few years with Mobile IPv4 [4]. These statically configured keys require no additional signaling to be established between a mobile node and a correspondent node, and thus offer immediate route optimization without extra setup latency and signaling overhead. Unfortunately, very late in the standardization process for Mobile IPv6, there was a demand from some members for the removal of this feature. In order to protect the widest possible consensus, the removal was agreed to with the understanding that a future specification would repair the lack.

Now, just such an alternative, low-latency security mechanism for protecting binding management messages (e.g., signaling related to route optimization) in Mobile IPv6 has been created and published as RFC 4449 [5]. It requires configuring a shared secret between the mobile node and the correspondent node, and as a result, the Return Routability tests can be avoided. It can also provide stronger assurance of the home address since it is assumed that the party performing preconfiguration would presumably attest to that address. However, there are obvious limitations in terms of scale and the trust placed in a mobile node not to misbehave due to the absence of a care-of address test. Given these considerations, the following is a scenario where preconfiguration can be applied:

Consider domains where both the mobile node and the correspondent node share the same trust, for instance, enterprises where a salesperson Bob on the road is talking to engineer Alice at a branch office. In a scenario such as this (and possibly similar ones), the correspondent node has a good reason to trust the mobile node. Specifically, it has a high degree of confidence that the mobile node will not launch flooding attacks against a third party as described in [2]. In addition, the enterprise must ensure that sufficiently long keys are used in configuration.

The actual mechanism for using the preconfigured key is very similar to the mechanism described above using Return Routability with the following considerations. In order to precompute the binding management key K_{bm}, the two nodes must share the key K_{cn} (see Section 8.4). This K_{cn} is used to generate the Home and Care-of Keygen Tokens. For each keygen token, there has to be a nonce to use. Since these nonces are preconfigured, the Nonce Indices (described in Section 7.7.2) are not present in the Binding Update message. There is an additional consideration regarding the use of Sequence Number for replay protection: the correspondent node must maintain the recently used value of Sequence Number in some stable storage. Otherwise, a malicious node could replay an old Binding Update binding the home address to an old care-of address. Other than that, the computation of keygen tokens using K_{cn} and K_{bm} itself is the same as it is for Return Routability which we described in the previous sections [5].

8.9 SUMMARY

We have discussed in detail the protocol, messages and underlying threats that guided the design of the Return Routability protocol. Perhaps it is important to understand the delicate nature of incorporating security into Internet protocol design. For purists, there can never be enough of it, while others may point out the demarcation between perception and reality. Pragmatic design has to balance these two situations and arrive at a (usually compromised) result which can be *deployed*. Since many factors are involved, especially in the current phase of the Internet, in deploying a fundamental functionality such as mobility, the task becomes harder.

While Return Routability is not perfect, it reflects the reality of protocol design that has a chance to be deployed. Even though it is considered a weak form of authentication, its principle of attributing address ownership based on the routability of the address and then allowing binding between addresses is good enough, all things considered on the current Internet. Yet, we do not have a good understanding of the lifetimes used in the protocol. For instance, the maximum lifetime for a home address ownership is 240 seconds. There is no strong empirical evidence why this has to be the case or why it could not be changed. Future research and deployment experience could help us arrive at a better understanding of such issues.

In this short description of preconfiguring keys, we have recapitulated the main points about how to preconfigure certain shared data between a mobile node and a correspondent node. The shared data is useful for creating a shared binding management key K_{bm}, but it can only be used when the care-of address supplied by the mobile node is not endangered. It would be possible to augment the procedures defined in this chapter with a minimal variation on the CoT. We may also expect to see new methods for sharing the preconfiguration data in order to widen the applicability of this technique for reducing the amount of data needed for mobility management between mobile nodes and correspondent nodes.

REFERENCES

1. D. Johnson, C. Perkins, and J. Arkko, "Mobility Support in IPv6," RFC 3775, Internet Engineering Task Force, 2004.

2. P. Nikander, J. Arkko, T. Aura, G. Montenegro and E. Nordmark. "Mobile IP Version 6 Route Optimization Security Design Background", RFC 4225, Internet Engineering Task Force, December 2005.

3. Krawczyk, H., Bellare, M. and R. Canetti. "HMAC: Keyed-Hashing for Message Authentication," RFC 2104, February 1997.

4. C. E. Perkins (Editor). "IP Mobility Support for IPv4," RFC 3344, Internet Engineering Task Force, August 2002.

5. C. E. Perkins. "Securing Mobile IPv6 Route Optimization Using a Static Shared Key," RFC 4449, Internet Engineering Task Force, June 2006.

6. National Institute of Standards and Technology, "Secure Hash Standard," FIPS PUB 180-1, April 1995, http://www.itl.nist.gov/fipspubs/fip180-1.htm

9

IP Security for Mobile Nodes and their Home Agents

The culture of the original Internet was one of trust. – Leonard Kleinrock

9.1 INTRODUCTION

In this chapter, we discuss how a mobile node and its home agent establish and maintain an IPsec security association. There are two key security problems in mobile node and HA interaction. First, a mobile node must be able to establish a security association with its home agent. The second problem concerns how IPsec itself is used to secure various types of traffic between the two nodes. The first problem is addressed by manual or dynamic keying. The second problem is addressed by means of specifying how to construct packets using both IPsec and Mobile IP. We address each of these separately.

As we explained earlier, the Mobile IP working group was chartered to specify the details of both key establishment and securing the messages used for effecting Binding Cache management. In the previous chapter, we discussed the Return Routability protocol for establishing the security association between a mobile node and an arbitrary correspondent node on the Internet for route-optimized communication. It turns out that IPsec can be assumed for the purpose of both key exchange and securing the binding updates from a mobile node to its home agent. This is primarily facilitated by the presumed trust model; it is anticipated that any Mobile IP deployment will ensure provisioning of necessary credentials as well as configuration of parameters necessary for the trust model to work in practice.

A companion protocol specifies the nits and details of how to use IPsec with Mobile IP. The interaction between these two protocols is "involved enough" that a separate specification was deemed necessary. This chapter illustrates the specification in [2], which is needed for implementing IPsec together with Mobile IPv6. We use some acronyms (such as CoA, HA, and HoA) defined earlier in the book.

9.2 ESTABLISHING A SECURITY ASSOCIATION BETWEEN A MOBILE NODE AND ITS HOME AGENT

As we saw in Chapter 3, an SA determines IPsec processing of inbound and outbound packets. So, a home agent needs to populate its Security Association Database (SAD) and Security Policy Database (SPD) with entries that determine processing rules for Mobile IPv6 signaling messages as well as for data traffic itself. These rules can be entered either manually or can be established by a keying protocol such as IKE.

An example of such a rule in the SPD database for outgoing packets at a mobile node is as follows:

```
IF src IP addr == HoA &&
   dst IP addr == HA  &&
   Protocol    == Mobility Header (MH),
Then Use SA1
```

For this rule, the entries in the SAD are as follows:

```
Selectors: HoA, HA, MH
Treatment: ESP in Transport mode, SPI-1, HA
```

The treatment above indicates that ESP should be applied in transport mode for all outbound packets that match the rule in SPD. The exact cryptographic algorithms and keys are indexed by SPI-1.

When such an SA establishment is performed by IKE, it is important to use the correct identities in each of the two phases. For instance, the identity of the mobile node in phase 1 cannot be the home address when preshared keys are used in Main mode because IKE uses the IP address of the exchange (i.e., the care-of address) as the identifier. The home address cannot be used as the identity in other modes as well if the home address is dynamically assigned or the secret between the mobile node and the home agent is provisioned based on some other identifier. For instance, the Network Access Identifier (NAI) [1] or the Fully Qualified Domain Name (FQDN), such as "user.home.net.", of the user may be used as an identity for phase 1. The

IKE phase 2 identifier has to be the home address since all the IPsec SAs will be keyed using home address as one of the fixed addresses. Using care-of address as the identity will not work since the home agent may not be able to determine who the mobile node is even if it presented a shared secret (i.e., the secret has to match a fixed identity).

Once an IPsec SA is established by means of IKE, Binding Updates can be sent using IPsec. When a mobile node moves, however, the existing SA at the home agent will have the outer tunnel IP address pointing to the mobile node's previous care-of address. The only way to update this is by rerunning IKE phase 2. Mobile IPv6 provides a method to update just the outer tunnel IP address of the existing SA by including a 'K' bit in the Binding Update. If the Binding Update processing is successful, then the home agent updates the tunnel outer IP address with the mobile node's current care-of address.

9.3 BINDING UPDATE AND IPSEC PROCESSING AT A MOBILE NODE

In this section, we illustrate how a packet flows from the Mobile IP module through IPsec and eventually into the device driver. This is a logical illustration which can be implemented in multiple ways. We use a Binding Update message from a mobile node to the home agent as an example packet.

Assume that a mobile node visits a new network, obtains a care-of address and then forms a Binding Update message. See Figure 9.1. This packet has home address as source IP address. Why? Because whether the packet should use reverse tunneling or be sent with Home Address option has not yet been decided at the time of constructing this packet. A mobile node's local policy settings determine this. In any case, the packet must contain a Mobility Header with the Binding Update message. It is also worthwhile to observe the Alternate CoA option. See step 1 in Figure 9.1. Regardless of whether reverse-tunneling is used or route optimization is used, the eventual packet must contain the care-of address as the source IP address. So, why is it necessary to include the Alternate CoA option? We will come back to this later.

This packet is passed to the IPsec module. There is no Home Address option present, again because no decision has yet been made about how to transmit the packet. Since a particular implementation can choose either one, the IPsec selectors cannot assume the presence of the Home Address option. Hence, the SPD entries only include the source IP address (HoA), the destination IP address (Home Agent) and the protocol type (Mobility Header). The packet matches the selectors in the database, but the IPsec module "recognizes" that the packet may not be fully formed, since either reverse tunneling or route optimization would add additional headers. This kind of recognition, which is specific to Mobile IP, can be done based on the protocol *type* value (i.e., Mobility Header Type for Binding Update). So, the IPsec module passes the packet back to the Mobile IP module. See step 2 in Figure 9.1.

The Mobile IP module determines, based on local policy, that route optimization is used. Hence, a destination option is added. But the address used in the destination option is the CoA. See step 3 in Figure 9.1. The reason for this can be understood by considering how a destination of this Binding Update actually processes this packet when it is received. IPv6 extension header processing requires that the destination option be processed before IPsec AH or ESP processing takes place. This means the destination (i.e., the home agent) will first process the destination option which will swap the home address in the destination option with the care-of address in the source IP address. So, by the time the packet is processed by the IPsec module on the home agent, the destination option would contain the care-of address and the source IP address would contain the home address. Hence, when the Mobile IP module on the mobile node adds the destination option prior to IPsec processing, it includes the care-of address.

The Mobile IP module passes the packet back to the IPsec module to add the ESP header. In step 2, we have already identified the SPI so that only an application of the ESP transform using the appropriate key and the algorithm needs to be done. The resulting packet now contains an ESP header (in transport mode) immediately after the destination option. See step 4 in Figure 9.1. At this point, the Binding Update is protected using the established SA, but the values in the destination option and the source IP address fields must be exchanged. Hence, the packet is passed back to the Mobile IP module, which swaps the two addresses so that the destination option now contains the home address and the source IP address is set to care-of address. This step, not shown in Figure 9.1, ensures that the packet is transmitted with a topologically correct source IP address. It also ensures that correct addresses are used when the destination processes the destination option.

Earlier, we deferred the reason for including the Alternate CoA in the Binding Update. An attacker could tamper with the care-of address present in the source IP address of the Binding Update. For instance, an attacker could try to redirect the mobile node's traffic to some other node (whose care-of address appears in the source IP address). With ESP, there is no protection for the fields in the IP header; they cannot be protected with an authentication field. [1] When an attacker succeeds in redirecting traffic, the traffic meant for the mobile node will be tunneled to an unsuspecting node when the home agent creates a Binding Cache Entry. There is a solution available for this problem in Mobile IP. Since Alternate CoA is encrypted by ESP, it cannot be tampered with. Hence, including Alternate CoA protects the Binding Update denial-of-service attacks generated by tampering with the source IP address.

9.4 BINDING UPDATE AND IPSEC PROCESSING AT A HOME AGENT

When the Binding Update packet arrives, the IPv6 module passes the packet for processing the Home Address destination option. See Step 1 in Figure 9.2. The Mobile IPv6 module swaps the address in the destination option with the source IP address. Since the next header in the destination option is ESP, the packet is passed on

[1]The AH provides such a protection for the IP header fields.

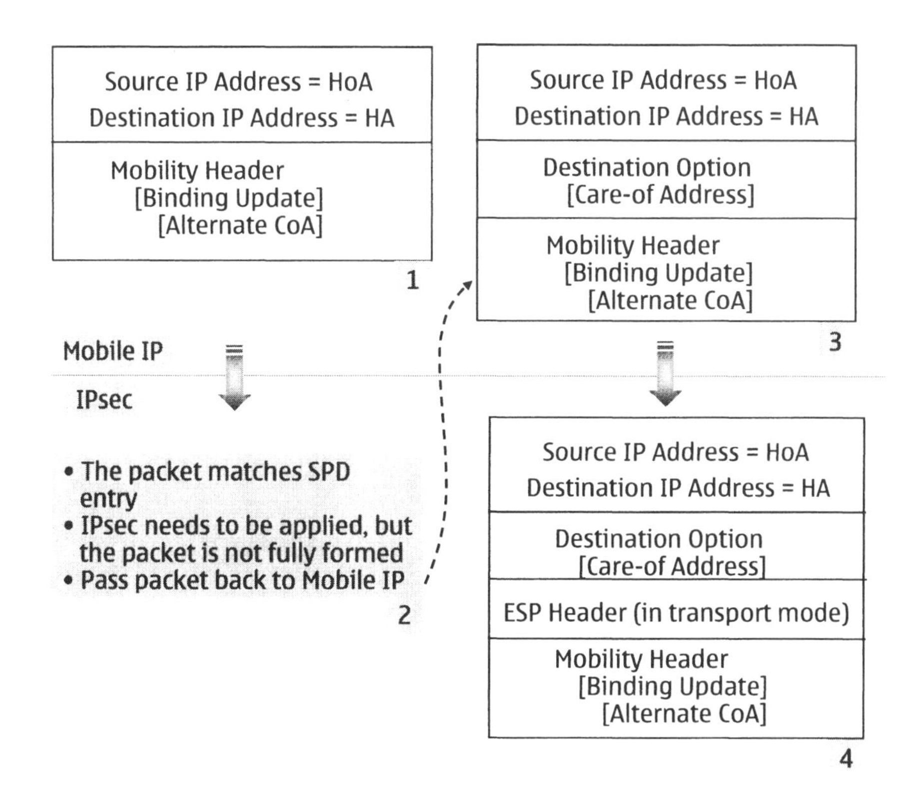

Fig. 9.1 Outbound Binding Update Packet at a Mobile Node

to the IPsec module. Using the SPI and the destination IP address, the ESP processing takes place, which decrypts the contents of the Mobility Header. At this point (step 3 in Figure 9.2), the selectors match the values in the SA maintained at the home agent. Since the client of IPsec here is the Mobility Header, the packet is passed back to the Mobile IPv6 module. The Binding Update is then processed. The IP address present in the Alternate CoA option is used as the care-of address when the Binding Cache Entry is created. In addition, if the 'K' bit is set, the home agent also updates the tunnel mode SAs with the new care-of address, so that a packet like HoT can be sent with protection to the mobile node's new location.

Let us discuss the purpose of the 'K' bit in the Binding Update a little further. The IKE phase 1 security association is established using the mobile node's care-of address as the endpoint. Furthermore, the tunnel mode IPsec SA for Return Routability and data packets at the home agent uses the care-of address as the tunnel end-point. When a mobile node moves to a new location and acquires a new care-of address, simply using the existing tunnel mode IPsec SA results in tunneling the HOT packet to the mobile node's previous care-of address even if a new Binding Cache Entry is created at the home agent, unless the corresponding IPsec SA is also updated. Hence, a home agent updates the outer tunnel address for the tunnel mode IPsec SA to the new care-of address. Although this ensures that the Return Routability protocol will work correctly, it does not change the IKE phase 1 SA itself, which also uses the care-of address as the SA endpoint. A new IKE phase 1 connection can be established by running IKE again. Mobile IP provides a mechanism to avoid having to rerun IKE by means of the 'K' bit; a mobile node includes the bit to indicate to the home agent that it has the capability to update the IKE endpoint and a participating home agent responds by updating the IKE SA and setting a corresponding 'K' bit in the Binding Acknowledgment. Why is IKE phase 1 connection survivability important? Well, without an IKE SA in place, any modifications to its child SAs (such as IPsec SA) are not possible with dynamic keying. In addition, it is wasteful to run IKE each time a mobile node handovers just to update the phase 1 connection SA. So, the 'K' bit is a useful means of updating the IKE SA using Mobile IP signaling, which has to take place upon every handover.

9.5 IKE, IPSEC AND MOBILE IPV6

It is quite possible that there is no SA at all when a Binding Update packet needs to sent. At this stage, an SA needs to be established, and IKE is typically the protocol of choice. Figure 9.3 illustrates the additional steps necessary. As shown, the very first Binding Update triggers IKE dynamic keying since no security association exists. Once IPsec SA is established by IKE, the Binding Update packet goes through the steps we have identified in Figure 9.1 and is eventually sent out.

Subsequently, the mobile node undergoes a handover. It is able to send a new Binding Update using the IPsec SA established earlier. When processing the Binding Update, the home agent modifies the outer tunnel address for the IPsec SA for Return Routability signaling (to accommodate processing the new care-of address). Furthermore, if both the mobile node and the home agent are capable of processing the 'K' bit, they rekey the IKE SA for previous care-of address with new care-of address.

This looks quite straightforward. However, it took meticulous engineering to make sure that parameters and additional fields necessary to interwork with IKE and IPsec are specified unambiguously for implementations.

9.6 SUMMARY

In this chapter, we have studied how Mobile IP, IKE and IPsec interact. Mobile IP relies on IPsec to secure the Binding Update and Binding Acknowledgment messages between a mobile node and its home agent. IPsec, in turn, relies on either manual keying or dynamic keying for establishing an SA. IKE is central to SA establishment. Hence, interaction between all three protocols has to be clear in order for Mobile

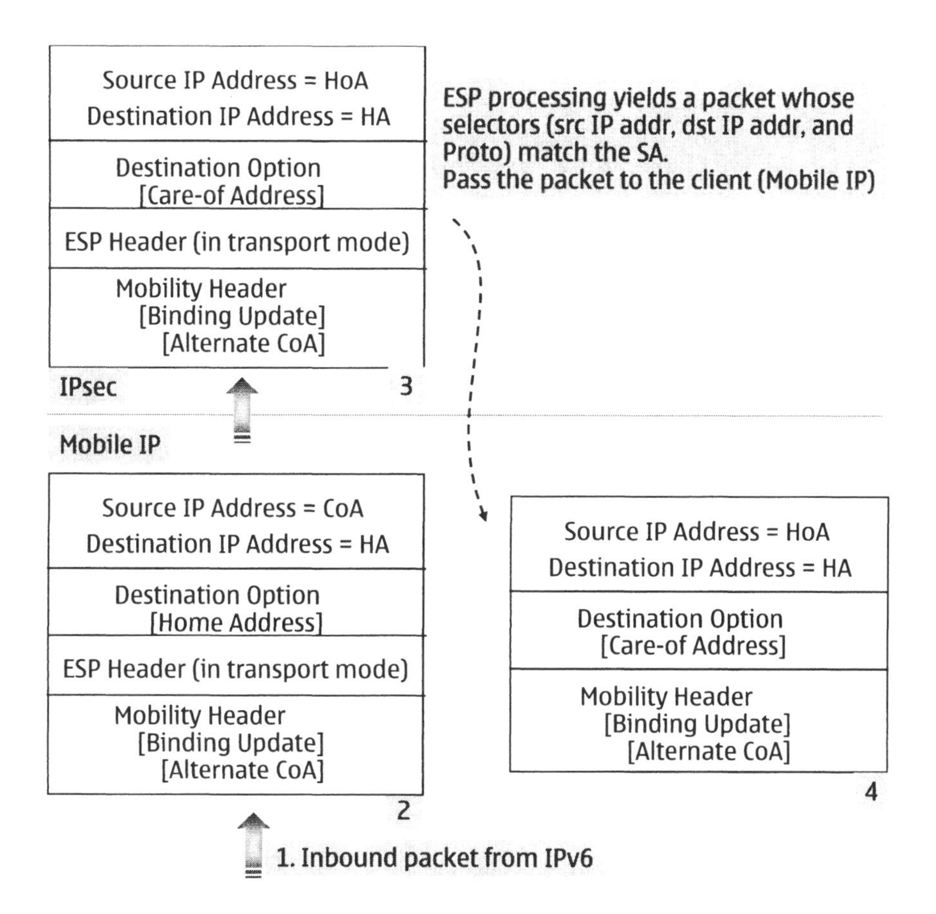

Fig. 9.2 Inbound Binding Update Processing at a Home Agent

IP signaling to work correctly. Furthermore, mobility brings some nuances to SAs and their semantics under changing IP addresses. Without any further protocol enhancements, change in IP address results in IKE signaling for rekeying which could be avoided. Such an enhancement is indeed provided by means of the 'K' bit in the Mobile IP Binding Update message.

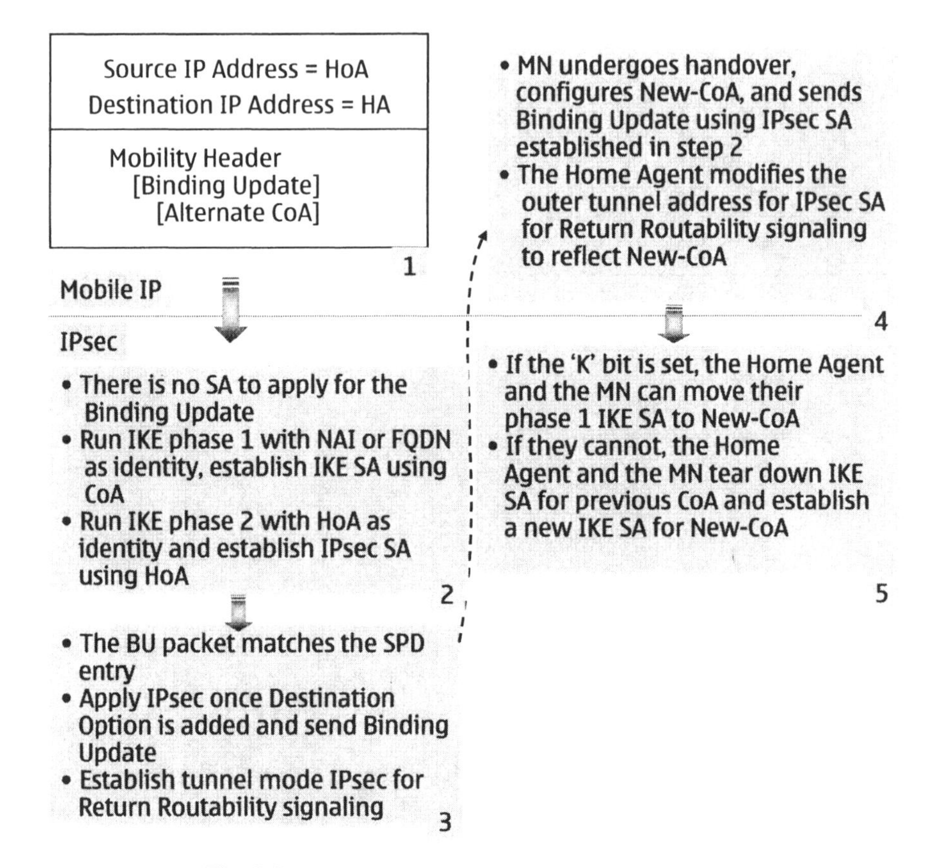

Fig. 9.3 IKE, IPsec and Mobile IP Processing

REFERENCES

1. B. Aboba and M. Beadles. "The Network Access Identifier," RFC 2486, Internet Engineering Task Force, January 1999.

2. J. Arkko, V. Devarapalli, and F. Dupont. "Using IPsec to Protect Mobile IPv6 Signaling Between Mobile Nodes and Home Agents," RFC 3776, Internet Engineering Task Force, June 2004.

10

Packet Handling

The first principle is that you must not fool yourself - and you are the easiest person to fool. – Richard Feynman

—

Mobile IPv6 specifies several mechanisms that have to be used to send and receive IPv6 packets to and from mobile devices. These mechanisms can best be understood as various forms of encapsulation, which are tailored to be minimally intrusive. In all cases, the IPv6 addresses visible to the routing infrastructure are required to be topologically correct using source and destination addresses that belong to the appropriate subnets from which the communicating IPv6 nodes are accessible. The signaling considered in the previous chapters has the purpose of informing the other nodes about the mobile node's topologically correct address (i.e., its care-of address). The techniques and protocols in this chapter utilize the address information which has been previously (securely!) provided by the mobile node to its home agent and its correspondent nodes.

There are several kinds of encapsulation specified in the base Mobile IPv6 document:

- IPv6-within-IPv6 encapsulation [1]. This is used by the home agent to deliver packets to the mobile node if the packets have arrived at the mobile node's home network. This is also used by the mobile node to send packets to a correspondent by means of its home network.

- *Routing Header Type 2* [2]. This is used by a correspondent node to indicate that traffic which is intended for the mobile node's home address must first travel to the care-of address. Of course, once the traffic arrives at the care-of address, it does not have to go much farther.

- *Home Address destination option* [2]. While not commonly described as an encapsulation, the use of this option causes the same effect. It is as if the existing IPv6 header for a packet were the outer header, and was encapsulating another inner header that was identical except for the replacement of the Source IPv6 address by the mobile node's home address. This option is used by the mobile node to enable the use of its (topologically correct) care-of address, fulfilling the restriction imposed by ingress-filtering routers in the IPv6 header while still supplying its home address to the applications hosted by the correspondent node.

- *IPsec encapsulation*. This can be used to provide one of two separate functions: either authentication, or authentication combined with encryption. As far as Mobile IPv6 is concerned, these IPsec methods are utilized only for signaling purposes – for example, with Binding Cache management messages. There is no specification within RFC 3775 that requires encrypting or authenticating normal data traffic.

Each of the data encapsulation methods is conceptually quite simple. Even so, there were numerous details involved in selecting the design currently represented in the protocol. Much of the discussion has centered around security details, and consequently certain restrictions have been made which were not originally envisioned during the initial design phases of Mobile IPv6. These restrictions, if they stand the test of time, are likely to have significant effects on the deployment of Mobile IPv6. For that reason, the restrictive rationale has to be understood and taken into account whenever designing and implementing Mobile IPv6. It is to be hoped that in the future people will discover ways to reduce the impact of the current restrictions or even better, that revised versions of the protocol document will eliminate some of the restrictions. This could easily happen if some of the threats which motivated certain design choices were understood not to be threats after all.

The encapsulation methods already mentioned above account for all the important traffic conditioning called for. Most of the rest of the specification deals with signaling protocols. However, in the future, we expect that Mobile IPv6 data will be more fully integrated with Binding Cache management. If mobility headers could be inserted in IPv6 data packets, then typically a device could expect some performance improvement by informing its communications partners (i.e., correspondent nodes) of mobility events without any significant impact on the data stream. As it is now, a mobile node has to be quite careful in deciding when it may start using its new care-of address; if data packets could contain the new care-of address along with the means to validate it, significant timing issues could be avoided entirely. This integrated approach, called "piggybacking," was part of the original design of Mobile IPv6, but had to be deleted because of design restrictions imposed by IPsec.

The rest of the chapter is organized as follows. First a brief overview of the entire data transport design of Mobile IPv6 will be presented. Then, the three major encapsulation techniques will be described in detail.

10.1 OVERVIEW

There are surprisingly many separate scenarios for data transport that are affected by the design of Mobile IPv6 and, usually, the use of the mobile node's care-of address for routing. Here are the major scenarios of interest that involve one of the three encapsulation methods:

1. Delivering packets from a mobile node directly to a correspondent node
2. Delivering packets from a mobile node to a correspondent node by way of a reverse tunnel
3. Delivering packets to a mobile node by way of its home network
4. Delivering packets to a mobile node directly, bypassing the need to be routed through the home network
5. Sending binding updates to the home agent

When a packet arrives at the home network, the home agent uses IPv6-within-IPv6 encapsulation to get it to the mobile node. This method of encapsulation is described in Section 10.2. If a correspondent node has the mobile node's care-of address, then it can send packets to the mobile node directly and bypass the home network. When the correspondent node does this, it also must use some variant of encapsulation for the same purpose of shielding the home address during the process of routing the packet. The method of encapsulation used by the correspondent node is the IPv6 routing header, which is comparable to the Loose Source Route option in IPv4 [8]. The routing header used for this purpose is defined to be slightly different than the original IPv6 routing header and is described further in Section 10.3.

These are the methods of encapsulation used to deliver packets to a mobile node. In order for a mobile node to deliver packets to a home agent or a correspondent node, it must make its home address known to the receiver while still causing that home address to be shielded from the routing infrastructure. As mentioned, the mobile node could do this by formulating a IPv6 packet header with its home address and then encapsulating that header by another IPv6 header containing its care-of address. But IPv6 headers are not so small – 40 bytes plus any extensions. So, it makes sense to economize and use the Home Address destination option for the purpose of supplying the home address to the receiver.

When a mobile node receives a packet that contains a Type 2 routing header, it can safely assume that the packet came from a correspondent node. Otherwise, if the packet is delivered using IPv6-within-IPv6 encapsulation, the mobile node can typically assume that the packet came from the home agent. Since the home agent sent the packet, the originating correspondent node is presumed to be lacking a Binding Cache entry for the mobile node. The mobile node can then use this as an indicator that the Return Routability procedure should be initiated in order to establish the necessary security association with the correspondent node and thus improve the performance of future packet deliveries.

If the correspondent node does not have any SA with the mobile node, then the mobile node is not supposed to send its home address using the Home Address destination option. This decision was made in order to prevent that option from being abused. When the discussion took place in the IETF working group, it was pointed out that the care-of address supplied by the mobile node isn't really verified by the correspondent node at all unless the Return Routability protocol has been carried out.

In this case, the correspondent would still deliver packet to the mobile node's home address, but the Home Address destination option was judged to have too much potential for evading the policies for ingress filtering that are common in today's Internet. One line of thinking suggests that a malicious node could conform to simplistic ingress-filtering policies by simply picking a topologically correct care-of address, but at the same time could appear to a correspondent node to be addressable anywhere within the Internet (i.e., at any arbitrary home address).

This line of reasoning is probably flawed, however, since the ability of a node to use its topologically correct care-of address means either (i) that it has gotten authorization to use that address or (ii) that there isn't much network administration to enforce such careful policies and so most likely no ingress filtering anyway. Moreover, the typical damage that could be caused by a malicious node using this strategy would be limited to aiming traffic at the address it claimed as its own topologically correct address, which is the same vulnerability that exists in today's Internet.

A more serious threat would involve a malicious node claiming to have the home address of some other mobile node. One can imagine a situation in which such a malicious node might try to "steal" traffic belonging to a victim mobile node by using its home address without authorization. If the malicious node could identify correspondent nodes that would be likely candidates for communications with the victim mobile node, that malicious node could try inserting a Home Address destination option and representing its own address as the care-of address. This strategy would fail in most circumstance and probably would fail in all circumstances of any importance. Usually if a correspondent node has any data for a mobile node that is worth stealing, the correspondent node will already have an SA with the mobile node, and will flatly reject a random attempt to supply a malicious care-of address. For cases when the real mobile node has already established, by way of Return Routability, a temporary SA with the correspondent node, the malicious node's attempt would be flagged and once again rejected.

Of course, it might be possible to get random web pages or other miscellaneous data misdirected to an unauthorized care-of address, but it seems unlikely that serious attackers would use such a method to collect such data, especially when so many other easier methods are available and the threat of exposure is quite high whenever care-of addresses are carefully administered.

Even though we could question relative possibilities of attacks above, a design that takes them into account conforms to the "Do No Harm" dictum better than otherwise. Since there is just not enough evidence one way or the other, perhaps it is better to err on the safer side, making sure that no new additional vulnerabilities are introduced. At least this line of thinking appears to be underlining the prevailing decision making process in IETF.

In the rest of this chapter, we provide details about the encapsulation formats, as well as operational details about the five scenarios enumerated at the start of this section.

10.2 IPV6-IN-IPV6 ENCAPSULATION

IPv6-in-IPv6 encapsulation refers to the process of inserting an IPv6 packet header before the bits of a fully formed IPv6 packet, treating the latter as payload for the encapsulating IPv6 header. Thus, there is no new packet format to be shown for such an IPv6-in-IPv6 encapsulation operation, since it just uses the already well-known IPv6 packet header as specified in the IPv6 specification [2]. This process of encapsulation is also known as *tunneling*, since the encapsulating header "protects" the inner header from having any effect on the routing of the packet until it is decapsulated. The one other piece of essential information is that the *Next Header* field of the encapsulating IPv6 header has to be set to 41 (decimal) to indicate that the next header is an(other) IPv6 header.

Schematically, the process of encapsulation can be illustrated as follows, using a diagram modeled after Figure 3 in [1]:

```
+---------+ - - - - - +-------------------------//--------------+
| IPv6    | IPv6      |                                         |
|         | Extension |              Original Packet            |
| Header  | Headers   |                                         |
+---------+ - - - - - +-------------------------//--------------+
```

Fig. 10.1 IPv6-in-IPv6 Encapsulation Schematic Diagram

Typically, the initial IPv6 header will itself use the value of 41 for the next header type, showing that the encapsulated payload is an IPv6 packet including the full IPv6 header. It is, however, possible to insert other IPv6 extension headers between the encapsulating IPv6 header and the encapsulated IPv6 packet, as shown in Figure 10.1. This could be done, for instance, in cases where the transmitter of the encapsulated packet needed to use fragmentation or needed to provide a source route for the delivery of the encapsulated packet.

There are many detailed considerations for the use of IPv6-in-IPv6 encapsulations. Since these are mostly of no major importance for Mobile IPv6, it is probably better not to discuss them at length in this book. It should be pointed out, however, that there is a destination option for use with IPv6 encapsulation that could be useful in certain circumstances with Mobile IPv6. It has to do with nested encapsulations.

Since the result of IPv6-in-IPv6 encapsulation is another IPv6 packet, the process of encapsulation could conceivably be iterated – either by the transmitting network node or by another node (e.g., a router) along the way towards the destination of the encapsulated packet. This is called *nested encapsulation* (or, *nested tunneling*). As with any potentially iterative process, there may be a danger that too many iterations will be performed, and with each iteration the size of the total packet grows by at least the size of an IPv6 header, which is 40 bytes. Eventually, this would cause fragmentation, followed by more and more congestion, and probably would only happen by mistake. Since each encapsulation step has the effect of prolonging the lifetime of the original packet, the effects of iterated packet enlargement are magnified in time as well as in space.

If it is desired to protect against the possibility of such a mistake, the transmitter can insert the *Tunnel Encapsulation Limit Option* header between the encapsulating IPv6 header and the IPv6 payload. Then any other node which might attempt to perform a further encapsulation of the packet is required to first check all the IPv6 headers to make sure that there are not too many encapsulating headers. This strategy is not perfect, and there are numerous ways that it can fail to protect against packets growing too large or traveling around the Internet for too long. For instance, an intermediate node could use a UDP encapsulation, which would defeat the checking process for the encapsulation limit option. Or an intermediate node could use encryption. Both of these situations are likely to happen more and more in the future Internet, and there are quite a few other ways that checking could fail or be infeasible to do.

10.3 ROUTING HEADER TYPE 2

The Routing Header type 2 is specifically designed for use by correspondent nodes using Mobile IPv6. The required functionality is almost the same as that of the general IPv6 Routing Header type 0: the recipient of the packet with the routing header is supposed to check whether it is an intermediate endpoint for one of the segments of the route indicated by the addresses contained within the routing header.

However, during the final phases of standardization, another requirement was imposed on the routing header which would not necessarily be relevant for other devices using the routing header for its more generally designed purposes. The additional requirement, which changes the header from type 0 to type 2, is that there can only be one intermediate node address in a routing header of type 2. Moreover, the destination IPv6 address for the packet has to belong to a network interface of the node owning the intermediate address. The total effect is that the two addresses involved both belong to the recipient, and no additional forwarding is allowed. The constraints, taken together, are meant to ensure that the recipient of the type 2 routing header cannot serve as a portal for uncontrolled traffic into its visited domain. One can imagine that future network administrators might choose firewall settings that would admit packets containing routing headers of type 2 while blocking packets containing routing headers of other types. At least that was the theory promulgated by the engineers championing the decision to choose a new routing header type.

The format of the Routing Header type 2 is as follows:

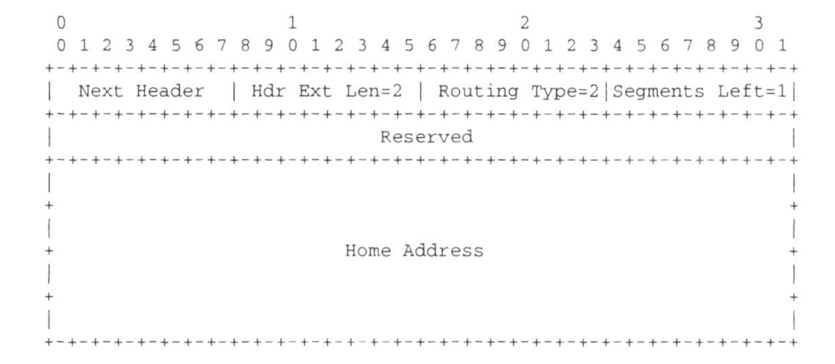

Fig. 10.2 Routing Header Type 2 Schematic Diagram

The fields of the routing header are as follows

- Next Header: used in the same way as any other IPv6 extension header to identify the type of the next IPv6 extension or protocol header following this routing header.

- Hdr Ext Len: 2, the length of the routing header in units of 8 octets (not including the 8 octets allocated for the initial fixed fields)

- Routing Type: 2, to signify that this routing header is type 2

- Segments Left: set to 1 to signify that there is only one more leg remaining in the routing path

- Reserved: 4 octets; sent as zero, ignored on reception

- Home Address: The home address of the mobile node.

Routing headers are subject to placement and ordering rules within the sequence of IPv6 extension headers. These rules are laid out in RFC 2460; briefly, a routing header is supposed to precede all other IPv6 extension headers, including the fragmentation and security and mobility headers, and any other destination options. The only exception to this is that hop-by-hop options, if present, would precede any routing headers; but hop-by-hop options are rarely used, especially in situations where a routing header of any type is likely to be in use. For instance, sending a jumbogram to a mobile node does not conform to the typical use cases motivating the creation of the jumbogram.

In case there are two routing headers of different types, the routing header of type 2 goes second, as a natural result of the intended use by a destination mobile node.

In order to understand the layout of the Routing Header type 2, it is best to understand the layout of the Routing Header type 0 from which it was derived. The more general type 0 routing header allows for a relatively large number of intermediate routing points, since there can be up to 255 segments and each segment is defined by the next intermediate forwarding IP address in the list of Addresses of the type 0 routing header. The type 2 routing header thus looks exactly like a type 0 routing header, but one with only one address allowed in the list of addresses of intermediate routing points.

Type 0 routing headers were originally designed to be the IPv6 analog of the Source Route Options specified for IPv4. There is a Loose Source Route option and a Strict Source Route option for IPv4; strict source routing requires that the option contain every intermediate routing point, whereas loose source routing does not have this requirement. In fact, a loose source route does not require that any of the intermediate routing points even be contiguous. As it has evolved, the Routing Header type 2 can be understood as a very strict variation of the IPv4 strict source route header.

There is one very important difference, however, between IPv4 routing headers and IPv6 routing headers. For IPv4, there is the implicit assumption that the receiver of the source routed packet should reverse the source route when trying to respond to the sender. This can be done simply by reversing the order of the IPv4 addresses that are provided, whether for a strict source route or a loose source route. Unfortunately, this behavior opens up a very nasty security exposure. A malicious node could insert itself as a putative intermediate routing point for such a source routed packet, and then it would have free access to data transmitted from the receiver of the source routed packet back towards the source. In fact, a malicious node can even originate the packet itself but make it seem as if some other unaware node were the originator. This is done by creating a fictitious source route option and putting the IPv4 address of the unaware node as the source IP address in the IPv4 header.

For IPv6 routing headers, there is no such assumption, and in fact no such reversal is recommended. The experience with IPv4 has caused most firewall administrators to disallow entry to source-routed IPv4 packets, and the designers of IPv6 wanted to make sure that IPv6 routing headers did not suffer the same fate. For Routing Header type 2, it is no hardship for the mobile node, of course, since it is presumed to own both (i.e., all) of the IPv6 intermediate and destination addresses for the source-routed packet.

One of the reasons for the original design using a type 0 routing header was to allow for nested and network mobility scenarios. If a correspondent node could be economically informed about a mobile node's presence within a mobile network, then packets could be delivered to the mobile node by way of, first, the care-of address of the mobile network and, second, by the care-of address of the mobile node itself. The proponents of type 2 routing headers did not deem this to be a sufficiently valuable resource, so it was not maintained. One can hope that in the future, when mobile networking is much better established and appropriate security mechanisms much are better understood, these restrictions can be removed.

10.4 ENCAPSULATING PACKETS TO THE MOBILE NODE

Unless already supplied with the care-of address of the mobile node, a correspondent node will use the home address of the mobile to initiate communications or to respond to communications from the mobile node. When the mobile node is away from its home network, the home agent will intercept such packets from the correspondent node and take steps to deliver the packet to the mobile node. This delivery proceeds by using the mobile node's care-of address as the destination address of a new packet which is constructed from the packet sent by the correspondent node. The new packet is created by using IPv6-in-IPv6 encapsulation, as described in Section 10.2. Once the encapsulation has created a new packet, the home agent delivers it to the mobile node without any other special considerations.

In order to receive packets from the correspondent node and subsequently carry out the encapsulation, the home agent has to be able to intercept packets which are routed to the mobile node. If the home agent is also a default router for the home network, and advertises connectivity to the home network to other neighboring routers, then it will receive packets for the mobile node as they arrive from correspondent nodes in the rest of the Internet. Whenever feasible, it makes a lot of sense to configure the home agent as the home network router for this reason, but Mobile IPv6 does not require this. Other configurations are allowable, so that the home agent could appear anywhere on the home network as long as it arranges to intercept packets destined for the mobile node. Moreover, even when the home agent is the only router for the home network, it still must take additional steps that will enable interception of all packets for the mobile node. This is required because other computers on the home network may try to send packets to the mobile node, as if it were still a neighboring node on the home network.

In order for the home agent to intercept packets that otherwise would be handled by local reception on the home network, the home agent has to do the same things that the mobile node would do if it were still located on the home network. In other words, the home agent has to act as a proxy for the mobile node and run the same protocols that the mobile node would use in order to receive these locally delivered packets. Those protocols are part of IPv6's Neighbor Discovery mechanisms.

The home agent fulfills this proxy function by causing all nodes on the home network to associate the mobile node's IPv6 address to the layer-2 address of the home agent. Whenever any node on the home network tries to associate the IPv6 address of the mobile node to a layer-2 address (e.g., MAC address), the home agent has to take the appropriate action and respond as if it were the mobile node. What this means is that the home agent has to respond with a Neighbor Advertisement message whenever any node on the home network issues a Neighbor Solicitation for the mobile node's IP address. Neighbor Advertisement and Neighbor Solicitation messages are described in detail in the IPv6 ND protocol document [6] which we reviewed in Chapter 2.

After the home agent arranges to intercept packets for the mobile node in this way, it also must perform all the other ND functions for the mobile node as if the mobile node were itself present on the network. In particular, the home agent has to participate in all Duplicate Address Detection (DAD) operations and Neighbor Unreachability Detection (NUD) operations on behalf of the mobile node.

When a mobile node is present on its home network, it does not need the home agent to do anything for it. The home agent and mobile node operate without any need for Mobile IPv6, and until the mobile node acquires a care-of address on a visited network, network-layer operation proceeds as usual. This means, in particular, that the mobile node will perform all local ND operations. Consequently, local neighbors of the mobile node are likely to have Neighbor Cache information allowing them to associate the IPv6 address of the mobile node to the mobile node's layer-2 address.

When the home agent first learns that a mobile node has moved away from the home network, it prods all the other nodes on the home network to update the relevant Neighbor Cache information for the mobile node. It does this by multicasting a Neighbor Advertisement message on behalf of the mobile node. The multicast message contains the layer-2 address of the home agent and amounts to a directive for the neighboring nodes to associate the mobile node's IPv6 address (i.e., home address) to the layer-2 address of the home agent.

When a mobile node moves back to the home network, the home agent has to stop intercepting packets and stop doing all other ND operations on behalf of the mobile node. In order for the home agent to change its mode of operation with respect to the mobile node, the mobile node has to take some immediate and recognizable action when it returns to its home network. Then when the home agent detects the presence of the mobile node, it immediately stops performing the proxy functions on behalf of the mobile node. In order to effect this change of state at the home agent, the mobile node sends a Binding Update to the home agent, with the care-of address set to the same address as the home address. This is called *returning home* or *binding de-registration.*

When a mobile node receives an encapsulated packet, it means that the correspondent node probably does not have available the current value of the mobile node's care-of address. Otherwise, the correspondent node would typically utilize a Routing Header (type 2) for delivery of the data to the mobile node instead of sending it to the home network for encapsulation by the home agent. This can serve as a clear hint to the mobile node to initiate route optimization with that correspondent. Route optimization may not be a default choice with *all* correspondents. A mobile node should have the choice of continuing to use the home network to communicate with a correspondent, for example, for *location privacy* reasons (which we discuss in Chapter 23). Indeed, this choice is left open for the mobile node.

10.5 REVERSE TUNNELING

In the original design of Mobile IPv6, a mobile node would use its care-of address for the Source IP address of outgoing packets and insert its home address in a destination option, which would only be interpreted by the receiver (i.e., the correspondent node). Doing so allowed the direct delivery of packets to correspondent nodes even when the correspondent node had to send packets back to the mobile node (transparently) by way of the home agent on the home network.

As more and more security requirements were mandated within the Mobile IPv6 working group, this simple and direct method of delivery was claimed to be less valuable compared to the protections offered by ingress filtering mechanisms [3]. To see the nature of the claimed vulnerability, imagine that a malicious node could use the Home Address destination option to masquerade as an IPv6 device on some other network, unrelated to the network suggested by the IPv6 address as indicated in the Source Address field of the IPv6 header. Then the receiver of the packet would typically respond to the home address, which could be in a completely different part of the Internet than the part patrolled by the ingress filtering routers. In other words, allowing the free and unencumbered use of the Home Address destination option was seen as a way to defeat ingress filtering.

When the mobile node cannot make use of the Home Address destination option, it has to tunnel packets back to the home network before they can be delivered to the correspondent node. The reverse-tunneled packets are encapsulated by an IPv6 header using the mobile node's home agent as the destination IPv6 address. Then the home agent will decapsulate these packets on behalf of the mobile node and complete the delivery to the correspondent node by forwarding them as usual.

In this way, ingress filtering mechanisms that may be in place for the visited network will be satisfied, since the outer packet will show the mobile node's care-of address as the source of the packet, and the receiver (i.e., the home agent) will not have any reason to respond to a potentially spurious care-of address. Likewise, the decapsulated packet emanating from the home network towards the correspondent node will adhere to all ingress filtering restrictions since it will appear to come from the mobile node's home address (and in fact, it does).

The mobile node can similarly employ reverse tunneling of packets from the visited network to the home agent for other reasons. For instance, the mobile node may wish to hide its care-of address from the correspondent node for some reason. By using reverse tunneling, the mobile node enlists the aid of the home agent, which decapsulates the tunneled packet and discards the encapsulating IPv6 header. Since the care-of address is only present in the encapsulating header, not in the encapsulated original packet, neither the correspondent node nor any other intermediate forwarding node will see the care-of address after the packet leaves the mobile node's home network.

10.6 DIRECT DELIVERY TO A CORRESPONDENT NODE

Suppose a mobile node has established a security association with a correspondent node (perhaps by using Return Routability, or by having some preconfigured key material as in Section 8.8). When this SA is available and the mobile node has sent a Binding Update to the correspondent node, then there is a sufficient level of trust between the two nodes that the mobile node can transmit packets to the correspondent node using the Home Address destination option. It is not required to supply authentication data when using this destination option; it is only required that the SA exists and the care-of address has been made known to the correspondent node. Since the correspondent node is thus protected against the possibility of sending data packets to the wrong IPv6 address, the danger of victimizing unaware nodes is eliminated and the direct method of delivery can be utilized without fear.

10.7 DELIVERING PACKETS TO A MOBILE NODE DIRECTLY

When a correspondent node wants to send packets directly to the mobile node, it utilizes another packet delivery mode: the correspondent node inserts a Routing Header (type 2) (see Section 10.3) after the IPv6 header. This bypasses the need for its packets to be routed through the home network.

It is expected that such improvements in routing will offer better performance for many time-critical applications such as VoIP and interactive conferencing. As mentioned previously, there is also hope that, in the future, routing headers might play some important role in efficiently handling mobile networks and nested mobile networks.

10.8 SENDING BINDING UPDATES

The Home Address destination option is crucial for one other kind of packet delivery specified by Mobile IPv6: the mobile node uses it to supply its home agent with its home address whenever it sends a Binding Update to the home agent. It also uses this destination option to provide its home address to a correspondent node as part of the Binding Update. However, as we saw in Chapter 9, Binding Update to the home agent can be reverse tunneled as well; there is no need for the destination option in this case.

10.9 INLINE SIGNALING OR PIGGYBACKING

In the original design of Mobile IPv6, a Binding Update message could be sent along with any IPv6 payload by simply including the necessary IPv6 extension header before the UDP or TCP header which precedes the application payload. In this way, smoother handovers and reduced jitter and bandwidth utilization could be achieved. Because of the way IPsec operations were interpreted, it was thought that the inclusion of a Mobility Header would preclude extending the IPsec header protections to any additional payload data. Thus, in the interests of trying to expedite the promulgation of the Mobile IPv6 standard, a decision was taken to avoid further elaboration on IPsec features that might lead to better coverage of packets that included both IPv6 extension headers and other payloads.

The perceived benefits of "piggybacking" a Binding Update, or "tagging along" data packets are perhaps somewhat affected by Return Routability procedure, the inclusion of which enforces at least one round trip of message exchange before a Binding Update (which can be piggybacked) can be sent. However, with preconfiguration of keys, the potential benefit originally envisioned remains. In any case, use of piggybacking requires modifications to IPsec processing rules under the strictest interpretation of RFC 2401 [5]. In order to address this concern, the designers worked on a new method to achieve inline signaling. The requirement is to use two new header types: one extension header called "Nonfinal Mobility Header" that can be placed before a transport header and one final Mobility Header type to enable use of IPsec for verifying authentication data.

The format of the Nonfinal and final Mobility Header are identical and are the same as those discussed in previous chapters. Since an extension header cannot be protected according to the strictest interpretation of RFC 2401 [5], and the final header type is considered as transport, there is no requirement for changing IPsec at all in any node, peer or intermediate. Since these two headers have an identical format, the effect on mobility implementation is minimal. The full details on the specification are outlined in [7], including the message formats and IPsec interworking.

10.10 SUMMARY

We have described different scenarios of data packet forwarding in Mobile IPv6. There are three primary methods of packet delivery: IPv6-within-IPv6 encapsulation, Home Address destination option, and Routing Header Type 2. All three achieve the same purpose of presenting the home address to the mobile node's peer while using the care-of address for routing purposes. The semantics used by each are obviously different, and the usage scenarios are different as well. For instance, IPv6-within-IPv6 encapsulation is used when a mobile node wishes to communicate with its correspondent by means of the mobile node's home network. The destination option and routing headers are used for route-optimized communication and signaling. Mobile IPv6 continues to be the only protocol that uses the IPv6 destination option. It is also one of the first protocols to adopt routing header usage, although the definition of the routing header used in Mobile IP is now far more restrictive than that of the type 0 IPv6 header.

We also briefly discussed piggybacking, which was discussed in great detail during the standardization. Even though the concept is somewhat simple, it unearthed numerous subtleties in a variety of issues ranging from IPsec processing to radio bearer management, combined use of voice, video and data and so on. Even though it is not part of the eventual standard, it is worth investigation as an exercise in networking in general.

REFERENCES

1. A. Conta and S. Deering. "Generic Packet Tunneling in IPv6 Specification," RFC 2473, Internet Engineering Task Force, December 1998.

2. S. Deering and R. Hinden. "Internet Protocol, Version 6 (IPv6) Specification," RFC 2460, Internet Engineering Task Force, December 1998.

3. P. Ferguson and D. Senie. "Network Ingress Filtering: Defeating Denial of Service Attacks Which Employ IP Source Address Spoofing," RFC 2267, Internet Engineering Task Force, January 1998.

4. D. Johnson, C. Perkins, and J. Arkko, "Mobility Support in IPv6," RFC 3775, Internet Engineering Task Force, 2004.

5. S. Kent and R. Atkinson. "Security Architecture for the Internet Protocol," RFC 2401, Internet Engineering Task Force, November 1998.

6. T. Narten, E. Nordmark, and W. Simpson. "Neighbor Discovery for IP Version 6 (IPv6)," RFC 2461, Internet Engineering Task Force, December 1998.

7. C. E. Perkins and F. Dupont. "Nonfinal Mobility Header for Mobile IPv6," draft-ietf-mobileip-piggyback-00.txt (work in progress), April 2002.

8. RFC: 791, INTERNET PROTOCOL, DARPA INTERNET PROGRAM PROTOCOL SPECIFICATION, September 1981, prepared for Defense Advanced Research Projects Agency Information Processing Techniques Office 1400 Wilson Boulevard Arlington, Virginia 22209, by Information Sciences Institute, University of Southern California, 4676 Admiralty Way Marina del Rey, California 90291

11

Movement Detection

Never confuse movement with action. – Ernest Hemingway

Movement detection is one of the crucial events that affects the performance of handover. Yet, it is not easy to do it quickly and reliably for several reasons. First, movement detection at the IP layer is dependent on reliably detecting the movement at the lower layer, namely the link layer. In wireless technologies, a link may be lost and regained, which does not mean movement. However, the IP operations are typically performed anyway when a link "comes up." Movement may actually happen from one link to another, for instance from one WLAN access point to another. However, that may not constitute movement to a new subnet. The challenge is to determine when a subnet change has actually taken place; this requires performing at least some operations, and these operations inevitably introduce delay. Second, multiple prefixes may be advertised by the same router on a single link, for instance to balance the traffic load on the network. So, hearing a new prefix does not mean a new subnet. Third, a router may use the same link-local address on multiple interfaces but advertise different prefixes. In this case, the mobile node needs to perform an IP handover, but it may not recognize that the router is still reachable at the previous address. Finally, there may be multiple routers on the same link advertising different prefixes. This does not mean that the mobile node has to perform a handover from one router to another, since its current access router may still be reachable.

As we can see, all these situations contribute to the complexity of reliably detecting movement to a new subnet. Hence, the Mobile IP specification defines a generic method using IPv6 router discovery and Neighbor Unreachability Detection (NUD) procedures, which we discussed in Chapter 2.

11.1 MOVEMENT DETECTION ALGORITHM

Overall, the guideline is to postpone an IP handover as long as possible until the current default router is no longer reachable. This ensures that the packet loss and signaling overhead that follow an IP handover are minimized. The tradeoff is delay associated with the movement detection procedure. Once movement to a new subnet is detected, the mobile node performs configuration of the link-local address (including the DAD), router discovery for new prefix(es) and new care-of address configuration. Subsequently, it registers the new care-of address with its home agent and correspondents.

Roughly speaking, movement detection mechanisms can be classified as either passive or active mechanisms. We discuss both types.

In a passive movement detection approach, a mobile node waits for Router Advertisements to see if it has undergone an IP handover. Mobile IP specifies a new Advertisement Interval Option for Router Advertisement. The Advertisement Interval specifies the maximum time in milliseconds between successive advertisements. If this interval expires without having received an advertisement from the router in question, the mobile node can ascertain that it has missed at least one Router Advertisement. If it misses a certain number of advertisements, the mobile node can determine that the current router is no longer reachable. The exact number of missed advertisements that determine an IP handover is subject to the internal policy of the mobile node. As can be seen, this approach can take a long time to detect movement even if the interval between the advertisements is kept fairly short.

In an active approach, the mobile node acts based on the hints it gets (e.g., from the link layer). Upon reception of a hint from the link layer that a new link is available, the mobile first sends a unicast Neighbor Solicitation message to the default router. If it does not receive a solicited Neighbor Advertisement from the router, the mobile node sends a multicast Router Solicitation. Even though the mobile node acts quickly in response to a link event, the ND operations contribute to the delay. When the mobile node has actually traversed subnets, its Neighbor Solicitation is not answered. So, it retransmits (typically up to three times) after a gap of 1 second. When it does not hear any advertisement, it sends a multicast router solicitation. The new router may not send a Router Advertisement as a response right away, which adds further delay.

In some sense, detection based on the NUD algorithm can be classified as a hybrid of the two approaches. It is the default algorithm used to reliably detect that a mobile node has crossed subnet boundaries. The entire NUD algorithm itself is fairly complicated; see Section 7.3 in [1]. We discuss it briefly here. The NUD algorithm requires bidirectional reachability between peers. This means a sender should be able to conclude that it is able to both send to and receive from a neighbor. Hence, an unsolicited message such as a router advertisement is not considered an indicator of bidirectional reachability since the receiver of such a message can only conclude that the path from the sender of the message to the receiver is working (but not necessarily the other way around).

NUD relies on either hints from the upper-layer protocols (such as TCP acknowledgments) or responses to solicitation to confirm bidirectional reachability. The algorithm itself is triggered when certain period of time expires after the last known confirmation of reachability of a peer. This is referred to as the STALE state. At this time, if the sender has no packets to send, nothing happens (hence the "passive" approach). The STALE state can transition to a PROBE state upon another time-out. At this time, the sender actively probes the peer with a Neighbor Solicitation message (hence the "active" approach). The operation at this time is similar to the active approach discussed in the previous paragraph; the node waits for Neighbor Advertisements and then multicasts a Router Solicitation. As we can see, NUD can incur substantial delay before the mobile node can go ahead and configure new addresses and register its new care-of address.

11.2 IP ADDRESS CONFIGURATION

Once it has detected movement to a new subnet, the mobile node configures a link-local address and performs DAD on that address. With the prefix information obtained in the Router Advertisement, it configures a new care-of address. If DAD succeeds for the link-local address, there is no need to perform it again on the care-of address as long as the same Interface Identifier is used in both the link-local and global addresses. The mobile node then sends a Binding Update to the home agent, possibly followed by the Return Routability protocol with its correspondents.

11.3 RETURNING HOME

A mobile node follows the same movement detection algorithm outlined above, but detects that it has moved to its home network based on the prefix it receives from the router (home agent). When this happens, the mobile node deregisters from its home agent by sending a Binding Update with Lifetime set to zero. Although the detection process itself is the same as for any network, there are some special considerations that apply pertaining to the ND operations. The mobile node may need to learn its home agent's link-layer address. In this case, it cannot send a Neighbor Solicitation using its home address as the source IP address since the home agent is defending that address. Hence, it is required to send the packet with source IP address set to the unspecified address (see Chapter 2) and the target address set to its home address. The home agent then multicasts the Neighbor Advertisement which contains its link-layer address. The mobile node can then send a Binding Update to deregister from the home agent.

Immediately after sending the Binding Update, the mobile node must be ready to respond to Neighbor Solicitations for its home address. This is necessary since the home agent will typically send out a solicitation for the home address before it can transmit the Binding Acknowledgment message to the mobile node.

We mentioned previously that the mobile node engages in address configuration once movement is detected. When returning home (with valid binding for the home address), however, doing DAD on the home address before it is deregistered would cause conflicts. The home agent would try to defend the home address for the mobile node. In addition, the home agent would defend the link-local address for the mobile node if the 'L' bit was set in the Binding Update message. Hence, the mobile node first deregisters itself from the home agent before performing the DAD operation on its addresses.

After it has deregistered itself from the home agent, the mobile node multicasts an unsolicited Neighbor Advertisement message informing its neighbors to update their neighbor cache entries with its link-layer address. The Target Address is set to its home address, and the message includes the Target Link Layer Address Option set to the mobile node's link-layer address. This allows nodes to start using the mobile node's link-layer address for the home address (as opposed to the home agent's link-layer address).

11.4 CHANGES TO NEIGHBOR DISCOVERY

The main change that helps movement detection is the allowance for sending unsolicited router advertisements at faster intervals. Neighbor Discovery protocol [1] specifies a minimum of 3 seconds between successive multicast router advertisements. Mobile IP relaxes this value to as little as 50 milliseconds where deployments can support it. Mobile IP also requires the ability to disable multicast router advertisements in some deployments where detection is based link indications followed by solicited advertisements.

In order to facilitate determination of the router's global address, a new 'R' bit is defined in the Prefix Information Option carried in the Router Advertisements. When set, the Prefix field contains the complete IP address of the router assigned to the sending interface. This helps disambiguate movement scenarios where the router uses the same link-local address on more than one interface. There is also an 'H' flag added to Router Advertisements to indicate that the router is also a home agent. For a visiting mobile node, this may be useful in obtaining a local home agent, although no such concept is formally defined in the base specification itself.

When configuring a new address by stateless autoconfiguration, ND protocol requires a random delay before sending a DAD probe. Since this could introduce additional delay, Mobile IP recommends performing DAD right away when implementations allow enough randomness to desynchronize the steps that happen prior to DAD [2].

As we conclude this chapter, we sum up the main points of this chapter: that movement detection is surprisingly complicated primarily due to deployment variations. Movement detection introduces delay, which is perhaps acceptable to portable scenarios. For instance, if Bob were to unplug his computer from his office Ethernet, and plug back in at a conference room, the delay incurred for such applications as e-mail may not be beyond some annoyance, if any. However, if Bob were talking using VoIP, surely, the call quality will suffer at every instance of handover, and movement detection delay only adds to the quality degradation.

In this chapter, we saw that the default approach in Mobile IP based on NUD can take a long time before the mobile node even begins configuring its IP address, which itself is a highly delay-prone operation. Hence, approaches using link-layer indications are useful. In the next part of this book, we will see how network neighborhood information can be used to offset this delay as part of fast handovers. We also studied some special actions necessary on the part of the mobile node when it returns home. Specifically, it has to use the unspecified address as the source address in its Neighbor Solicitation when it needs to learn the link-layer address of the home agent. We also studied some enhancements to ND protocol to enable Mobile IP functions such as the allowance for discovering the home agent's global IP address as well as the knowledge of availability of the home agent itself.

REFERENCES

1. T. Narten, E. Nordmark, and W. Simpson. "Neighbor Discovery for IP Version 6 (IPv6)," RFC 2461, Internet Engineering Task Force, December 1998.

2. S. Thomson and T. Narten. "IPv6 Stateless Address Autoconfiguration, RFC 2462, Internet Engineering Task Force, December 1998.

12

Dynamic Home Agent Discovery

No computer has ever been designed that is ever aware of what it's doing; but most of the time, we aren't either. – Marvin Minsky

12.1 MOTIVATION

The Mobile IPv6 protocol allows a mobile node to discover its home agent without relying on static configuration. This is useful since the home network may undergo reconfiguration or renumbering that results in a "change of guards" so that a previously configured home agent may not be available for the mobile node. It is also often mentioned that dynamic home agent discovery is useful in balancing the load across multiple home agents in certain deployments. In addition to the mobile node discovering one or more home agents, the home agents can discover each other as well. Finally, home network prefix discovery is closely related to home agent discovery in which a mobile node can learn or refresh its knowledge of prefixes that are currently being supported.

This short chapter describes the essential parts of the Mobile IPv6 specification which deal with home agent discovery and prefix discovery.

12.2 HOME AGENT AND PREFIX DISCOVERY

12.2.1 Dynamic Home Agent Discovery

The mobile node may choose to perform home agent discovery at any time, and the events that cause it to perform the discovery operation are not specified. For instance, a configuration profile may force the mobile node to perform discovery after a predefined time period has expired, or a mobile node may determine that its current home agent is not functioning, or a mobile node needs to discover a home agent upon power-up. Regardless, the mobile node sends am ICMP Dynamic Home Agent Address Discovery (DHAAD) Request message to the well-known Mobile IPv6 Home-Agents anycast address [2]. An anycast address, as we mentioned in Chapter 2, is a reserved address assigned to all routers acting as home agents, but only one of them will receive the packet by virtue of anycast routing. Even though it is a reserved address, the mobile node needs to be configured with the subnet prefix in order to construct this address. The format of the DHAAD Request message follows the ICMP format, with Type being set to 144, and is identical to the DHAAD Reply message shown in Figure 12.1 except for the Home Agent Addresses field. The Identifier field is chosen randomly, to thwart any profiling. The source IP address of the message is the mobile node's current location where it is reachable (e.g., the care-of address), but it is not necessary that the home agent have a binding for that address since the mobile node probably has not established such a binding. The destination IP address is the anycast address just discussed.

As a response, a home agent sends the DHAAD Reply message shown in Figure 12.1.

As with ICMP messages, the Identifier field contains the value copied from the corresponding (DHAAD) Request message. The Home Agent Addresses field contains one or more Home Agent Addresses. The order is based on the "preference" that each home agent would have advertised on the home link (see Section 12.3). The mobile node may attempt to register with any of the home agents whose address is supplied in the DHAAD Reply message using the Binding Update message. The mobile node should try reaching the home agents according to the order specified in the DHAAD Reply message until it receives a Binding Acknowledgment. Recall that the Binding Update needs to be secured using the security association, which may either exist or needs to be established using a protocol such as IKE, as discussed in Chapter 9.

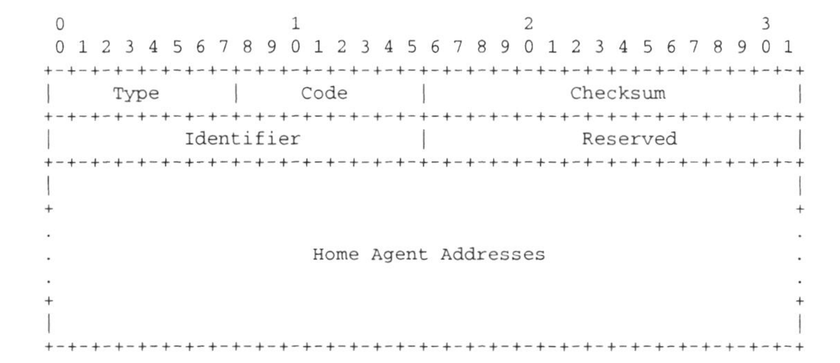

Fig. 12.1 Dynamic Home Agent Address Discovery Reply Format

12.2.2 Mobile Prefix Discovery

In addition to discovering the home agent address, a mobile node needs to learn the home prefix and the associated lifetime in order to configure a home address. And an existing home network prefix may become invalid after some time. The Mobile Prefix Solicitation and Mobile Prefix Advertisement messages can be used to learn new prefixes or refresh an existing prefix, but not, interestingly, to learn the prefixes the first time; the Mobile IPv6 *Bootstrapping* protocol now being designed in the IETF provides extensions to the IKE version 2 (IKEv2) protocol [3] to discover a single home network prefix when the mobile node has none to begin with. [1]

Both Prefix messages use ICMP, just like the DHAAD messages. The Mobile Prefix Solicitation message includes the home address destination option and is sent to the home agent's address discovered in the previous step. The message should also be sent with IPsec ESP protection. The home agent is required to respond to the solicitation with a Mobile Prefix Advertisement message shown in Figure 12.2. The home agent may also send the advertisement unsolicited if the preferred or valid lifetime, or the state of flags associated with the prefix in the home address of a valid Binding Cache entry has changed.

The 'M' and 'O' flags refer to the *Managed* and *Other* stateful configuration indication in [4]. These flags are set if the home agent supports DHCPv6 [1]. The message carries the Prefix Information Option, which consists of all the supported home network prefixes, their lengths, and the associated preferred and valid lifetimes. The destination IP address of the advertisement message is set to the source IP address of the solicitation message, or the home address in the binding cache entry if the message is sent unsolicited. In either case, the message includes a Routing Header type 2 and is sent protected using IPsec ESP.

[1]Without bootstrapping, the mobile node needs to be statically configured with the home network prefix and prefix length, which is quite feasible.

```
 0                   1                   2                   3
 0 1 2 3 4 5 6 7 8 9 0 1 2 3 4 5 6 7 8 9 0 1 2 3 4 5 6 7 8 9 0 1
+-+-+-+-+-+-+-+-+-+-+-+-+-+-+-+-+-+-+-+-+-+-+-+-+-+-+-+-+-+-+-+-+
|     Type      |     Code      |           Checksum            |
+-+-+-+-+-+-+-+-+-+-+-+-+-+-+-+-+-+-+-+-+-+-+-+-+-+-+-+-+-+-+-+-+
|          Identifier           |M|O|        Reserved           |
+-+-+-+-+-+-+-+-+-+-+-+-+-+-+-+-+-+-+-+-+-+-+-+-+-+-+-+-+-+-+-+-+
|          Options ...
+-+-+-+-+-+-+-+-+-+-+-+-+-+-+-+-+-+-+-+-+-+-+-+-+-+-+-+-+-+-+-+-+
```

Fig. 12.2 Mobile Prefix Advertisement Format

When a mobile node receives a Mobile Prefix Advertisement message, it first checks that the source IP address is that of the home agent with which it has registered its primary care-of address. If not, the source IP address should be that of a home agent it recognizes. Additional validation steps include verifying IPsec, the presence of Routing Header type 2 and the ICMP Identifier that matches the one it sent in the solicitation. If the validation fails, the mobile node discards the message. If the message is considered valid, the mobile node has to process the prefix information present and may need to reconfigure its home address.

12.3 DISCOVERY OF OTHER HOME AGENTS

A router acting as a home agent learns of other home agents on the link by means of unsolicited multicast router advertisements which are sent periodically. When the 'H' bit is set in a router advertisement, the router is acting as a home agent. Each home agent creates a Home Agent List of all other routers acting as home agents. A Home Agent List is a data structure where each entry includes the source IP address of the other home agent, preference value, and the lifetime. The source (link-local) IP address is extracted from the IP header, and the preference and lifetime values are extracted from the Home Agent Information Option when present. Each entry also contains the global IP address, which is extracted from the Prefix Information Option present in the router advertisement. The format of the Home Agent Information Option is shown in Figure 12.3. Higher values mean higher preferences in ordering the home agent address in the DHAAD Reply message. If this option is not included in the Router Advertisement, default for preference is zero. Similarly, the Lifetime specifies how long this router acts as a home agent in units of seconds; the default is the same as that of *Router Lifetime* in the Router Advertisement message, with a maximum of 18.2 hours.

If the home agent lifetime value (in a Router Advertisement) is zero for a home agent address already present in the Home Agent List, the receiving home agent immediately deletes the entry. Otherwise, the receiving home agent updates the values for the preference and lifetime values for an existing entry in the list. If no entry exists, the receiver creates one. As we noted earlier, a home agent that responds to the DHAAD Request message uses the information present in the Home Agent List in order to construct the DHAAD Reply message to the mobile node.

```
 0                   1                   2                   3
 0 1 2 3 4 5 6 7 8 9 0 1 2 3 4 5 6 7 8 9 0 1 2 3 4 5 6 7 8 9 0 1
+-+-+-+-+-+-+-+-+-+-+-+-+-+-+-+-+-+-+-+-+-+-+-+-+-+-+-+-+-+-+-+-+
|     Type      |    Length     |           Reserved            |
+-+-+-+-+-+-+-+-+-+-+-+-+-+-+-+-+-+-+-+-+-+-+-+-+-+-+-+-+-+-+-+-+
|     Home Agent Preference     |      Home Agent Lifetime      |
+-+-+-+-+-+-+-+-+-+-+-+-+-+-+-+-+-+-+-+-+-+-+-+-+-+-+-+-+-+-+-+-+
```

Fig. 12.3 Home Agent Information Option Format

In summary, there are three parts to home network discovery: home agent address, prefix information and the discovery of other home agents. Only the last is exclusively between home agents. In order to learn the home agent's address, the home network prefix must be configured in the mobile node. Only then can it reach a home agent using the anycast address and formulate a home address. Subsequently, it can learn new prefixes using the Mobile Prefix Solicitation and Mobile Prefix Advertisement messages.

Some of the configuration can be improved by means of a bootstrapping protocol. The current design thinking of such a protocol would involve configuring the domain name of a home agent and discovering the home agent by DNS. Subsequently, the mobile node can use extensions to the IKEv2 protocol to learn a home network prefix and configure a home address. Finally, it can register that home address with its home agent. Even with such a protocol, the prefix solicitation and advertisement messages are useful to be "in sync" with the home network configuration dynamics.

REFERENCES

1. R. Droms, J. Bound, B. Volz, T. Lemon, Perkins, C., and M. Carney. "Dynamic Host Configuration Protocol for IPv6 (DHCPv6)," RFC 3315, Internet Engineering Task Force, July 2003.

2. D. Johnson and S. Deering. "Reserved IPv6 Subnet Anycast Addresses," RFC 2526, Internet Engineering Task Force, March 1999.

3. C. Kaufman (Editor). "Internet key Exchange (IKEv2) Protocol," RFC 4306, Internet Engineering Task Force, December 2005.

4. T. Narten, E. Nordmark, and W. Simpson. "Neighbor Discovery for IP Version 6 (IPv6)," RFC 2461, Internet Engineering Task Force, December 1998.

13

Network Mobility

There is no reason anyone would want a computer in their home. – Ken Olson, founder, chairman and president of DEC, 1977

—

13.1 INTRODUCTION

In the previous chapters, we discussed the mobility of a mobile node in great detail. Beginning from the basic concepts of mobility to multiple ways of supporting mobility to IP mobility using Mobile IP, we have studied the functional elements and their behaviors corresponding to protocol specifications. In this chapter, we consider network mobility, in which an entire network moves from one location to another on the Internet. As an example, consider Alice taking a network in her car on her vacation; it should be possible for her vehicular network to maintain seamless connectivity with her home network. We consider how Mobile IP can be extended so that individual mobile nodes do not have to run the Mobile IP protocol when the network they are attached to moves.

Let us consider applying Mobile IP directly when an entire network moves. Examples of networks that move include personal vehicular networks, networks on transportation systems such as a train and airline systems, and personal area networks. When the network attaches itself to a new point of attachment, each node has to perform Mobile IP operations. That is, each node has to first detect that it has moved to a new network, configure a new IP address and send Binding Update to its home agent. This could constitute a significant signaling overhead. In addition, there is a provisioning overhead of security association establishment between the home agent and the mobile nodes. Furthermore, all hosts need to be aware of mobility. It can be argued that these considerations are the direct result of deploying Mobile IP. Although this is accurate, there are benefits that can be derived from considering the entire network of nodes as a unit. Obviously, signaling can be assigned to a single node on behalf of the mobile network. And, in deployments such as transportation networks, provisioning an SA between the home agent and the mobile nodes is a burden that can be avoided by concentrating the mobility functions on a single node. Perhaps equally important, except for a single node assigned to handle mobility operations, the rest of the nodes need not be concerned with mobility at all. This allows nodes to function as if they are "virtually home."

The Network Mobility (NEMO) group in IETF [11] and the NEMO Basic protocol [2] have adopted the model of a *Mobile Router* which handles mobility operations on behalf of the entire moving network. In the following sections, we describe the model and the protocol in detail.

13.2 NEMO MODEL AND TERMINOLOGY

We illustrate the NEMO model using Figure 13.1.

A *Mobile Network Node* (MNN) is any node in the network that moves as a unit. It is either a generic IPv6 host which moves with the rest of the MNNs or Mobile IP capable. Indeed, it can also be a Mobile Router, which we discuss next.

The new element in the figure is the Mobile Router which performs multiple operations. First, it acts like any other Mobile IP host performing movement detection, care-of address configuration and Binding Update operations (among others) on the *egress interface*. This allows it to create a bidirectional tunnel with the home agent for all communication. Second, it negotiates a prefix list with the home agent, which uses the list to capture packets arriving for the MNNs that share the common prefix and forward them to the Mobile Router. The prefix list negotiation could be done manually, via an extension to the Binding Update or by other means, such as DHCP prefix delegation (which we discuss in Section 13.5). Finally, the Mobile Router advertises one or more prefixes on the *ingress interface* towards the Mobile Network attached. It does not advertise those prefixes on the egress interface when attached to a visited network.

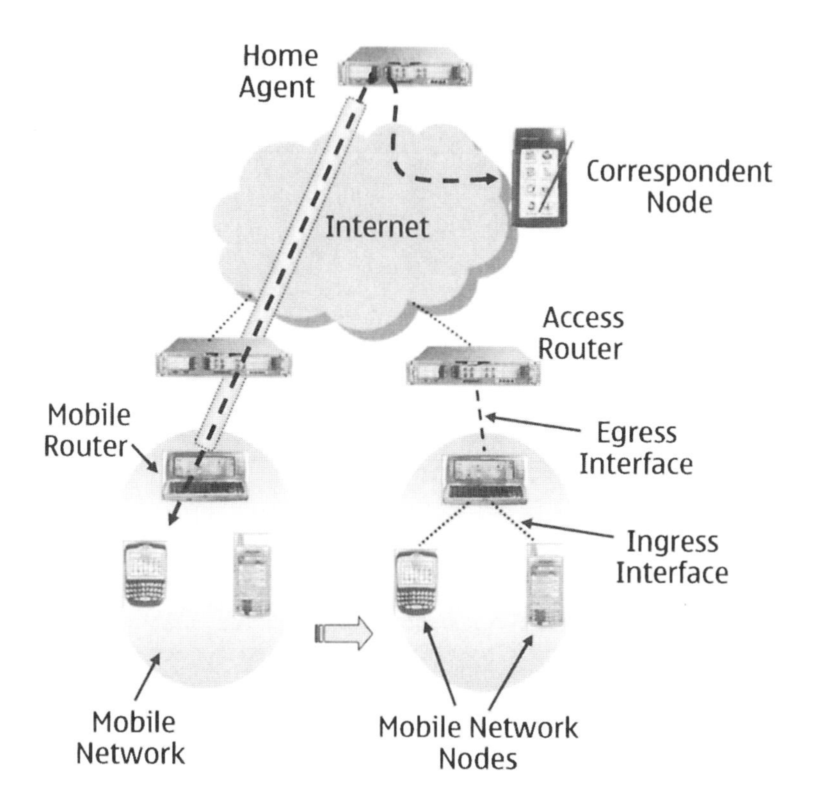

Fig. 13.1 NEMO Reference Model

The home agent is the same as in Mobile IP with the following enhancements. It maintains a binding between the Mobile Router's home address and the care-of address. In this sense, it treats the Mobile Router as any other Mobile Node, and creates binding between the two addresses as well as a bidirectional tunnel. However, it uses the bidirectional tunnel to forward *any* packet whose address matches the mobile network prefix(es) belonging to the Mobile Router. The mobile network prefixes are indexed by the Mobile Router's home address and are maintained as part of the binding cache data structure in a *Prefix Table*.

Even though the NEMO Basic protocol model appears to be simple, many complex scenarios are possible. For instance, a Mobile Network in Figure 13.1 can consist of not just the fixed hosts, but also mobile nodes. When a MNN is mobile, it could be a Local Mobile Node (LMN) or a Visiting Mobile Node (VMN). An LMN is typically a mobile node whose home network belongs to the mobile network; in some cases, an LMN could be a Mobile Router itself. A VMN, on the other hand, has a home network that does not belong to the mobile network. A VMN could also be a Mobile Router, and derives visited address from the mobile network. As opposed to the fixed nodes, both the LMN and VMN are able to change their point of attachment while maintaining existing sessions. As we noted earlier, a MNN can be a Mobile Router itself with its own network of fixed hosts and mobile nodes. Hence, nested mobile networks are possible. Nevertheless, the network as a whole should be able to maintain connectivity to the Internet as well as the existing sessions as it moves on the Internet.

13.3 NEMO BASIC PROTOCOL

Conceptually, the NEMO Basic protocol is simple. A special node (Mobile Router) acts both as Mobile IP Mobile Node as well as a router for the network it serves. This way, the Mobile Network Nodes do not see mobility at all. We consider the protocol operation in detail.

As a mobile node, the Mobile Router performs all the functions of a Mobile IP host. This includes the functions when at home. However, it is a default router for all the nodes in the mobile network at all times. When at home, it advertises the same mobile network prefix, but that prefix is either aggregated by the home agent or is exchanged in a routing protocol with the home agent.

When the mobile network moves and attaches to a new subnet, the Mobile Router configures a care-of address using the visited network prefix and sends a Binding Update to its home agent. If the Mobile Router wishes to provide connectivity to the nodes in its mobile network, it indicates this by means of an 'R' flag in the Binding Update. This allows the home agent not only to set up a Binding Cache entry for the Mobile Router's home address and care-of address, but also to set up a tunnel for the entire prefix belonging to the mobile network.

The mobile prefixes are supplied in a new Mobility Header option in the Binding Update in the so-called *Explicit Mode*. In the *Implicit Mode*, the set of mobile network prefixes is already preconfigured on the home agent corresponding to the Mobile Router. In such a case, the Binding Update does not include the mobile prefix option. In either case, for each of the prefixes, the home agent creates a forwarding table entry to the Mobile Router's care-of address. The forwarding table entry uses the tunnel interface, matching the prefix in the destination address of incoming packets to the Mobile Router's care-of address.

The Binding Update is secured just as in Mobile IP using the security association set up between the Mobile Router and the home agent. Presumably, the configuration on the home agent extends the SA to include packet forwarding to a set of nodes not just a single node. Exactly how this is decided and how the Mobile Router is entrusted with the task of forwarding to a set of nodes is not specified. In fact, the home agent may have no specific relationship with any of the nodes in the mobile network.

After setting up the forwarding table entry for the mobile network prefix, the home agent sends a Binding Acknowledgement with an 'R' bit set back to the Mobile Router. Subsequently, the Mobile Router sets up a reverse tunnel back to the home agent for all the mobile network prefixes. Any packet with a source IP address using the mobile network prefix is sent back to the home agent using this tunnel. The home agent decapsulates the tunneled packet and forwards the inner packet to the correspondent. The bidirectional tunnel between the Mobile Router and the home agent uses the IP-in-IP encapsulation.

A correspondent's packets to a MNN arrive at the home agent for one of the two reasons. The mobile network prefix is aggregated at the home agent (e.g., the home agent may have provided a prefix chunk that it owns to the Mobile Router), which advertises the aggregated prefix to the external networks. Or the home agent advertises the mobile network prefix as one of the networks to which it is capable of routing. The latter allows the mobile network to be an intra-domain network on its own. In such a case, the Mobile Router would run a routing protocol with the home agent exchanging the mobile network prefix(es). When attached to a visited network, the Mobile Router runs the routing protocol inside the bidirectional tunnel. The Mobile Router must not run a routing protocol with its visited point of attachment.

The Mobile Router maintains an additional data structure in its Binding Update List called Prefix Information, which includes what is sent in the Binding Update to the home agent. If no prefixes are included in the Binding Update, this field is set to null. The data structure also includes the 'R' flag as well as the fields maintained by the Mobile IPv6 protocol itself. If a Mobile Router receives a BAck message in which the 'R' flag is not set, it concludes that its current home agent no longer supports forwarding for the mobile network. It may subsequently attempt to dynamically discover a home agent that supports network mobility.

When at home, a Mobile Router may respond to Router Solicitations on the link attached to the home link. However, the Router Lifetime should be set to zero so that no hosts (other than the MNNs) configure it to be their default router. Generally speaking, the Mobile Router is not supposed to act as a router on the egress interface. It is recommended to behave as a host on the egress interface, especially when attached to a visited network where it must process a Router Advertisement in order to configure a care-of address, and subsequently send a Binding Update.

The home agent maintains an additional data structure in its binding cache called the Prefix Table, which includes all the mobile network prefixes associated with a specific Mobile Router's home address. The table is populated when the home agent processes the Binding Update with 'R' flag set. For each received packet, the home agent verifies if the destination address matches any of the prefixes in the Prefix Table. If it does, it forwards the packet to the Mobile Router's care-of address using the tunnel established after processing the Binding Update.

13.4 NEMO ROUTE OPTIMIZATION

The NEMO Basic protocol provides always-on connectivity to nodes in the mobile network without even requiring the nodes to be mobility-aware. All communication must go through the Mobile Router and its home agent. Compared to Mobile IPv6, no route-optimized communication is possible between a MNN and its correspondent using the NEMO Basic protocol. We look at what it means to provide route optimization in a NEMO in this section.

The first question that arises is, how does a MNN even know what is route optimization since its IP address never changes. This is because the Mobile Router never relays a visited network router advertisements. [1] So, even if a MNN is IP mobility-capable, it may never be able to determine that it could use route optimization. Given this, one could argue that the burden of providing the route optimization rests on the Mobile Router. On the other hand, we could construct scenarios where a Mobile Router could inform its nodes about the visited network prefix information. In such a case, if a node is fixed, it can configure a new address and communicate directly with its correspondents. However, any existing communication (that uses its home IP address) will break. A Mobile IP node could establish a route-optimized communication with its correspondents using the Return Routability protocol. In order to extend this to nested mobile networks, an iterative prefix delegation process from the visited network router to all the Mobile Routers in a mobile network is necessary [5, 8].

Assuming that the Mobile Router itself is in charge of providing route optimization, many challenges arise. First, how does a Mobile Router determine the correspondents with which to perform route optimization? Arguably, it would have to inspect the destination IP address in each of the packets being sent by the MNNs and initiate Return Routability with them. In a sense, it would have to act as a proxy [1] for each of the MNNs and perform the Return Routability operation and Binding Update. We leave the exact details of the protocol changes necessary, as well as the security and trust models to use, as an exercise for the reader.

[1] Nodes could overhear such a Router Advertisement, in which case the NEMO Basic protocol does not describe any specific behavior.

Another challenge to route optimization is that the correspondents may not be mobility-aware, like many nodes in the mobile network itself. In such a case, the Mobile Router needs a counterpart proxy in the correspondent's network to terminate Return Routability signaling. Based solely on the destination IP address in an IP packet, how does a Mobile Router securely determine its counterpart (which is called the *Corresponding Router* in [7])? One approach, presented in [10], uses an ICMP message sent to a well-known anycast address containing the correspondent's IP address and expects the Corresponding Router capable of terminating the RR signaling to respond. However, anycast address security is not well understood on the Internet, and approaches using anycast between arbitrary nodes face the challenges of trust models and verifying the authenticity of responders.

If the nontrivial discovery and security issues are addressed, a Mobile Router could establish a tunnel with a Corresponding Router using Return Routability signaling extensions and forward packets, just as it does with its home agent. This approach would be transparent to end hosts but requires significant enhancements, not the least of which is being able to scale to an arbitrary number of IP sessions traversing all across the Internet. As a comparison, the virtual private networks dynamically establish tunnels between end hosts and special-purpose VPN gateways to facilitate secure remote access. However, securely instantiating tunnels (for multiple mobile nodes) with arbitrary correspondent routers has no precedence on the Internet today. So, this problem continues to be challenging.

Finally, route optimization has to reckon with location privacy, a topic we will cover in Chapter 23. For now, we can define it as the ability of onlookers and correspondents to determine that a mobile node, or a mobile network itself, has roamed on the Internet. When a node can inspect two different addresses, one for optimal routing and another for keeping a connection persistent, it can also determine that the mobile node has roamed. This side effect of mobility may be undesirable for some users who may wish to keep their roaming information private. This mobile node location privacy problem is compounded when we consider mobile network mobility. One could argue that even defining the problem for a mobile network is challenging. We encourage the readers to consider this problem once they familiarize themselves with the mobile node location privacy problem.

13.5 PREFIX DELEGATION AND MANAGEMENT

One of the problems we have not specifically discussed is the actual mechanism used to provide a mobile network prefix to a Mobile Router. This is generally referred to as the *Prefix Delegation* problem. The requirements for such a delegation are specified in [6]. RFC 3633 [9] specifies one way of achieving prefix delegation using the DHCPv6 [3] protocol. This DHCP-based prefix delegation approach is adapted for NEMO in [4]. We will not discuss the protocol itself any further, encouraging the interested readers to consult the references.

13.6 SUMMARY

In this chapter, we have described the protocol operations necessary to address mobility on the Internet for an entire mobile network. We have discussed the NEMO Basic protocol, which assumes that only a Mobile Router is charged with handling mobility operations and that the rest of the nodes in the network are effectively mobility-unaware. We have seen how, with simple extensions to the Mobile IPv6 protocol, an entire mobile network is able to maintain always-on connectivity to the Internet. However, this model also has its limitations. Specifically, no explicit support is provided to allow nodes to differentiate between attachments to the home network and to a visited network; indeed, this is seen as one of the strengths of NEMO Basic approach. There are numerous proposals to provide route-optimized communication between nodes in a mobile network and their correspondents. We have provided brief descriptions of such approaches, leaving the comprehensive treatments to the references at the end of this chapter. Nevertheless, we recognize that optimizations beyond the NEMO Basic protocol are clearly more challenging than is the case with Mobile IPv6. We encourage readers to consult the relevant references for exploring such challenges further.

Exercises

13.1 How will a mobile network be able to learn that it has roamed if the Mobile Router always advertises the home prefix? What are the advantages of discovering the visited network prefix? What are the disadvantages? What should the Mobile Router do, if anything? (Is it necessary for the MNNs to know that their default router is a Mobile Router?).

13.2 Assume that a node in the mobile network configures a care-of address and sends a Binding Update to the home agent which happens to be the same for the Mobile Router and the MNN. How does packet forwarding work? Draw the IP headers for packets traversing in either direction.

13.3 Consider the scenario in which the mobile network becomes disconnected from the Internet. How could the MNNs communicate with each other in a disconnected mobile network? What protocol actions are necessary?

13.4 One of NEMO's basic design goals was to support mobility for nodes that are not mobility-aware or mobility-capable. This is achieved by designating the Mobile Router to perform all mobility operations which are otherwise expected of the mobile network nodes. Given this model, is route optimization for all nodes in the mobile network really practical? There are various approaches to support route optimization [7]. Compare and tabulate the tradeoffs between complexity, transparency (from mobility), and the ability to support arbitrary topologies. Is the problem solvable from a deployment perspective?

REFERENCES

1. C. Bernardos, M. Bagnulo, and M. Calderon. "MIRON: MIPv6 Route Optimization for NEMO," 4th Workshop on Applications and Services in Wireless Network, August 2004.

2. V. Devarapalli, R. Wakikawa, A. Petrescu, A., and P. Thubert. "Network Mobility (NEMO) Basic Support Protocol," RFC 3963, Internet Engineering Task Force, January 2005.

3. R. Droms, J. Bound, B. Volz, T. Lemon, Perkins, C., and M. Carney. "Dynamic Host Configuration Protocol for IPv6 (DHCPv6)," RFC 3315, Internet Engineering Task Force, July 2003.

4. R. Droms and P. Thubert. "DHCPv6 Prefix Delegation for NEMO" (work in progress), draft-ietf-nemo-dhcpv6-pd-01, February 2006.

5. K. Lee, J. Park, and H. Kim. "Route Optimization for Mobile Nodes in Mobile Network based on Prefix Delegation," 58th IEEE Vehicular Technology Conference, vol 3, pp 2035-2038, October 2003.

6. S. Miyakawa and R. Droms. "Requirements for IPv6 Prefix Delegation," RFC 3769, Internet Engineering Task Force, June 2004.

7. C. Ng, et al. "Network Mobility Route Optimization Solution Space Analysis" (work in progress), draft-ietf-nemo-ro-space-analysis-02.txt, February 2006.

8. E. Perera, R. Hsieh, and A. Seneviratne, "Extended Network Mobility Support," (work in progress), draft-perera-nemo-extended-00, July 2003.

9. O. Troan, and R. Droms, "IPv6 Prefix Options for Dynamic Host Configuration Protocol (DHCP) version 6," RFC 3633, Internet Engineering Task Force, December 2003.

10. R. Wakikawa, S. Koshiba, K. Uehara, and J. Murai, "ORC: Optimized Route Cache Management Protocol for Network Mobility," 10th International Conference on Telecommunications, vol 2, pp 1194-1200, February 2003.

11. "Network Mobility (NEMO)" Working Group, Internet Engineering Task Force, http://ietf.org/html.charters/nemo-charter.html.

Part III

Advanced Mobility Protocols

Unified communications networks are not exactly a new idea. Even the legacy voice telephone networks have worked extensively on video telephony technology, even though it never achieved commercial success. Now with IP, unified networks with various forms of applications are increasingly becoming a reality, although much work continues to be done in commercializing applications with various business models. This is particularly the case in VoIP, which is finding its way to the Internet core, as well as access networks such as DSL and cable modem networks. Along with VoIP, video in the form of IPTV is seen as a major application.

While applications continue to mature for commercialization on the Internet, mobility is bringing the so-called *quad play*. Users increasingly want their applications "on the go," which means that applications such as VoIP are expected to provide the same level of convenience offered by the erstwhile *circuit-switched* mobile voice technology. Two developments related to access technologies are significant in this context. First, cellular data networks, most notably the CDMA EV-DO and WCDMA HSDPA networks, are being deployed to support not only data applications, but also VoIP. And second, the IEEE 802.11 networks, otherwise known as the WLANs, are being adopted by multiple market segments including business, cities, airports, hotels, and convention centers. Given the emergence of VoIP support in the core networks, perhaps service providers and enterprises now see it as natural to offer VoIP in the access networks to achieve the economies of scale that can be achieved with converged networks.

A major piece of the puzzle in the converged networks involves mobility management to support real-time applications. In the next few chapters, we focus on two key problems in *real-time mobility management*: how to quickly establish IP connectivity in handovers and how to make handovers smooth for applications. All services of interest to us depend upon Internet connectivity. Thus, reestablishing routing paths to the Internet (i.e., IP connectivity) in the presence of user mobility becomes a crucial problem. Once a mobile node establishes basic IP network connectivity, steps can be taken to make sure that transport protocols, such as TCP and RTP, do not suffer performance degradation due to mobility. Our goal is to describe these routing and transport enhancements to provide IP mobility management especially for, but not limited to, real-time applications. In addition, we discuss a variation of Mobile IPv6, called Hierarchical Mobile IPv6, that allows a mobile node to use a visited network home agent, which has the benefits of reducing signaling and allowing a mobile node not to reveal its actual care-of address to its correspondents; an extension of this is the topic of *location privacy*, which we discuss in a subsequent part of this book. In addition, we walk through the source code of an implementation of fast handovers in Chapter 15. We hope that these chapters provide insights and practical guidelines on how IP can be used to provide advanced mobility management.

14

Fast Handovers

The biggest difference between time and space is that you can't reuse time. – Merrick Furst

—

In this chapter, we introduce the basic concepts behind fast IP handover, which allows a mobile node to quickly regain its IP connectivity in order to send and receive its payload immediately following a handover. Fast handovers are necessary for real-time applications such as VoIP. Since delay and packet loss are significantly reduced, they assist traditional TCP applications as well. We describe the Fast Handovers protocol specification [11] itself in Chapter 15.In Chapter 16, we describe context transfer, a technique to smooth the effect of routing change on transport protocol performance by eliminating the need to re-establish any network-resident contexts, such as access control and QoS. Together, the protocols provide necessary ingredients to build effective system solutions for supporting a variety of applications in a mobile networking environment.

14.1 SNAPSHOT OF A MOBILE NODE'S ACTIONS

Consider a mobile node that attaches to its access network, typically through a wireless link, using an *access point* or a base station. The access point provides link connectivity, and the access router provides IP connectivity. Although the access point and access router are separable functional entities, they both could be integrated in a single physical device. Since we are interested in IP layer mobility, let us focus on the Access Router, using Figure 14.1 as a reference in the following discussion.

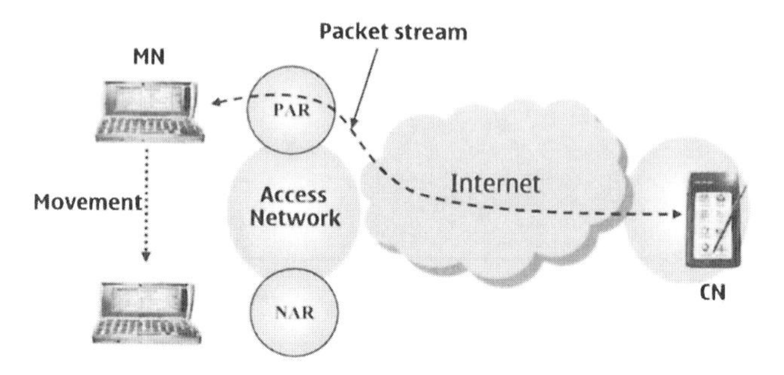

Fig. 14.1 Mobility Reference Diagram

The actions performed by a mobile node can be summed up by the following:

while (true) do

{

1. Establish link connectivity
2. Establish IP connectivity
3. Request network resources
4. Run application(s)
5. Undergo handover
6. Go back to step 1

}

Once a mobile node powers up, it first establishes link connectivity. It then performs Neighbor Discovery operations in order to establish IP connectivity. In IPv6, as we saw in previous chapters, these operations involve configuring a link-local address, which allows on-link communication only [5] using a well-known network prefix (`FE80::`) and the mobile node's interface identifier. The mobile node then has to ensure that its new link-local address is unique. We also saw that address autoconfiguration [19] allows a node to configure its IP address without requiring any stateful inspection on the network side. This process simplifies address assignment but requires each node to ensure the uniqueness of its address by performing Duplicate Address Detection (DAD). After configuring its link-local IP address, the mobile node performs router discovery [16], allowing it to identify its default router as well as to create a globally routable IP address. During this process, the network may require that the mobile node present its credentials for network access authentication and authorization [1]. The successful authentication results in establishing a network access control state or *context* on the access router. After these operations, an IPv6 mobile node is capable of sending and receiving IP packets using its new IP addresses.

The operations for IPv4 are similar, except that the address configuration is mostly performed through the Dynamic Host Configuration Protocol (DHCP).

After a mobile node establishes IP connectivity, it runs some applications, such as VoIP with its correspondent node. An application such as VoIP typically requires some QoS support from the network. This functionality, which reserves desirable *forwarding treatment* to certain distinguished packet streams, again necessitates the establishment of some context for the particular packet stream. This context, for example, can include packet classification, packet metering and packet marking parameters [4]. In addition to QoS, a lower-speed wireless link makes it highly advantageous to implement IP and transport header compression functionality in order to use the expensive, limited bandwidth resources efficiently. The header compression process is stateful, requiring establishing and maintaining context on the access router.

These examples show that context establishment is often necessary once basic connectivity is established. Except for the network access authentication context (established when a mobile node first attaches to the network), the remaining contexts are typically specific to each packet stream. For now, we will not consider how the contexts are established, i.e., the specific signaling methods used.

Now, suppose that the mobile node undergoes handover from its current access router to another. In Figure 14.1, the mobile node moves away from the Previous Access Router (PAR) to the New Access Router (NAR). It is important to separate the algorithm used for selecting a target router from the actual handover mechanics itself, since multiple approaches to target router selection can be conceived. We identify a few such approaches later on. For now, let us assume that either the mobile node or its default router know which target router the MN will hand over to. During this handover process, the mobile node has to relinquish its current link and establish a new one. Then the whole process above repeats.

With this background, we can formulate the problems we need to address to provide real-time mobility management:

1. How to enable the mobile node to *send* and *receive* IP packets immediately after regaining link connectivity. We call this the *fast handover* problem.

2. How to maintain uninterrupted access to network resources so that the disruption caused by handover for transport protocols is minimized. We call this the *smooth handover* problem.

The fast handover problem is about expeditious routing of IP packets, whereas the smooth handover problem has to do with cushioning the impact of handover on transport protocol performance. Context transfer is a solution to the smooth handover problem, which we will consider in Chapter 16.

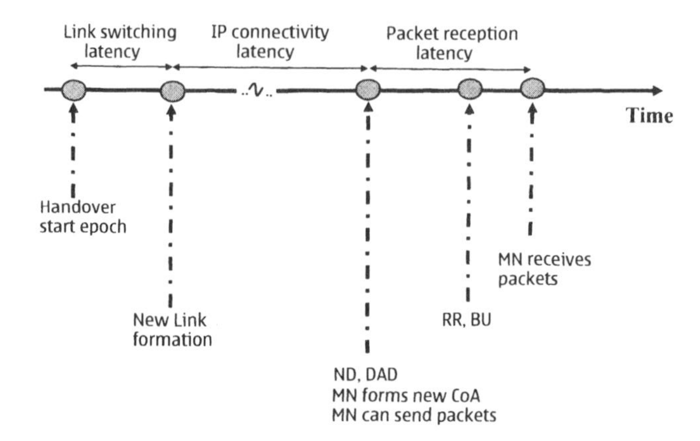

Fig. 14.2 Handover Delay Timeline

14.2 ENABLING FAST HANDOVERS

There are two design points in the fast handover problem. First, the latencies due to subnet movement detection, IP address acquisition and configuration subsequent to handover have to be reduced. This design point addresses the problem of how quickly the mobile node can transmit IP packets. We call this *connectivity latency* improvement. Second, the latency in forwarding IP packets to the mobile node's new IP address must be reduced. Observe that packets continue to arrive at the mobile node's previous IP address until the correspondent node is notified, typically through the Return Routability procedure followed by the Binding Update [9]. These packets have to be rerouted to the mobile node's new IP address until the location update becomes effective. We call this design point *reception latency* improvement. Both connectivity latency and packet reception latency follow *link switching* from the previous to the new link which introduces its own latency. A timeline illustrating these latencies is shown in Figure 14.2. [1]

14.2.1 Connectivity Latency Bottlenecks

A mobile node needs to know its router's advertised network prefix in order to configure a globally unique IP address. With IPv6 stateless address autoconfiguration, the mobile node also needs to perform DAD to ensure address uniqueness. Both of these operations, which we called ND procedures earlier, contribute to the connectivity latency. We looked at the delays involved in address autoconfiguration in detail in Section 2.6.2 in Chapter 2. Before it can even perform the ND operations however, the mobile node has to determine that it has moved to a new subnet.

[1]Since the connectivity latency far exceeds the other two, it is shown with a "break in time".

In order to understand the effect of these operations, consider the default latencies associated with them. The Mobile IP protocol specifies mechanisms for movement detection. It relies on mobility-friendly routers performing unsolicited Router Advertisements more frequently but with at least 30 ms gap, and on lower layers and Neighbor Discovery providing information about subnet change. Even with more frequent Router Advertisements, it correctly recognizes the difficulty of accurately determining subnet movement detection for multiple reasons we outlined in Chapter 11. The "hybrid approach" based on the Neighbor Unreachability Detection (NUD) can lead to variable delay in determining that subnet change has occurred, up to hundreds of milliseconds.

With a confirmed handover to a new subnet, and a unique IP address, the mobile node can send packets. The DAD latency dominates IP address configuration, averaging over a second. Until DAD completes, an application such as VoIP continues to generate packets which are dropped by the operating system due to lack of network interface support.

Even after a mobile node has configured an IP address, it cannot use that address right away with a correspondent without first performing the Return Routability procedure. This is because the correspondent node will simply drop any packet containing the Home Address Option if the source address of the packet is not found in its Binding Cache. Since the mobile node has just configured a new care-of address, it must first convince the correspondent that it is reachable at the new care-of address and then bind the home address to that care-of address. This operation takes at least 1.5 Round-Trip Time (RTT) to the correspondent (assuming that the MN optimistically sends packets immediately after transmitting a Binding Update without waiting for a Binding Acknowledgment), and also contributes to the *effective connectivity latency*.

14.2.2 Reducing the Connectivity Latency

The key concept behind reducing connectivity latencies is to enable a mobile node to learn about its neighborhood before handover takes place. Since a mobile node can detect the possibility for radio communication with adjoining networks, it can also obtain the IP configuration information about them even before it actually attaches to them. Such IP configuration information should allow a mobile node to quickly detect movement to a new subnet and allow fast IP address formulation. This means the mobile node would know its default access router and a prospective IP address prior to leaving the PAR.

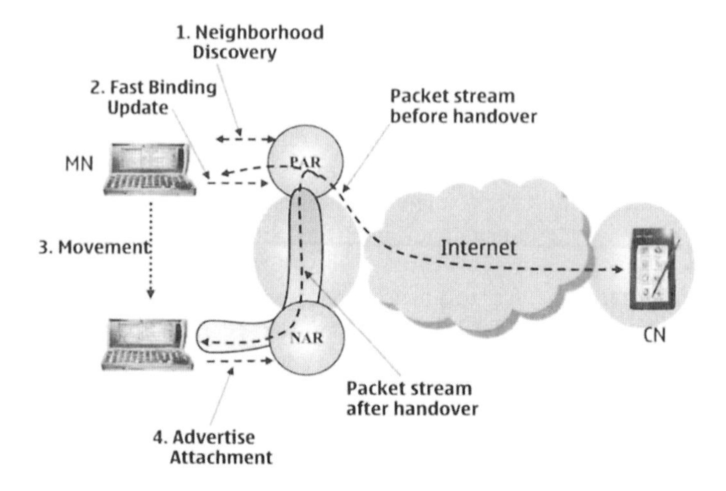

Fig. 14.3 Fast Handover Protocol

As earlier, when a mobile node's network interface is enabled, the link-layer operations establish a communication link with the network. For instance, in IEEE 802.11 systems, the wireless interface establishes *association* with an appropriate access point according to the IEEE 802.11 system specification. Once the link is available, the IP stack performs router discovery to learn the default router's parameters and the routing prefix for address configuration. [2]. When establishing link connectivity, however, the mobile node may be able to discover all the available radio access points or base stations in the neighborhood. This discovery produces information such as the access point identifier, the operating channel, available maximum bandwidth and so on, for each access point. Using the list of access point identifiers, the mobile node proceeds to discover the subnet information corresponding to the individual access points. We call this *resolving an access point.* Essentially, the mobile node supplies the link-layer identifiers of one or more access points and requests its default router to resolve them to subnet-specific information in a *Router Solicitation for Proxy Advertisement* (RtSolPr) message. This is like performing reverse address translation but for a node on a different link. The default router then provides in a *Proxy Router Advertisement* (PrRtAdv) the subnet information including the prefix, IP address and link-layer address of the router serving the access point whose identifier the mobile node has supplied. The mobile node then builds a tuple [Access Point ID, Subnet-information] for its neighborhood, constituting *neighborhood discovery* (contrast with ND!). This is illustrated in Figure 14.3.

[2]When router discovery indicates stateful address assignment, the mobile node has to perform DHCP to configure an IP address for its interface.

The result of neighborhood discovery is that the mobile node has a map of all the available radio access points and the corresponding IP configuration information to use should the mobile node attach to any of them. Such a map can have multiple entries in a WLAN environment, and fewer entries in a Wireless Wide Area Network (WWAN). In some networks, the network discovery may not be done after powering up and may be done only during a handover. An existing map may be updated at any time. In any case, the neighborhood map provides the information the mobile node needs to avoid the latencies due to movement detection and IP configuration. The neighborhood map allows a mobile node to quickly detect movement because link connectivity with an access point whose identifier is mapped to a different subnet prefix than the one the mobile node is currently using implies subnet traversal. The map enables fast IP configuration because once movement to a particular subnet is determined or impending, the mobile node will know the default router's IP and link-layer addresses to use, as well as the prospective new IP address it can use.

The mobile node may perform neighborhood discovery at any time. It is quite useful, however, to do it when an interface becomes available since the handover process itself, which can take place any time once the interface is enabled, can be decoupled from the process of learning about the neighborhood. This is a important for several reasons. First, the handover mechanics are highly delay-sensitive, and hence it is crucial to ensure that the number of operations needed is relatively few. Second, learning about the neighborhood does not necessarily constitute an immediate handover, which may take place (or not) at any time in the future. Finally, handover also can imply managing state information, such as creating and managing a route table entry, defending an address for the mobile node, and so on, all of which are clearly not performed until the actual handover takes place. For these reasons, it is important to decouple neighborhood discovery, which is a form of *handover planning*, from handover execution.

A wireless interface is usually capable of *scanning* for available radio access points at any time. Indeed, it is important to keep the neighborhood map up-to-date. Link-specific technologies determine when the link layer performs such operations. For instance, the radio layer continuously monitors the signal strength with its current access point in both the cellular systems and in WLAN. In either case, new access points may be discovered, although the actual discovery process may vary in cellular systems and WLAN. When a new access point is added to the map, the mobile node should resolve that access point to obtain the subnet-specific information. This keeps the map up-to-date.

The process of scanning for access points needs careful engineering in WLAN. Many wireless cards may automatically follow the scan operation by link handover. Also, the default scan operation itself takes a considerable amount of time. These operations can cause handover delay leading to poor performance for real-time applications. We look at this in detail in Chapter 20.

We saw above that the mobile node's default access router provides neighborhood information when solicited by a mobile node with an access point identifier. We did not describe how an access router is able to construct its neighborhood map. The simplest solution is to statically configure the map, just as system administrators configure their IP subnets and make this database available on all the access routers. When the map can also contain more dynamic information, such as access router capabilities and their instantaneous availability, a new protocol called *Candidate Access Router Discovery* (CARD) has been specified in the IETF [13]. In essence, the CARD protocol addresses the problem of discovering the capabilities associated with the candidate routers so that a mobile node can make its own decision on target router selection. For this, the protocol defines a set of messages for the mobile node to exchange with its default router. These messages allow it to gather an ordered list of capabilities associated with the router connected to an access point. We may observe that such an exchange is an enhancement to the basic access point resolution we described earlier, and hence could be implemented using the same set of message carriers that the fast handover protocol defines. Indeed, the CARD protocol allows piggybacking of all the capability exchange messages over existing RtSolPr and PrRtAdv messages.

The CARD protocol also defines messages for the access routers to exchange capabilities. Exactly when the routers exchange their capabilities is dependent on whether a router has information currently available when a request arrives from a mobile node. This also depends to a certain extent on the mechanism used by the access routers to discover their adjacency. The CARD protocol does not mandate any single method for routers to discover each other and subsequently populate the neighborhood map, but it does describe two approaches.

In the so called *centralized approach*, a server maintains topology information about all the access routers and their access link topology. The access routers perform registration with the server in order to enlist themselves as capable of providing handover support. So, when a request arrives from a mobile node to discover capabilities associated with an access point, the access router first contacts the *CARD server* to resolve the access point identifier to its access router. With a valid IP address from the CARD server, the access router then contacts the adjacent router to discover its capabilities using the message structure the CARD protocol defines.

In a *de-centralized approach using mobile node handover*, the access routers build their map using information provided by mobile nodes that undergo handover between them. So, when a mobile node hands over to a new router, it supplies the previous access point identifier and the IP address of the previous access router to its new access router. The new access router verifies the authenticity of the information provided by the mobile node by contacting the previous access router and supplying it with the mobile node's IP address. If the previous access router approves of the supplied information, the new access router adds an entry for it in its neighborhood map and the two newly discovered routers proceed to exchange their capabilities.

It is worthwhile to point out that what an access router is capable of offering is not the same as what it can offer for a mobile node that is about to undergo handover. In order for an access router to assure a mobile node that the candidate router can indeed offer the advertised capability once the mobile node attaches to that candidate router, the capability information has to be fairly up-to-date. Of course, the access router can immediately query the candidate router, but such a straightforward approach can unduly increase the signaling load as a function of the number of queries, which do not necessarily mean a certain handover. One could argue that an access network link-state routing algorithm, such as Open Shortest Path First (OSPF), can provide the necessary updates. In fact, the QoS-extended OSPF or QOSPF [2] is such a protocol. However, little deployment experience exists, and relatively little attention has been paid to designing routing algorithms for access networks that must support mobility.

It is also worthwhile to observe that adjacency discovery is relatively infrequent, while capability exchange needs to be done fairly often. As a matter of principle, the two can be separated. For instance, adjacency discovery can be attempted when a router bootstraps, while capability exchange can take place between adjacent routers using a suitable mobility-aware routing algorithm.

14.2.3 Bottlenecks in Reception Latency

There is latency in receiving packets from correspondent nodes after the handover because the correspondents' Binding Cache need to be updated with the mobile node's current IP address, the configuration of which itself depends on the connectivity latency. Before the correspondents are updated, the mobile node's home agent needs to be updated. This Binding Update procedure, which we studied in detail in Chapter 7, precedes the Return Routability procedure, which we studied in Chapter 8. Here, we briefly review the Binding Update procedure.

Once a mobile node detects movement to a new subnet and is able to configure a new care-of address, it first sends a Binding Update to its home agent so that any packets arriving at its home network can be forwarded immediately. Next, or simultaneously, it sends the Home Address Test Init (HoTI) message to its correspondent by reverse tunneling through its home agent, and the Care of Test Init (CoTI) message directly to the correspondent. When the corresponding Home Address Test (HoT) and Care of Test (CoT) messages arrive, the mobile node can secure and send a Binding Update message to its correspondent. The mobile node may then optimistically start sending packets without waiting for the Binding Acknowledgment. Even so, roughly 1.5 RTT to the correspondent is necessary before the mobile node can update the binding at its correspondent. This delay can be easily over 100 ms on the Internet[3], and is in addition to the link switching latency. All this time, packets keep arriving at the mobile node's previous care-of address. Note that Binding Update is per correspondent, which only increases the latency when multiple correspondents are involved.

[3]The delay is worse over WWAN such as EV-DO. The time to reach a server located immediately behind such a network averages 100 ms.

The reception latency reflects the Mobile IPv6 signaling delay. It is in addition to the connectivity latency, which reflects the overall handover delay as seen by the IP stack.

14.2.4 Reducing Reception Latency

The approach taken to reduce reception latency is to set up temporary forwarding from the mobile node's previous access router to its new care-of address. Clearly, this forwarding has to be carefully timed with the mobile node's movement. Whether this forwarding can be set up before the mobile node leaves its previous link or only after the mobile node establishes a new link is a good design point to consider. We treat each case separately.

14.2.4.1 Establishing Forwarding Prior to Handover: Assume that a mobile node engaged in a VoIP session has to undergo handover. How the mobile node arrives at this conclusion is worthy of discussion. Typically, handover is necessitated by the falling signal strength and quality of the existing wireless link. This can happen because of user movement. However, in systems such as WLAN, it can also happen due to congestion since there is no admission control. There can also be interference caused by other nodes and devices operating in the same frequency but with much higher power levels (such as microwave ovens). These are the basic considerations that can drive the quality of the signal level low enough that a handover becomes necessary. There can also be other reasons. For instance, a mobile node may discover that a different access point can serve its needs better; e.g., better QoS. The existing network may decide to force the mobile node to handover to a particular access point for load balancing reasons. In such cases, a mobile node may decide to undergo handover even if its existing signal quality and strength are acceptable.

There is considerable interest in standardizing a set of link layer triggers or L2 triggers for initiating handovers. Presumably, such triggers would notify the mobility software in the IP stack when certain events of interest take place. For instance, a mobile node would benefit from the knowledge that its "Link" has just come "Up." A "Link Going Down" trigger can help a mobile node to prepare for a handover when the trigger is delivered sufficiently in advance. A "Link Down" trigger may be less useful from a handover planning perspective, but nevertheless may be useful in reacting to available alternatives. In addition to these basic indications, the IEEE 802.21 working group [8] is defining a set of services that a mobile node can also use in a query-response format to discover available network services. In some ways, there is a common purpose between this effort and the CARD protocol objectives.

It is also instructive to consider the triggers based on their origin. See Figure 14.4. Those triggers which originate from the host itself receive immediate attention. Such host-internal triggers typically emanate as responses to changing network conditions. Examples include dropping signal quality and strength of the existing radio link. Such a trigger can be supplied to a suitable mobility management module that can then take action. These triggers are local by nature and by themselves cannot make handover decisions. In contrast, network-generated triggers could be explicit indicators of handovers. To illustrate how, a small digression to traffic patterns in wired and wireless subnets is informative.

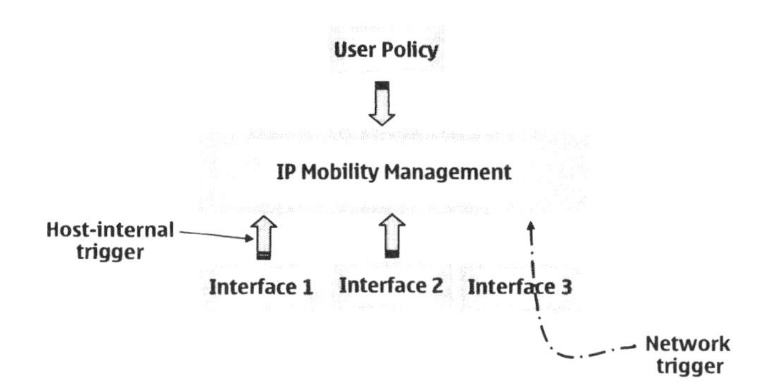

Fig. 14.4 Various Handover Triggers

As radio networks continue to proliferate, traffic management becomes an important topic. Unlike wired networks, where traffic does not reflect host mobility patterns for obvious reasons, wireless networks can create traffic imbalances across different access links and subnets primarily due to the convenience they offer for users to roam freely. This node mobility can create unpredictable traffic patterns and sometimes can cause traffic hot spots. In order to maintain an acceptable traffic load to continue serving real-time applications, a network node such as a router may need to command one or more mobile nodes to handover to neighboring access points. Once it receives such a command, a mobile node may be required to handover or else stand to lose service.

Using network-generated triggers for effective traffic management is a relatively new concept. Indeed, mobility here is viewed as a tool to manage resources in a wireless network, rather than reacting to mobility as a natural user phenomenon.

The third category is user-generated triggers, and this too has only just begun to receive attention. As the network diversity grows with an abundance of wireless networks available, a user may be presented with a tool that monitors the available networks and notifies the user when changes are seen. In turn, the user may effect handover by forcing the mobile node to change its attachment. Such user-initiated handovers present interesting challenges. For instance, a user may switch from WLAN to a cellular network forcing all or some of the connections to undergo handover with the associated performance implications. However, suitable presentation tools need to be developed before users will routinely perform such operations.

Both network-initiated and user-driven handovers represent a deviation from the traditional mobility model in which a handover is a necessity to maintain connectivity and service. Both these modes, on the other hand, are representations of network traffic management and user convenience as much as they also represent the need for connectivity.

Resuming the topic of establishing forwarding prior to handover, the mobile node formulates a tentative new care-of address once the decision to undergo handover to a particular access point is taken. It constructs such an address based on the subnet information it has collected using neighborhood discovery. If the target access point has not been resolved, the mobile node needs to query its access router; however, well-engineered deployments will minimize the likelihood of this happening. The mobile node then sends a a *Fast Binding Update* (FBU) message to the previous access router (see Figure 14.3) requesting it to forward incoming packets to the newly formulated care-of address. Soon afterward, the mobile node should terminate its link connection with the previous access router and establish a new link with the new access router. We will describe the FBU message in detail in Chapter 15.

When the previous access router processes the FBU message, it immediately sets up a forwarding entry in its route table. Hence, any incoming packets will be tunneled to the new care-of address which is valid on the new access router. So, what follows is Neighbor Discovery protocol in action at the new access router: When packets arrive for the new care-of address, the new access router performs Neighbor Solicitation for the new care-of address. If the mobile node is present by then, it will answer the Neighbor Solicitation. If it is not present, then the incoming packets beyond a small buffer are typically dropped. The new router attempts ND after a default time out of 1 second. However, the mobile node may send an unsolicited Neighbor Advertisement right after establishing a new link. This forces the new access router to perform ND again as soon as a new packet arrives and delivers the packet to the mobile node.

In the above scenario where the new access router does not have explicit knowledge of the mobile node's handover, the packet delivery to the mobile node takes place depending on link switching delay, the time when the new access router receives tunneled packets and the ND delay. The best-case scenario is when the mobile node is already present when the new access router performs ND for the very first packet. This is possible, but the delay between the previous and new access routers will be typically less than the link switching delay, making it less likely for the mobile node to be present "just in time." When this is not feasible, the likelihood of packet loss increases at the new access router. The average-case scenario is when the mobile node attaches and sends an unsolicited announcement to the new access router about its presence on the new link regardless of the Neighbor Solicitation from the new router. If the new access router has attempted neighbor resolution earlier, then it records the mobile node's IP and link layer address but performs reachability confirmation; it attempts to receive a Neighbor Advertisement for a direct solicitation. This is done, according to ND [16], because an unsolicited announcement from the mobile node does not ensure that forwarding on the path from the router to the mobile node is necessarily working. So, there is additional signaling before the router confirms reachability, although ND, interestingly, allows a router to forward any queued packets while reachability confirmation is in progress.

If the new access router has not attempted neighbor resolution earlier, for instance no packets have arrived for the mobile node, then it discards the announcement from the mobile node. In this case, the router would do neighbor resolution when a packet arrives. Of course, the worst-case scenario is when the unsolicited announcement is lost, leading to the default resolution, which has an interval of 1 second between retries.

It is perhaps useful to investigate why the default parameter settings in ND contribute to delay during handover. The protocol was designed to work when new connections or sessions are being set up for an address when there are no packet bursts from the application. For new sessions or sessions resuming after an idle time, TCP transmits data using "Slow Start." The UDP-based protocols are typically isochronous, transmitting data at a fixed rate and interval. But, when a protocol such as the Session Initiation Protocol (SIP) is used, there is ample time to first set up a session, and then data transmission ensues. In either of the transport scenarios, there is usually no packet burst to cause the ND buffer holding packets awaiting address resolution to overflow, nor is a packet stream unaccompanied by prior signaling. The initial signaling provides plenty of time for ND to resolve an address while keeping a small queue to buffer packets.

For handover connections or sessions, there is an *existing* packet stream. The handover packet stream arrives without prior "heads up" signaling. A typical ND buffer may not be able to accommodate a burst. Moreover, the ND protocol cannot distinguish handover streams from packets that initiate a new session even if it were to offer additional buffering support for handover streams. This means if *all* address resolution queries are associated with larger buffers, they *can* open up a new Denial-of Service attack where a router is bombarded with large amount of data immediately following the queries for valid addresses that cannot be resolved (because no one owns those addresses).

A slightly enhanced design would make the previous access router send a message about the impending handover to the new access router once the FBU message is processed. Such a *Handover Initiate* (HI) message allows the new access router to allocate necessary buffers beyond what is typically allocated for ND. In the majority of implementations, the ND buffer is kept small to avoid any Denial-of Service attacks just described. The HI message serves as an indication to buffer incoming packets for the mobile node's new care-of address. [4] It also requests the new access router to defend the new care-of address until the mobile node announces itself on the new link. As a result of undertaking such a proxy defense, the new access router can also verify if a node is already using the address the mobile node wishes to use.

We have not specifically considered the issue of address collisions so far. What if there is a node on the new access router's link which is already using the new care-of address that the mobile node wishes to use? In the absence of any explicit signaling between the routers, the new access router will incorrectly forward packets to an unsuspecting node. The possibility of such an address collision is very small, 1 in 2^{64}, since the Interface Identifier field of the IPv6 address is 64 bits long. But a mobile node can abuse the allowance offered by the minuscule probability of collision by intentionally configuring an existing address which it has knowledge of. In such a case, merely assuming that there will not be collision because the probability is small might be somewhat simplistic. In deployments where this is a concern, the HI message provides an opportunity for the new access router to ensure that such misdirection of packet streams does not occur. The new access router can inspect its forwarding tables or access control lists for possible collision with the new care-of address supplied in the HI message. In such deployments, one can expect that the router maintains a guarded list of admitted nodes and hence can look for possible collisions. When there is a conflict, it is reported in a *Handover Acknowledge* (HAck) message back to the previous access router as part of the handover status information. The previous access router then informs the mobile node about the status of FBU processing in a *Fast Binding Acknowledge* (FBack) message. When there is a conflict, the mobile node has to reconfigure a new care-of address. When HI and HAck messages are used, the previous access router can be made to establish forwarding only after it receives a confirmation from the new access router, thus avoiding any possibility of traffic misdirection.

[4]We will see in Chapter 16 that the HI message can be used for context transfer purposes as well.

The foregoing description is broadly categorized as *predictive handover*. It can work with neighborhood discovery (i.e., RtSolPr and PrRtAdv messages) and fast binding update (i.e., FBU message). The former is typically done ahead of the actual handover, whereas the fast binding update is done only at the time of handover. There are certain advantages to augment this basic signaling with inter-access router messages (HI and HAck) as we discussed above.

Furthermore, receipt of an FBack on the previous link is not required for predictive handovers. A mobile node can leave its link immediately after transmitting the FBU message. It only has to announce its attachment to the new access router, and receive arriving and buffered packets including the FBack. For the most part, FBU sent this way is processed by the previous access router and a forwarding path is established even before the mobile node regains its connectivity. For the times when the previous access router fails to process FBU sent from the previous link, for instance when it is lost, the protocol switches to the reactive mode, which we discuss in the next section. So, essentially, a single message during the actual handover is sufficient to effect routing update. This is important since the handover process gets to be less reliable as the number of messages increase.

Knowledge of neighborhood topology and availability of suitable Application Programming Interfaces (APIs) from the link layer are essential for executing handovers rather than simply undergoing handovers. As we saw, a mobile node may soon be faced with the task of choosing network(s) from among available ones based on its user's preferences, in addition to the objective of maintaining connectivity. So, mobility management will not only have to address fast connectivity establishment, but also respond to user inputs and even network commands. We anticipate that algorithms for multiple interface management for mobility purposes will provide a rich arena for research. In the following section however, we will go over routing update subsequent to handover, which is a necessity since prediction, by definition, can fail.

14.2.4.2 Establishing Forwarding Subsequent to Handover: Sometimes, it may not be feasible to establish a forwarding path before the mobile node leaves its previous access router even if it has built a neighborhood map. The mobile node may be forced to attach to a new access point, which may be connected to a new subnet, due to abrupt changes in its link conditions. Or the Fast Binding Update message may get lost before reaching the router. Because of these reasons, forwarding update can only be effected once the mobile node regains link connectivity. Until a new forwarding entry is created, packet loss can occur.

Even though forwarding cannot be changed early enough, the mobile node can still benefit from reducing the movement detection and IP address configuration latencies. Since the mobile node has knowledge of the neighborhood, it can detect movement to a new subnet, and determine the new access router address and the prospective IP address to use. Since the address collision probability is very small, the mobile node can send an FBU immediately to the previous access router to establish a forwarding entry. However, as we mentioned earlier, some deployments may want to consider accidental or malicious use by acquiring another node's IP address as the care-of address. Safeguards are necessary and should be designed; deployments may choose to enable them or not, depending on their requirements.

The mobile node may need to send an FBU from the new link even if an FBU has been successfully processed previously. For instance, when the mobile node decides to leave its previous link right after transmitting the FBU in order to maximize the overlap of its link switching with the IP messaging operations, it may not receive an FBack on the previous link. Without receiving FBack, the mobile node cannot be certain that the FBU has been processed successfully, however. Hence, it resends the FBU even if it was able to send the FBU from the previous link. If it never sent the FBU previously, it sends it after reclaiming link connectivity. Sending FBU from the new link constitutes the *reactive handover*, regardless of whether it has actually been processed by the previous access router earlier. If an FBack has not been received, the mobile node cannot be certain if its FBU has been processed. On the other hand, when the FBU has been processed (but the mobile node has no knowledge of it) we would like the mobile node to receive packets right away. This brings us to the discussion of announcing the mobile node's attachment.

14.2.4.3 Announcing Attachment to the New Link: In order to receive packets as soon as it regains new link connectivity, the new access router must learn the mobile node's attachment as soon as possible. We discussed at length in previous sections how a mobile node might be able to circumvent connectivity latency. Equipped with those enhancements for movement detection and new care-of address configuration, the mobile node can almost immediately announce its attachment to the new access router so that arriving and any buffered packets can be delivered to the mobile node right away. In [11], this announcement is done using the *Fast Neighbor Advertisement* (FNA) message, which includes the mobile node's new care-of address and its link-layer address.

The FNA serves two purposes. First, it informs the new access router about the mobile node's attachment. When forwarding has been established before the mobile node's attachment to the new link, packets are already flowing to the new care-of address, and some may even be buffered at the new access router. These packets can be delivered to the mobile node as soon as FNA is processed by the new access router. This packet delivery takes place according to the Neighbor Discovery protocol which we have studied earlier (see for instance, Chapter 2 and Chapter 11). Let us look at just some important steps.

When a router has not resolved an IP address to its layer 2 address, it maintains the corresponding Neighbor Cache entry in INCOMPLETE state. When it hears an unsolicited advertisement such as FNA, the router *probes* the reachability of the address while changing the Neighbor Cache entry to STALE. This means it sends another Neighbor Solicitation, but attempts to deliver the packet while probing for reachability. In the predictive handover case, this is the likely mode of packet delivery since the mobile node cannot be assumed to be always present to answer the very first call for address resolution.

As an optimization, RFC 4068 suggests optimistically setting the state to REACHABLE when the router hears the FNA message. This would not be strictly compliant with the ND protocol requirements; however, the constraints imposed by wireless network and real-time applications are far more recent than the original design of the ND protocol. Nevertheless, the fast handover protocol can provide the desired performance in delivering packets to the mobile node even with the original ND protocol as we discussed in the previous paragraph. It would have to incur the additional signaling for reachability confirmation with the original ND protocol, even on links where such probing may not always necessary however.

This optimization was originally conceived to be used with the HI and HAck messages between the access routers. Recall that the new access router begins to defend the new care-of address once it is deemed acceptable. It does this by maintaining a "proxy" state associated with the address. So, when it hears the FNA message, it changes the state to REACHABLE and begins forwarding buffered and arriving packets to the mobile node.

When the new access router receives an FNA message, it may have no entry for the new care-of address in its Neighbor Cache at all. For instance, no packets may have been received previously for the new care-of address because the binding has not been established at the previous access router. In this case, the ND protocol would simply ignore the announcement message. Once the binding is established, packets will begin arriving, and the new access router performs address resolution by sending the Neighbor Solicitation message for the new care-of address. When the mobile node responds with a Neighbor Advertisement, the Neighbor Cache entry is immediately established in REACHABLE state followed by packet forwarding. As with the predictive handover, this scenario also works seamlessly with the base ND protocol.

Indeed, because the fast handover protocol provides the desired level of delay and packet loss performance without requiring any changes to the base ND protocol, the need for a special message (i.e., FNA) itself has been debated extensively. It has often been suggested that an unsolicited Neighbor Advertisement message with the 'O'verride bit set to zero should be sufficient for the purpose. The original FNA message was just that, with a different Type number to distinguish it from its ND cousin, in order to provide some features (such as improved buffering) specifically for fast handovers. That design had to be consolidated with the other purpose of the FNA message: provide some mechanisms to protect against address conflicts.

Our description of announcing attachment above has not considered the possibility of another node being present on the new access router's link with new care-of address as its IP address. This is justifiable because of the probability of a random collision in extremely small, 1 in 2^{64}. We could consider this sufficient to ignore address collisions for all practical purposes. However, as we discussed in Section 14.2.4.1, collisions can happen, and sometimes they may not be random. Deployments may demand some form of protection. We also saw in Section 14.2.4.1 that the access routers can exchange signaling to reduce the possibility of such collisions. However, what if a deployment did not allow the use of such communication? For instance, two different administrative domains may not support inter-access router communication. In such cases, it should still be possible for the new access router to avert any misdirection of traffic.

Specifically, when a mobile node sends FBU from the new access router's link immediately following the FNA message, it has no knowledge of the uniqueness of new care-of address. Or, it may accidentally or with malicious intent choose an existing address as its new care-of address. If the new access router detects collision when processing the FNA message, it cannot stop a binding being established at the previous access router because of the FBU message, unless the two messages are somehow synchronized. This is where encapsulation is handy and useful. If we allowed the FBU to be encapsulated within FNA, then the new access router could drop FBU when it detects a collision for the new care-of address present in FNA. There was one problem in realizing this: the original ICMP format of FNA did not allow another packet to be encapsulated. Primarily because of this reason, the format of FNA was changed to Mobility Header, and allowance made to include another packet header in the Next Header field. All in all, such encapsulation was recommended to be used (where address collisions were a concern) but not mandated. While this design allowed an important feature, the downside is that the ND state machinery needed to be accessed by the mobility software processing the FNA message (just like in setting the Neighbor Cache state directly to REACHABLE above).

In order to allow the fast handover protocol to work with the least amount of interfacing with the ND protocol, the FNA message has to be made almost identical to the unsolicited Neighbor Advertisement message. This has the consequence of removing the provision for encapsulation. So, we will be obliged to allow FBU and FNA to operate independently, which means we can only consider recovering from traffic misdirection rather than trying to avoid it in the first place. It remains to be seen which of these options will be preferred.

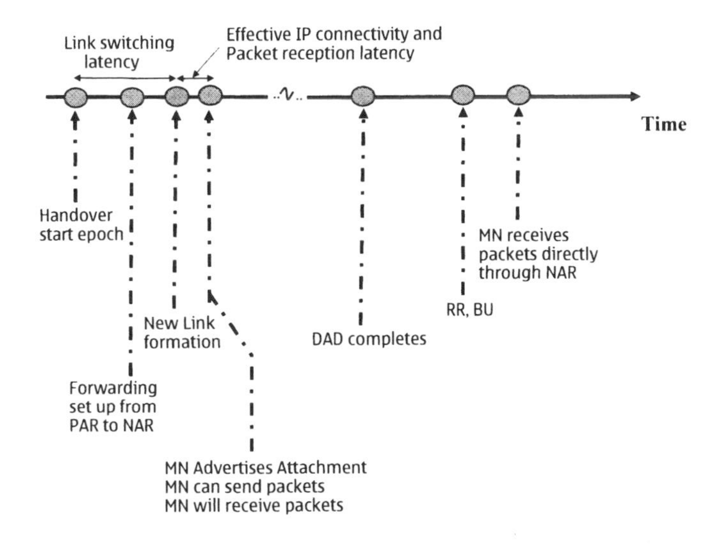

Fig. 14.5 Improved Handover Delay Timeline

Assuming that there is no encapsulation of FBU within FNA, let us consider the scenario where the router detects an existing entry for the new care-of address. It should immediately send a Router Advertisement to the mobile node indicating a collision. The new access router may also supply a different care-of address to use; a special case is when the previous care-of address is to be used. By this time, packets may already be arriving for the unsuspecting node since the FBU is processed independently. Only after the mobile node sends another FBU with a collision-free address, this misdirection will stop. There are different considerations based on the whether there is already an entry or not and the order of response if there is no existing entry. We leave this as an exercise for the reader.

In order to be able to send an FNA and perhaps an FBU as well, soon after movement to a new subnet is detected, the IP stack can benefit from a "Link Up" indication from the link layer. The IP stack can then verify if the new link also means a new subnet using the neighborhood map. If the mobile node has indeed moved to a new subnet, the IP stack has to verify if it has received FBack for the new address it can use. If not, an FBU has to be sent. In any event, an FNA has to be sent to announce attachment and, in the event that no FBack has been received, to confirm the tentative address.

14.2.4.4 Concluding Thoughts on Reception Latency: As we conclude the description of improving reception latency, we make several observations. First, in order for the packets to follow the mobile node as it undergoes handover, the forwarding path should preferably be set up before the mobile node leaves its existing link. This requires some handover planning, which already exists in cellular systems. Even in a WLAN, basic control logic exists for when to effect a handover. These operations have to be closely integrated with IP operations in order to design systems for supporting real-time packet-data applications. Such integration is a function of attentive system design of the link layer and the IP stack, as well as modifications of handover operations performed at the link layer. For instance, a system could be designed to provide a "Link Up" trigger to the IP stack, re-designing the link layer to not perform a comprehensive scan of neighborhood access points during an impending handover. This needs more attention and some modifications to the current design. We will address this issue in more detail in Chapter 20, when we explore implementing fast handovers in a WLAN.

Second, as forwarding is set up before the mobile node moves out of coverage, buffering support for handover packet streams is extremely useful. Such buffering, when complemented with buffering at the host itself (e.g., of, acknowledgment packets), can effectively reduce packet loss and hence improve the performance of streaming and TCP-based applications. The usefulness of buffered packets for interactive real-time applications such as VoIP has been a subject of some debate. Conventional wisdom would dictate that delivering packets beyond their play-out deadline is as good as discarding them. While this might be true for traditional circuit-switched voice where there is effectively no jitter, applications in a packet-switched network need to be prepared for network jitter. Even though the IP networks are being enhanced to offer differentiated forwarding services based on application characteristics, applications still need to be *elastic* in handling variable delay, and this applies to VoIP players. Indeed, most VoIP players use jitter buffers for cushioning.

There is also adaptive play-out based on elongating and compressing silence between talkspurts [18]. The essential idea is to postpone playing the beginning of a talkspurt until at least a few packets are in the player's buffer, depending on the play-out deadline which tracks the current network delay. Similarly, an existing talkspurt can be artificially extended when there are too many packets in a burst. These considerations lead us to argue that a mobile node should be in charge of deciding whether to use or discard packets delayed by the handover, rather than the network itself deciding to drop the packets.

When forwarding cannot be set up before the mobile node leaves its link, the previous access router can keep forwarding packets on the previous link, where they will be lost. Where such unpredictable handovers are anticipated, the the previous access router could maintain a copy of each packet forwarded in a small buffer. So, when the forwarding request arrives subsequent to handover, the previous access router can forward the buffered packets prior to delivering incoming packets. Such a scheme, while not perfect, could still reduce the amount of packet loss. The experimental evidence for instrumenting buffer size, when to activate the buffer and so on, is yet to be determined however.

Finally, the design for improving reception latency is intertwined with the problem of ensuring address uniqueness. This is because it is not feasible to require a mobile node to remain on the previous link to receive confirmation of the uniqueness of its proposed address. Such a constraint may not always be possible to enforce, and it also prevents the possibility of overlapping the link switching delay with IP handover messages. Hence, the mechanism should allow optimistically using the new care-of address while ensuring that proper hooks exist to verify uniqueness, especially when accidental or malicious abuse is a concern. The FNA message serves this dual purpose of both announcing and confirming the new CoA. In addition, when an FBU needs to be sent from the new link, encapsulation allows the new access router the option of dropping it if address conflict does occur.

Together, the above solutions for connectivity and reception latency design points allow a mobile node to quickly establish IP connectivity at the NAR. Figure 14.5 shows the handover delay with the improvements. As soon as the new link is established (see Section 14.2.1), the mobile node can start sending *and receiving* IP packets immediately. This is the essence of fast IP handover. The mobile node may then send a normal Binding Update [9] to its mobility agent and the correspondent node(s) so that they can begin packet transmission directly to the mobile node's new IP address.

14.2.4.5 When a Neighborhood Map Is Unavailable: In some cases, a mobile node may not be able to construct the neighborhood map. The protocol work being developed in the DNA working group of the IETF [6] can be useful in such cases. A mobile node that implements the DNA movement detection protocol (being worked on) can detect movement across subnet boundaries from a router that also implements a corresponding protocol. [5] The mobile node then runs the DNA router discovery protocol to obtain new access router information from a co-operating access router. Subsequently, it runs the optimistic DAD protocol [15] to configure its new IPv6 address. At this point, the mobile node can send an FBU message to its PAR. In this way, the reactive mode of fast handovers takes place.

The design principle behind fast handovers is to support handover planning while accommodating even the cases where no neighborhood map is available. In such a situation, *all* signaling, including movement detection, IP configuration and FBU, takes place after the mobile node regains link connectivity. Since this signaling is on the critical timepath, reliability concerns also increase in addition to the time taken to perform the signaling operations. Hence, performing all signaling subsequent to handover will be inferior to performing signaling necessary to announce attachment alone. Nevertheless, this may still be needed under some operating conditions, and it is still better than having no support at all.

[5] All but one movement detection protocol requires no changes to an access router.

14.3 UNDERSTANDING THE IMPLICATIONS OF PACKET REROUTING

In the past few sections, we have studied how to improve connectivity latency and reception latency. We have studied how the knowledge of the neighborhood can be used to effectively eliminate delays due to movement detection and IP configuration. We have also considered how forwarding update can be synchronized with the mobile node's handover to overcome reception latency. In this section, we will consider some implications of these techniques.

The process of binding the previous care-of address to the new care-of address has to address the problems that the basic Mobile IPv6 protocol is faced with: A bogus node can *steal* traffic if it succeeds in binding an innocent node's IP address to itself. It can also *flood* traffic to some other innocent node by redirecting traffic. Finally, a genuine mobile node itself can cause flooding by accidental or intentional redirection. All these concerns are valid for fast handovers, just as they are for Mobile IP since the routing change is initiated by a host. The problem of establishing a security association between a mobile node and any arbitrary access router is more general than provisioning an SA between a mobile node and its home agent, where natural trust exists between the two. Nevertheless, the FBU, which causes forwarding change, must be protected using an appropriate security mechanism between a mobile node and its access router. Exactly how this security is provisioned is largely dependent on what mechanism is used to allow a mobile node to attach to the router itself in the first place. This is because the process of access control typically also establishes secret keys necessary to protect signaling. For instance, if a AAA infrastructure is used, then suitable keys can be derived for use between a mobile node and its access router using such an infrastructure [7, 3, 17].

If an access router can trust a mobile node by virtue of some mechanism, for example, using the AAA infrastructure, then perhaps it can trust the signaling messages that the mobile node sends as long as the messages are appropriately protected. So, when the initial authentication succeeds, the access router stores the security parameters including the session key(s), the algorithm to use for ciphering, and so on, and the mobile node's IP address in its database for future use. As long as correctly formed and secured signaling messages arrive from the same IP address, the access router is assured that it is dealing with the same mobile node. Specifically, when a secure FBU message is processed, the access router is certain that traffic stealing cannot take place since the IP address (i.e., the previous CoA) and other parameters could only be encrypted by the node which has the same session key as the access router. No other node can possess both the IP address and the session key. The FBU message has a format similar to that of the Mobile IP Binding Update message. So, it can be protected by IPsec [9], or Binding Authorization Data (BAD) option.

Another example of secure access is SEND, which protects ND signaling between hosts and routers using Cryptographically Generated Addresses (CGAs). The mobile node and the router can derive and exchange a shared secret and protect it using the SEND public keys, although whether a public key could be used for encrypting even such infrequent handover key exchange is being debated. Deriving keys for handover purposes using access authentication is being developed in the IETF working groups. For instance, the MIPSHOP group is specifically addressing this problem for SEND and AAA.

Although it is possible to thwart traffic stealing by requiring secure FBUs, flooding can still take place. A genuine mobile node could accidentally choose a new care-of address (i.e., address collision happens), or it could even intentionally redirect traffic to an unsuspecting node. To see this, it helps to consider two scenarios under which an FBU is sent. First, when a mobile node sends an FBU from its previous link (see Section 14.2.4.1), if the previous access router starts forwarding packets immediately to new care-of address, then there is a possibility that the traffic could be misdelivered. So, the previous access router should exchange the HI message providing the new care-of address (and the mobile node's MAC address that will be used on the new link) to the new access router, which should verify if the proposed address is already in use. The address uniqueness test, if done using DAD, is prohibitively delay-prone. Instead, the new access router may consult its forwarding or access control entries to scan for collisions. Should any conflicts be found, it is conveyed back to the previous access router, which establishes a forwarding entry only after the HAck message is received. So, the possibility of flooding is avoided. The second scenario is when the mobile node sends an FBU from the new link. As we described earlier, a mobile node sends an FBU from the new link even if it sent one from the previous link if it has not received a response for the FBU message in the form of FBack. When sent from the new link, the original design allowed encapsulation of FBU in an FNA message to prevent misdelivery if there is any collision. This is specified in [11], but there is also interest in allowing FBU to be sent independently and recovering from collisions should there be any. Also, the look-up of router's database entries is not perfect; a "miss," no matter how small, could happen. So, a mobile node should be required to perform DAD in any event.

The new access router may want to ensure that it is not offering connectivity to a rogue node. To do this, it can utilize the trust it shares with the previous access router to assume that the mobile node is trustworthy. If the previous access router trusted the mobile node, for example by means of some authentication mechanism, then that knowledge is sufficient for the new access router to offer connectivity during handover. The mobile node may be required to perform authentication for other reasons, but such a procedure can take place at a less critical time. The HI message can be used to transitively derive trust between the mobile node and the new access router. For this reason, the new access router may wait until it processes the HI message before approving the forwarding entry it creates for the mobile node when processing the FNA message.

14.3.1 Avoiding Address Collisions Alltogether

One way to address vulnerability to flooding is to always bind the previous care-of address to a well-known address, such as the new access router's address. This way, the previous access router would tunnel incoming packets for the previous care-of address to the new access router's IP address. This is possible with the existing design of the fast handover protocol, which allows this as an option. An implication is that a unique routing entry needs to be created at the new access router when the forwarding tunnel terminates at the new access router's IP address. To see this, observe that the destination address in the inner packet would be previous care-of address, which would have a topologically inconsistent prefix on the new access router's interfaces. So, the new access router would have to create a *host route* entry to forward the packet on its link and not back towards the previous access router. Such entries are not uncommon. For instance, Mobile IPv4 supports them when a mobile node has to use the foreign agent's IP address as its care-of address. Host routing is used in Mobile IPv4 primarily because of the scarcity of freely available addresses. Another implication is that the new access router has to determine how to route packets that the mobile node sends with the previous care-of address as the source IP address. Ingress filtering routers discard packets with topologically inconsistent IP addresses as a means of avoiding source address spoofing and Denial of Service attacks. Typically, the new access router would set up a route table entry to tunnel packets back to the previous access router when it creates a host entry. As before, any such tunnel creation should be performed with care and security. Such bidirectional tunneling support for the previous care-of address is a reasonable approach for IPv4, where the addresses are not abundant, and there is higher likelihood of address collisions due to a smaller host ID. IPv6 does not have address space constraints. However, some implementations may choose to continue supporting previous care-of address for handover traffic. Hence this option is maintained as part of the fast handover specification.

14.4 SUMMARY

In this chapter, we considered the problem of supporting handovers with minimal latency and packet loss. Essentially, the problem for IP fast handovers is to reduce the time taken to regain IP connectivity and, subsequently, receive IP packets. We saw that movement detection, router discovery, and IP configuration contribute to connectivity latency, and that subsequent location update contributes to packet reception latency. We also studied how neighborhood discovery can provide a mobile node with sufficient information to eliminate connectivity latency. With the help of a suitable mobility management module and with assistance from the link layer, predictive handovers also enable packet forwarding for the new address before the mobile node leaves its link, thus reducing packet reception latency to the time necessary for the mobile node to announce itself to the new access router. When prediction is not possible or when it fails, knowledge of the neighborhood still eliminates connectivity latency, allowing forwarding set-up to take place reactively. We also studied the implications of packet forwarding to an address without first confirming its uniqueness, introducing the basic concepts and principles. In the next chapter, we will consider how the protocol actually works by tracing through its steps and operations.

Exercises

14.1 The centralized approach to CARD introduces a server as an additional entity for discovering network adjacency. While this might be disadvantageous, there are benefits, especially in a network with no inherent mobility support. List these advantages.

14.2 In the decentralized approach to CARD, can two access routers be truly adjacent if the mobile node "jumped" an intermediate router at the edge? Does this matter from the perspective of protocol correctness? Also in the decentralized approach to CARD, what happens if a mobile node hands over, but communicates to the new access router after the time when the entry at PAR for the mobile node has expired? Why could the adjacency entries not be stored as hard state?

14.3 Routers typically exchange prefix information using a protocol such as RIP [14]. Design extensions that allow physical neighborhood discovery in addition to IP prefix reachability.

14.4 Design a multicast-based adjacency discovery for IPv6.

14.5 We have seen that DAD introduces undesirable latency during handover. We have considered solutions to "sideline" DAD. Nevertheless, it is important to perform DAD. Explain why DAD is a good thing to do.

14.6 What is the probability of detecting address collision with the router accessing its tables? Assume that at any given time, m out of n nodes have traffic from the router. Is the answer the same if we consider the traffic *to* the router?

14.7 In the predictive handover scenario, what actions must the new access router take after processing the HI message, assuming that there is no address collision? Now consider address collision. If the mobile node leaves its previous link without receiving FBack, what happens when processing FNA?

14.8 Assume that there is no entry for the new care-of address on the new access router's Neighbor Cache, but there is a node on its link with that address. Also assume that there is no binding at the previous access router. Now consider the mobile node on its link which sends an FBU immediately after sending the FNA. What happens to the Neighbor Solicitation from the router when packets arrive for new care-of address?

14.9 What is the effect of a completely loss-free handover on existing TCP connections on the new link? What should be the starting state for the TCP stream undergoing handover?

14.10 Consider a mobile node undergoing handover to two routers in succession, moving from PAR to NAR1 and then to NAR2 in rapid succession before completing Mobile IPv6 operations on NAR1. Extend the protocol to handle this *three-party* handover.

14.11 As a special case of the previous problem, a mobile node may move back swiftly to PAR after visiting NAR. What protocol operations are necessary?

REFERENCES

1. Wireless IP Architecture Based on IETF Protocols, 3GPP2 P.R0001, Version 1.0.0, July 2000.
2. G. Apostolopoules et al. "QoS Routing Mechanisms and OSPF Extensions," RFC 2676, Internet Engineering Task Force, August 1999.
3. N. Asokan, P. Flykt, C. E. Perkins, and T. Eklund, "AAA for IPv6 Network Access" (work in progress), draft-perkins-aaav6-03.txt, March 2001.
4. S. Blake et al. "An Architecture for Differentiated Services," RFC 2475, Internet Engineering Task Force, December 1998.
5. S. Deering and R. Hinden, "Internet Protocol, Version 6 (IPv6) Specification," RFC 2460, Internet Engineering Task Force, December 1998.
6. "Detecting Network Attachment (DNA)" Working Group, Internet Engineering Task Force, http://ietf.org/html.charters/dna-charter.html
7. S. Glass et al. "Mobile IP Authentication, Authorization, and Accounting Requirements," RFC 2977, Internet Engineering Task Force, October 2000.
8. The IEEE 802.21 group. http://www.ieee802.org/21
9. D. Johnson, C. Perkins, and J. Arkko. "Mobility Support in IPv6," RFC 3775, Internet Engineering Task Force,June 2004.
10. J. Kempf (Editor). "Requirements for Layer 2 Protocols to Support Optimized Handover for IP Mobility" (work in progress), draft-manyfolks-l2-mobilereq-00.txt, July 2001.
11. R. Koodli (Editor). "Fast Handovers for Mobile IPv6," RFC 4068, July 2005, Internet Engineering Task Force.
12. R. Koodli and C. E. Perkins. "Fast Handover and Context Transfer in Mobile Networks," ACM Computer Communication Review, Special issue on Wireless Extensions to the Internet, October 2001.
13. M. Liebsch, A. Singh, H. Chaskar, D. Funato, and E. Shim. "Candidate Access Router Discovery," RFC 4066, Internet Engineering Task Force, July 2005.
14. G. Malkin. “RIP Version 2,” RFC 2453, Internet Engineering Task Force, November 1998.
15. N. Moore. "Optimistic Duplicate Address Detection for IPv6," RFC 4429, Internet Engineering Task Force, April 2006.
16. T. Narten, E. Nordmark, and W. Simpson, "Neighbor Discovery for IP Version 6 (IPv6)," RFC 2461, Internet Engineering Task Force, December 1998.
17. C. Perkins and P. Calhoun. "Authentication, Authorization, and Accounting (AAA) Registration Keys for Mobile IPv4," RFC 3957, Internet Engineering Task Force, March 2005.
18. R. Ramjee, J. F. Kurose, D. F. Towsley, and H. Schulzrinne. "Adaptive Playout Mechanisms for Packetized Audio Applications in Wide-Area Networks," in the Proceedings of the IEEE INFOCOM, 1994.
19. S. Thomson and T. Narten, "IPv6 Stateless Address Autoconfiguration," RFC 2462, Internet Engineering Task Force, December 1998.

15
Fast Handovers Protocol

Any computer project will take twice as long as you think it will even when you take into account Hofstadter's law. – Douglas Hofstadter

We studied the basic concepts of fast handovers in the previous chapter. Specifically, we studied the bottlenecks causing delay and packet loss during a handover and how the design addresses movement detection, IP configuration and location update procedures to effectively support real-time applications such as VoIP during a handover. In this chapter, we will study the protocol specification [4] in much greater detail. As with IETF protocols, terms such as "MUST", "SHOULD" and "MAY" have specific formal definitions [1]. For most practical purposes, these terms translate into "Required", "Recommended", and "Optional" respectively.

15.1 NEIGHBORHOOD PREFIX DISCOVERY

15.1.1 Sending Router Solicitation for Proxy Advertisement

A mobile node can perform neighborhood prefix discovery at any time after it has discovered neighborhood access points or base stations. A convenient time to do both neighborhood access point discovery and prefix discovery is right after an interface is enabled and the mobile node has finished router discovery. As we saw in the previous chapter, a mobile node uses the *Router Solicitation for Proxy Advertisement* (RtSolPr) for neighborhood prefix discovery. Neighborhood access point discovery is done using the link-layer specific operations, such as a "scan" procedure in IEEE 802.11 systems. The format of the RtSolPr message is shown in Figure 15.1.

```
 0                   1                   2                   3
 0 1 2 3 4 5 6 7 8 9 0 1 2 3 4 5 6 7 8 9 0 1 2 3 4 5 6 7 8 9 0 1
+-+-+-+-+-+-+-+-+-+-+-+-+-+-+-+-+-+-+-+-+-+-+-+-+-+-+-+-+-+-+-+-+
|     Type      |      Code     |          Checksum             |
+-+-+-+-+-+-+-+-+-+-+-+-+-+-+-+-+-+-+-+-+-+-+-+-+-+-+-+-+-+-+-+-+
|   Subtype     |    Reserved   |          Identifier           |
+-+-+-+-+-+-+-+-+-+-+-+-+-+-+-+-+-+-+-+-+-+-+-+-+-+-+-+-+-+-+-+-+
|   Options ...
+-+-+-+-+-+-+-+-+-+-+-+-
```

Fig. 15.1 Router Solicitation for the Proxy Advertisement (RtSolPr) Message

The packet is sent just like a Router Solicitation message with the source IP address set to the sending interface address and the destination address set to the access router's address or the all routers multicast address. The Type field is specified in [3], the Code is 0 and Checksum is the ICMPv6 checksum. The Subtype value distinguishes the message from other similar messages. The Identifier is used to match the message with a corresponding reply from the router. The Reserved field is usually set to zero, and ignored by the receiver.

Some readers may have observed that the format of the RtSolPr is similar to the Router Solicitation message [5]. Indeed, the purpose is to be able to reuse existing design and software as much as possible. There are some differences, however. All the enhanced mobility protocols are assigned a single ICMP Type number, and the RtSolPr message is distinguished by the Subtype field value of 2. Also, the Identifier provides a means to match queries with replies.

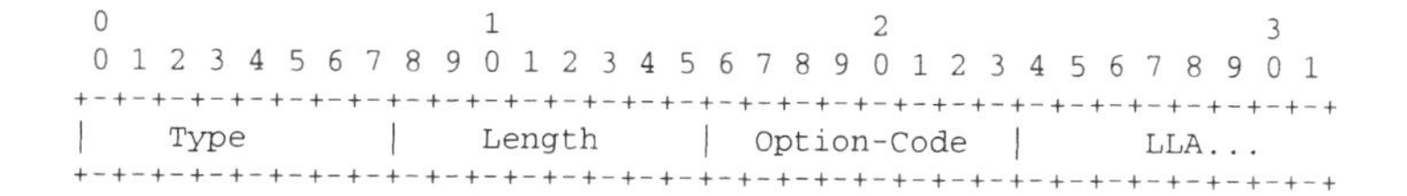

Fig. 15.2 Link-Layer Address Option

The message must contain at least one option that includes the access point identifier in order to be able to discover the neighborhood prefix(es). The generic option format is shown in Figure 15.2, and the header fields are described following the figure.

The options have a fixed format containing the Type, Length and Option-Code fields, followed by the data specific to the option. The LLA option is used by many protocol messages and its fields have the following values:

- Type: Specifies the type of the option carried in the message. Set to 19.
- Length: The size of this option in 8 octets including the Type, Option-Code and Length fields.
- Option-Code:
  0 wild card requesting resolution for all nearby access points
  1 link-layer address of the new access point
  2 link-layer address of the mobile node
  3 link-Layer Address of the NAR (i.e., Proxied Originator)
  4 link-Layer address of the source of the RtSolPr or PrRtAdv message
  5 the access point identified by the LLA belongs to the current interface of the router
  6 prefix information not available for the access point identified by the LLA
  7 fast handovers support not available for the access point identified by the LLA
- LLA: The variable-length link-layer address

So, a mobile node requesting resolution of a single access point would include Option-Code 1, and it should also supply another LLA option (with Option-Code 2) to avoid Neighbor Solicitation by the router (to discover the mobile node's link-layer address if a neighbor entry has expired).

There is no length field for the LLA itself. The implementations need to determine this based on the particular link layer on which the protocol is run. For instance, the length of LLA would be 6 bytes over Ethernet.

There is also a Mobility Header LLA (MH-LLA) option defined for use with messages such as FBU. We refer the reader to [4] for the format details.

It is possible to include requests for additional information about the target access points by defining new option types.

In the following, we include excerpts of RtSolPr implementation at www.fmipv6.org. In order to understand the source code, knowledge of the C programming language is necessary. For brevity, we do not include all the lines in the code. We also do not include all the referencing functions. Those interested in further implementation details are recommended to obtain the software from the above website.

```
/**
 * Sends an RTSOLPR message requesting information on the specified list
 * of access points.
 *
 * @return an errorcode (-ENOMEM or -EINVAL)  or 0 upon success
 */

int car_disc_send_rtsolpr(msg_transaction * rtsolpr_tran)
{
1 struct rtsolpr_descriptor * rtsolpr_desc =
      (struct rtsolpr_descriptor * )rtsolpr_tran->msg_descriptor;
2 struct iovec iov[MAX_AP_NEIGHBOURHOOD_SIZE + 2];
3                        //all APs + LLA + icmp6 hdr
4 ll_address lla;
5 struct list_head* ap_list_entry;
6 int ret = 0;
7 int iov_ind = 1;
8 struct in6_addr* src = &rtsolpr_tran->src;
9 struct in6_addr* dst = &rtsolpr_tran->dst;

10 FMDBG4("Entry. AP_LIST size is %d\n", list_size(rtsolpr_desc->ap_list));

11 //do not retransmit if we are not on the same link any more;
12 if(!md_is_coa_in_use_by_iface(&rtsolpr_tran->src,
                                   rtsolpr_tran->if_index) ){
              FMDBG2("%s is not valid any more -
                         aborting RtSolPr retransmission.\n",
                         inet6_ntoa(&rtsolpr_tran->src));
13 //tell the transaction manager that it needs to end our transaction
14 return -1;
15 }
```

The source code in lines 1 - 9 is declaration and initialization. The `iovec` structure defines a buffer which can be filled with data and passed to the kernel later on. The `iov_ind` keeps track of the current position for `iov` array. The line 10 simply prints the size of the Access Point (AP) list (i.e., indicates how many APs are in the list). It is common to use a macro such as "FMDBG4" [1] for testing and debugging purposes. Lines 11 - 15 verify if the current care-of address has changed on the interface. If so, we should not send the RtSolPr message. This is important since the mobile node may have roamed by the time control has reached this portion of the code, or the mobile node may have roamed after transmitting RtSolPr once but before receiving the PrRtAdv.

[1]The numeral 4 in string "FMDBG4" indicates the debugging level. Only certain levels may be enabled to selectively test and debug software.

```
16 memset(iov, 0, 2*sizeof(struct iovec));
17 /* init the size of the io vector to (AP list size) + 3:
18  * an entry per AP + the icmp6_hdr + the src lla + 1 ap addr in case the
19  * list didn't contain anything.*/

20 /* create the icmp6 message*/
21 if( fmip6_icmp6_create(iov, ICMP6_EXPERIMENTAL_MOBILITY,
                        FMIP6_RTSOLPR_CODE, FMIP6_SUBTYPE_RTSOLPR,
                        rtsolpr_tran->transaction_id) == NULL )
22 {
23              FMDBG("fmip6_icmp6_create failed.\n");
24              return -ENOMEM;
25 }

26 /* Get L2 address of the sending interface */
27 if ((lla.length = get_l2addr(rtsolpr_tran->if_index, lla.address)) < 0)
28 {
29              FMDBG("get_l2addr failed with %d\n", lla.length);
30              return -EINVAL;
31 }

32 /* Create Source Link-layer Address option */
33 /* When known the LLA SHOULD be included (6.1.2) */
34 if (lla.length > 0 && fmip6_icmp6_opt_create(&iov[iov_ind],
                              FMIP6_OPT_LINK_LAYER_ADDRESS,
                              FMIP6_OPT_LINK_LAYER_ADDRESS_OPTCODE_SRC,
                              lla.length,
                              lla.address) == NULL) {
35              FMDBG("Failed to create an L2 Address option");
36              free_iov_data(iov, iov_ind);
37              return -ENOMEM;
38 }
39 iov_ind ++;
```

Line 16 initializes the memory used to create the RtSolPr packet. Lines 21-25 are for creating the ICMP packet with FMIPv6 type definitions; of particular interest are the Type, Code, and Subtype fields defined in Figure 15.1. The rest of the fields in the RtSolPr are filled inside the function `fmip6_icmp6_create()`. The RtSolPr packet should contain the source link-layer address option so that the router can respond with a PrRtAdv immediately without having to perform a Neighbor Solicitation. So, first the LLA is obtained in lines 27 - 31. Subsequently, the Source Link-Layer option is appended to the ICMP packet just created (in lines 21 - 25) by the code in lines 34 - 38.

```
40 /* Create a "New Access Point Link-layer Address" option */
41 /* for each AP address present in the ap_lla_list */
42 /* We MUST have at least one of these so put a wild card
43  * if the list is empty. */

44 if(!list_empty(rtsolpr_desc->ap_list)){
45             FMDBG2("AP list size(%d) - sending contents in an RtSolPr\n",
                                   list_size(rtsolpr_desc->ap_list));
46             list_for_each(ap_list_entry, rtsolpr_desc->ap_list) {
47                      ap_descriptor* ap;
48                      ap = list_entry(ap_list_entry, struct ap_desc, list);


49                      //add to the rtsolpr
50                      if (fmip6_icmp6_opt_create(&iov[iov_ind],
                              FMIP6_OPT_LINK_LAYER_ADDRESS,
                              FMIP6_OPT_LINK_LAYER_ADDRESS_OPTCODE_NAP,
                              ETHER_ADDR_LEN,
                              &ap->l2_addr) == NULL) {
51                                FMDBG("Failed to create LLA option.");
52                                free_iov_data(iov, iov_ind);
53                                return -ENOMEM;
54                      }
55                      iov_ind++;
56             }
57       }
58        else{
59             FMDBG2("AP list is empty - sending a wild card\n");
60             if (fmip6_icmp6_opt_create(&iov[iov_ind],
                        FMIP6_OPT_LINK_LAYER_ADDRESS,
                        FMIP6_OPT_LINK_LAYER_ADDRESS_OPTCODE_WILDCARD,
                        ETHER_ADDR_LEN,
                        &any_ll_addr) == NULL) {
61                         FMDBG("Failed to create LLA option.\n");
62                         free_iov_data(iov, iov_ind);
63                         return -ENOMEM;
64             }
65             iov_ind++;
66       }
```

A New AP LLA option is created for each AP found in the AP list. This is done in lines 44 - 57. The AP list itself is created elsewhere and is specific to each link layer. For instance, it can be created using the *scan* operation in IEEE 802.11. The New AP LLA option is appended to the source LLA option created earlier.

A mobile node may choose to include a wild card instead of one or more APs in the RtSolPr message. See Figure 15.2. This is supported by the code in lines 58 - 64.

```
67 FMDBG3("Sending RtSolPr with iov_ind=%d, src=%s, \n", iov_ind, inet6_ntoa(src));
68 FMDBG3("dst=%s\n", inet6_ntoa(dst));

69 ret = icmp6_send(rtsolpr_tran->if_index, 255, src, dst, iov, iov_ind);
70 if( ret < 0){
71             FMDBG("Failed to send rtsolpr. error=%d!\n", ret);
72             perror("icmp6_send: ");
73 }

74 free_iov_data(iov, iov_ind);
75 FMDBG4("Exit!\n");
76 return ret;
77       }
```

Finally, the RtSolPr packet is sent. Subsequently, the memory associated with the `iov` array of structures is freed, and control returns to the calling function.

15.1.2 Sending a Proxy Router Advertisement

As a response to RtSolPr, the access router sends a *Proxy Router Advertisement* (PrRtAdv) message. This message provides an ordered response to each of the target access points listed in the RtSolPr message.

```
 0                   1                   2                   3
 0 1 2 3 4 5 6 7 8 9 0 1 2 3 4 5 6 7 8 9 0 1 2 3 4 5 6 7 8 9 0 1
+-+-+-+-+-+-+-+-+-+-+-+-+-+-+-+-+-+-+-+-+-+-+-+-+-+-+-+-+-+-+-+-+
|     Type      |     Code      |           Checksum            |
+-+-+-+-+-+-+-+-+-+-+-+-+-+-+-+-+-+-+-+-+-+-+-+-+-+-+-+-+-+-+-+-+
|   Subtype     |   Reserved    |           Identifier          |
+-+-+-+-+-+-+-+-+-+-+-+-+-+-+-+-+-+-+-+-+-+-+-+-+-+-+-+-+-+-+-+-+
|   Options ...
+-+-+-+-+-+-+-+-+-+-+-+-+-
```

Fig. 15.3 Proxy Router Advertisement (PrRtAdv) Message

- IP Fields:
  - Source address: An IP address assigned to the sending interface. Neighbor Discovery protocol [5] requires this address to be the link-local address.
  - Destination Address: Copied either from the source IP address in RtSolPr or from the address the access router is instructing to handover.
- ICMP fields:
  - Type: The Experimental Mobility Protocol Type. Different Types for IPv4 and IPv6. See [3]
  - Code: varies from 0 to 4
  - Checksum: The ICMPv6 checksum
- Subtype: 3 for IPv6
- Reserved: set to zero by the sender and ignored by the receiver.
- Identifier: Copied from Router Solicitation for Proxy Advertisement or set to Zero if unsolicited.

There are multiple possibilities with each access point whose resolution is requested.

If the access router does not recognize the LLA of the target AP, it cannot provide prefix information. Nevertheless, providing an indication as such allows the mobile node to know the router supports the fast handover protocol. So, the mobile node can still send an FBU message once it is able to complete its IP configuration on the new link.

If the target AP is attached to the same router's interface, no further protocol action is necessary. But the mobile node still needs to be informed with an appropriate indication.

If the router has knowledge of prefix information about an access router connected to the target AP, it must supply it as a [AP-ID, AR-Info] tuple, where AP-ID is the AP identifier, AR-Info is an access router's link-layer and IP addresses, and the prefix valid on the interface connected to the access point in question.

If the access router has knowledge that the router connected to the target AP does not support fast handover, then it has to indicate it to the mobile node so that the mobile node falls back to base Mobile IPv6 operations.

If a mobile node requests resolution of all available APs by providing a wild card, then the router supplies as many [AP-ID, AR-Info] tuples as would fit inside a PrRtAdv message. The format of the PrRtAdv message is identical to that of the RtSolPr message; only the options are different. Each [AP-ID, AR-Info] tuple consists of the following options:

1. AP LLA option
2. AR LLA option
3. AR IP Address option
4. AR Prefix Information, only if the prefix used in the AR IP address is different from that advertised on the subnet to which the AP is attached
5. New CoA option, which may be present if PrRtAdv message is sent unsolicited. This is for the scenario in which the handover is initiated by the network.

We have already seen the LLA option format. The IP Address option format is shown in Figure 15.4.

- Type: Specifies the type of the option carried in the message. Set to 17.
- Length: The size of this option in 8 octets including the Type, Option-Code and Length fields.
- Option-Code:
  1 Old Care-of Address
  2 New Care-of Address
  3 NAR's IP address
- Prefix Length: An 8-bit unsigned integer, indicates the number of leading bits in the address that form the prefix.
- IPv6 Address: The IPv6 address in question

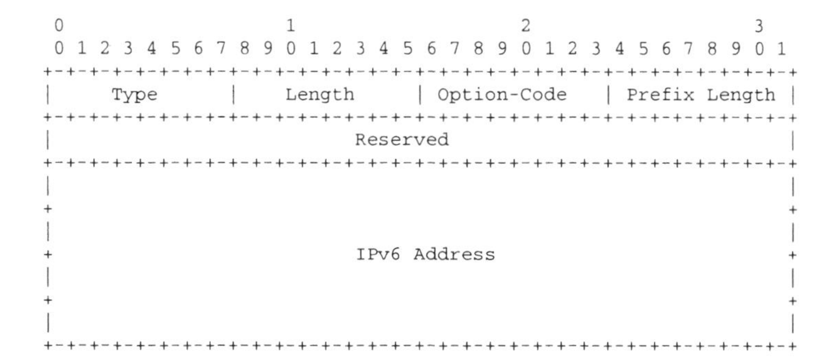

Fig. 15.4 IPv6 Address Option

A PrRtAdv can have multiple Code values:

1. A value of 0 means that at least one [AP-ID, AR-Info] tuple is present. A mobile node can thus make use of this information for movement detection and new care-of address formulation. The LLA option for the AP has the Option-Code of 1, and the LLA option for the new router has the Option-Code of 3.

2. A value of 1 means that the message is sent unsolicited, acting as an indication that the mobile node should undergo handover. See below for further description.

3. A value of 2 means that no prefix information is present. However, the LLA of the attachment point is returned, and the Option-Code values provide further details.

   (a) an Option-Code value of 5 means attachment to the target access point does not require a new care-of address

   (b) an Option-Code value of 6 means the previous router does not have prefix information corresponding to the AP in question. The mobile node may still send an FBU from the new link.

   (c) an Option-Code value of 7 means the new router does not support fast handover. In this case, the mobile node should stop fast handover operations.

4. A value of 3 means that only a subset of new router information corresponding to the requested access points is present in the message.

5. A value of 4 means that the neighborhood information is sent unsolicited but the message is not a handover indication as in the case of Code value 1.

A PrRtAdv must not be treated as a means to ascertain default router settings. Such an operation must only be initiated once the mobile node receives a Router Advertisement on the new link. The mobile node continues to use its current default router until it discovers a new one on the new link. Moreover, the new care-of address formulated using the PrRtAdv message is tentative until it is confirmed to be unique on the new link unless it is supplied by the new router in the HAck message.

Supplying Prefix Length makes it possible for the mobile node to also determine the Prefix from the IP address. So, a separate option for Prefix alone is not necessary. Furthermore, when the new router's address is configured using its LLA as input to the Interface Identifier, the IPv6 address option could also supply the LLA of the router. However, routers typically do not perform address autoconfiguration.

A PrRtAdv message can be sent unsolicited. In some scenarios, the router itself can send a handover indication to the mobile node without the mobile node engaging in the operations first. See Section 14.2.4.1. A router uses the Code value of 1 in an unsolicited PrRtAdv message. In this message, the router could also include a specific new care-of address to use in addition to the [AP-ID, AR-Info] tuple. As a response, the mobile node should send an FBU and handover to the new link within a short period of time. Otherwise, the router may choose to terminate forwarding for the mobile node. Until the FBU is received however, the router should continue forwarding packets to the mobile node.

Note that the Prefix of the new router does not have associated lifetime as in [10]. The expectation here is that when an advertised prefix changes, the access routers would exchange the new prefix information, for example, using a protocol such as CARD. The same holds true for other information such as the access router's LLA and IP address.

In the following, we review the code for sending the PrRtAdv on the previous access router.

```
/**
 * Send PrRtAdv message.
 * @return Exit code, can be:
 * - -EINVAL
 * - -ENOMEM
 * - 0 when PrRtAdv message is sent.
 * @author Martin Andre
 */

static int fmip6_ar_send_prrtadv(
int src_ifindex,
const struct in6_addr *dst,
struct list_head *nap_lla_list,
int is_ni_hover,
uint16_t id)
{
1 struct in6_addr src;
2 struct fmip6_prrtadv *prrtadv;
3 struct iovec iov[50]; // XXX eeek, this is burk
4 ll_address lla;
5 uint8_t prrtadv_code = 255;
6 int iov_ind = 1;
7 struct lla_list *nap_lla_entry = NULL;
8 struct list_head *lp = NULL;
9 car_descriptor * nar = NULL;
10      struct fmip6_opt_ip_addr ip_addr_opt;

11 if (nap_lla_list == NULL) {
12          FMDBG("nap_lla_list is NULL\n");
13          return -EINVAL;
14 }

15 memset(iov, 0, 50*sizeof(struct iovec));

16 /* setting src address to lladdr */
17 if (get_linklocal_addr(src_ifindex, &src) < 0) {
18          FMDBG("get_linklocal_addr error\n");
19          return -1;
20 }

21 /* Get L2 address of the sending interface */
22 if ((lla.length = get_l2addr(src_ifindex, lla.address)) < 0) {
23          FMDBG("get_l2addr error\n");
24          return -EINVAL;
25 }

26 /* Create Source Link-layer Address option */
27 /* This option SHOULD be present */
28 if (lla.length > 0 && fmip6_icmp6_opt_create(&iov[iov_ind],
29          FMIP6_OPT_LINK_LAYER_ADDRESS,
30          FMIP6_OPT_LINK_LAYER_ADDRESS_OPTCODE_SRC,
31          lla.length,
32          lla.address) == NULL) {
33              FMDBG("Failed to create Source Link-layer Address option\n");
34              free_iov_data(iov, iov_ind);
35              return -ENOMEM;
36 }
37 iov_ind ++;
```

The function begins with various declarations in lines 1 - 10. In lines 16 - 25, the code gets the LLA of the sending interface. This is necessary in constructing the source LLA option. In line 28, the source LLA option is created.

```
38 /* Support for wildcard */
39 /* If the first element contains only 0, then this is a wildcard
40  * so we replace the list (containing the wildcard) with a list
41  * containing all known AP */

42 list_for_each(lp, nap_lla_list) {
43         nap_lla_entry = list_entry(lp, struct lla_list, list);
44         break;
45 }

46 if (lladdr_is_any_addr(&nap_lla_entry->lla))
47 {
48        /* First entry is a wildcard: fill LLA list with all known AP */
49        FMDBG2("Found a wildcard\n");
50        struct list_head *car_list_entry = NULL;
51        car_descriptor* car;

52        list_del(&nap_lla_entry->list);
53        free(nap_lla_entry);

54        list_for_each(car_list_entry, &car_list) {
55            car = list_entry(car_list_entry, car_descriptor, list);
56            nap_lla_entry = malloc(sizeof(*nap_lla_entry));
57            if (nap_lla_entry != NULL) {
58               memset(nap_lla_entry, 0, sizeof(*nap_lla_entry));
59               memcpy(&nap_lla_entry->lla.address,
60               &car->nap_lla.address,
61               car->nap_lla.length);
62               nap_lla_entry->lla.length = car->nap_lla.length;
63               list_add_tail(&nap_lla_entry->list, nap_lla_list);
64                 }
65            }
66         }
```

This is laborious code for handling the wild card, where a mobile node requests all available information.

```
67 list_for_each(lp, nap_lla_list) {
68         nap_lla_entry = list_entry(lp, struct lla_list, list);

69         nar = fmip6_car_list_find_nap_lla(&car_list, &nap_lla_entry->lla);

70         if (!nar) { // NAR's LLA not found in CAR list
71            if (prrtadv_code == 255) //prrtadv_code is in initial state
72                prrtadv_code = FMIP6_PRRTADV_NORTINFO;
73            else if (prrtadv_code == FMIP6_PRRTADV_MNTUP)
74                prrtadv_code = FMIP6_PRRTADV_LIMRTINFO;
75         }

76         /* Determine LLA option code */
77         uint8_t lla_opt_code = get_lla_opt_code(nar);

78         /* Create New Access Point Link-layer Address option */
79         /* We're not aware of AR Prefix infos */
80         /* This option MUST be present (copied from the RtSolPr) */
81         if (fmip6_icmp6_opt_create(&iov[iov_ind],
82             FMIP6_OPT_LINK_LAYER_ADDRESS,
83             lla_opt_code,
84             nap_lla_entry->lla.length,
85             nap_lla_entry->lla.address) == NULL) {
86                FMDBG("Failed to create New AP Link-layer Address option\n");
87                free_iov_data(iov, iov_ind);
88                return -ENOMEM;
89         }
90         iov_ind ++;

91         if (lla_opt_code == FMIP6_OPT_LINK_LAYER_ADDRESS_OPTCODE_CURROUTER
                || lla_opt_code == FMIP6_OPT_LINK_LAYER_ADDRESS_OPTCODE_NOPREFIX
                || lla_opt_code == FMIP6_OPT_LINK_LAYER_ADDRESS_OPTCODE_NOSUPPORT)
92          {
93           if (prrtadv_code == 255) //prrtadv_code is in initial state
94               prrtadv_code = FMIP6_PRRTADV_NORTINFO;
95           else if (prrtadv_code == FMIP6_PRRTADV_MNTUP)
96               prrtadv_code = FMIP6_PRRTADV_LIMRTINFO;
97              continue; //Analyze next NAP's LLA since no other options are required
98          }

99          if (prrtadv_code == 255) //prrtadv_code is in initial state
100             prrtadv_code = FMIP6_PRRTADV_MNTUP;
101         else if (prrtadv_code == FMIP6_PRRTADV_NORTINFO)
102             prrtadv_code = FMIP6_PRRTADV_LIMRTINFO;
```

The code in lines 67 - 148 is within a loop. For each AP entry in the AP list, the code builds the required set of options. If no NAR entry is found for a particular AP LLA, the Code value is set to either no prefix information (FMIP6_PRRTADV_NORTINFO) or limited prefix information
(FMIP6_PRRTADV_LIMRTINFO) in lines 68 - 75. Next, the option-code for the New AP LLA is obtained in line 77. Various option-codes are enumerated in [4] [2]. Also, a New AP LLA option with the obtained code is created in lines 81 - 89. Based on the New AP LLA option-code, the Code value for the PrRtAdv message is changed accordingly in lines 91 - 102.

[2]The same option is used in both RtSolPr and PrRtAdv messages.

```
103         /* Create New Router Link-layer Address option */
104         /* This option MUST be present when Code is
105          * FMIP6_PRRTADV_MNTUP or FMIP6_PRRTADV_NI_HOVER */
106         if (fmip6_icmp6_opt_create(&iov[iov_ind],
107             FMIP6_OPT_LINK_LAYER_ADDRESS,
108             FMIP6_OPT_LINK_LAYER_ADDRESS_OPTCODE_NAR,
109             nar->nar_lla.length,
110             nar->nar_lla.address) == NULL) {
111              FMDBG("Failed to create New Router Link-layer Address option\n");
112              free_iov_data(iov, iov_ind);
113              return -ENOMEM;
}
114         iov_ind ++;

115         /* Create New Router's IP Address option */
116         /* This option MUST be present when Code is
117          * FMIP6_PRRTADV_MNTUP or FMIP6_PRRTADV_NI_HOVER */
118         memset(&ip_addr_opt, 0, sizeof(ip_addr_opt));
119         ip_addr_opt.fmip6_opt_prefix_len = nar->nar_prefix_len;
120         memcpy(&ip_addr_opt.fmip6_opt_ip6_address,
                    &nar->nar_addr,
                    sizeof(nar->nar_addr));
121         if (fmip6_icmp6_opt_create(&iov[iov_ind],
122             FMIP6_OPT_IP_ADDRESS,
123             FMIP6_OPT_IP_ADDRESS_OPTCODE_NAR,
124             sizeof(ip_addr_opt),
125             &ip_addr_opt) == NULL) {
126               FMDBG("Failed to create New Router's IP Address option\n");
127               free_iov_data(iov, iov_ind);
128               return -ENOMEM;
129             }
130         iov_ind ++;

131         /* Create New Router Prefix Information option */
132         /* This option MUST be present when Code is
133          * FMIP6_PRRTADV_MNTUP or FMIP6_PRRTADV_NI_HOVER */
134         memset(&ip_addr_opt, 0, sizeof(ip_addr_opt));
135         ip_addr_opt.fmip6_opt_prefix_len = nar->nar_prefix_len;
136         memcpy(&ip_addr_opt.fmip6_opt_ip6_address,
                    &nar->nar_prefix, sizeof(nar->nar_prefix));
137         if (fmip6_icmp6_opt_create(&iov[iov_ind],
138             FMIP6_OPT_NEW_ROUTER_PREFIX_INFO,
139             FMIP6_OPT_NEW_ROUTER_PREFIX_INFO_OPTCODE,
140             sizeof(ip_addr_opt),
141             &ip_addr_opt) == NULL) {
142               FMDBG("Failed to create New Router's Prefix Information option\n");
143               free_iov_data(iov, iov_ind);
144               return -ENOMEM;
145             }
146         iov_ind ++;

147         nap_lla_entry = NULL;
148     }
```

Following the New AP LLA option, the New AR LLA option (lines 106 - 114), the New AR IP Address option (lines 118 - 129), and the New AR Prefix Information option (lines 134 - 145) are created in order.

```
149 /* Support for Network Initiated Handover */
150 if (is_ni_hover) prrtadv_code = FMIP6_PRRTADV_NI_HOVER;

151 prrtadv = (struct fmip6_prrtadv *)
152       fmip6_icmp6_create(iov, ICMP6_EXPERIMENTAL_MOBILITY, prrtadv_code,
                             FMIP6_SUBTYPE_PRRTADV, id);

153 if (prrtadv == NULL) {
154         FMDBG("Failed to create a PrRtAdv message\n");
155         free_iov_data(iov, iov_ind);
156         return -ENOMEM;
157     }

158 int ret = icmp6_send(src_ifindex, 255, &src, dst, iov, iov_ind);
159 if( ret < 0){
160       FMDBG("Failed to send PrRtAdv. error=%d!\n", ret);
161       perror("icmp6_send: ");
162     }
163 else
164        FMDBG2("PrRtAdv sent\n");

165 free_iov_data(iov, iov_ind);

166 return 0;
}
```

Line 150 sets the PrRtAdv Code to network-initiated if an appropriate configuration variable is set. In line 151, the PrRtAdv message is created. [3] Lines 158 - 164 deal with the actual transmission of the packet. We have only shown the sending of the RtSolPr and PrRtAdv messages. Of course, there is processing of the RtSolPr message at an access router before it sends the PrRtAdv message, which the mobile node itself has to process.

15.2 FORWARDING SETUP

15.2.1 Fast Binding Update

At some point when a mobile node decides to undergo handover, or when it has realized that it has changed its subnet router, it sends a Fast Binding Update message in order to request a binding of its previous care-of address with its new care-of address. Such a binding allows the mobile node to receive packets at its new location. The FBU message also allows the mobile node to reverse tunnel its packets to the previous access router. The format of the FBU message is shown in Figure 15.5.

[3] In this case, all the options have been created first before the message has been built.

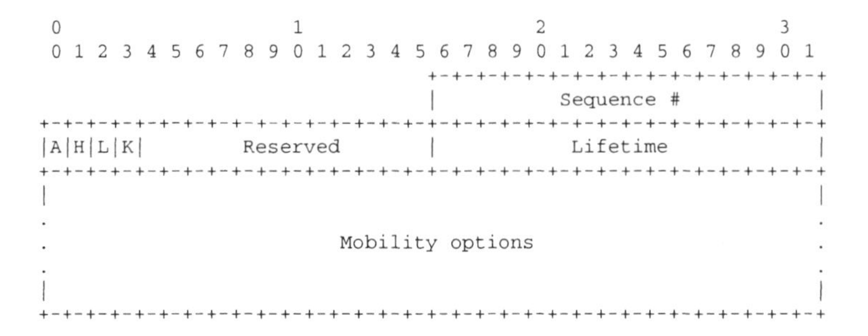

Fig. 15.5 Fast Binding Update (FBU) Message

- Source address: The Previous CoA or New CoA
- Destination address: The IP address of the Previous Access Router
- 'A' flag: MUST be set to one to request PAR to send a Fast Binding Acknowledgment message.
- 'H' flag: MUST be set to one. See [2].
- 'L' flag: See [2].
- 'K' flag: See [2].
- Reserved: This field is unused. MUST be set zero.
- Sequence Number: See [2].
- Lifetime: The time in seconds for which the sender wishes to have a binding between the previous CoA and the new CoA.
- Mobility Options: MUST contain alternate CoA option set to new CoA when an FBU is sent from PAR's link.

This format is identical to the format defined in [2]. Only the usage of certain fields is semantically different. For instance, the source IP address is either the previous care-of address or new care-of address depending on the link from which the FBU is sent. Only when a mobile node determines that it has changed its subnet once it associates with a new AP does it use the new care-of address for its FBU, assuming that it has not received an FBack already. When it uses the previous care-of address as the source IP address, the mobile node needs to indicate to the previous router the IP address to which its packets need to be rerouted. Normally, this address is contained in the source IP address itself. However, since the FBU is sent proactively, the new care-of address is included as an alternate care-of address under the Mobility Options. When the FBU is sent from the new link, this alternate CoA option is not necessary. However, implementations may still choose to include it anyway. The "Home Address" is always the previous care-of address.

The 'H' flag is set to 1 in order to request the previous router to act as a (temporary) home agent. This ensures that the previous router also defends the previous care-of address for Lifetime seconds. The 'A' flag is also set to request the previous router to send an acknowledgment (FBack).

Some may observe that the fast handover protocol is an access network protocol rather than an end-to-end protocol. That is, the protocol operates between a node and its access router. However, the access routers also tend to be home agents (although not necessarily for the same set of mobile nodes). So, the software needed to process the FBU is already present on such routers. Only the neighborhood discovery operation introduces additional messages that the protocol must support. And some networks may choose to use Handover Initiate and Handover Acknowledge messages when using the protocol together with the Context Transfer protocol, discussed in Chapter 16.

The Fast Binding Update is often sent as a response to some handover event, such as a change in link conditions. The exact mechanism used to integrate such changes with the protocol is specific to the link itself as well as the particular implementation. In the following, we review the code and design followed by the fmipv6.org implementation. The overall design is based on "observing" a particular link metric against a configurable threshold and dispatching an event whenever something interesting takes place. As a response to the event, handover actions are performed, including the transmission of FBU.

```
/**
 * The function follows quality over a link and notifies a registered callback
 * when signal goes beneath the threshold that the callback has been registered
 * with.
 *
 * currently the method would periodically send an ioctl asking the interface
 * for the current link state. We should rather register a threshold with
 * SIOCSIWTHRSPY and get notified on rtnetlink. when status goes too low.
 */
static void fmip6_mn_iw_link_qual_observe()
{
1 struct list_head * listener_list_entry, *n;

2 //only do anything if the listener list is not empty
3 if(list_empty(&link_listener_list))
4         return;

        //try to obtain a lock here in order to make sure that any scanning is over
        //and that we'll be sending real values here. if we don't get the lock
        //return immediately so that we don't block the event thread and get inside
        //a deadlock.

5       if(pthread_rwlock_trywrlock(&fmip6_iw_scan_lock) == EBUSY
6       || pthread_rwlock_trywrlock(&fmip6_iw_listener_list_lock) == EBUSY){
7           FMDBG4("Failed to acquire a lock (so else might be
                    scanning). Aborting!\n");
8           return;
9       }

10      list_for_each_safe(listener_list_entry, n, &link_listener_list) {
11          iw_link_listener * listener;
12          int sig_level;
13          listener = list_entry(listener_list_entry, iw_link_listener, list);

14          if(listener->if_index == 0){
            //No need to add code for wildcards here as this is only temporary anyway
            //Quality events should be generated from the driver with the IWTHRSPY thinny
15            FMDBG("We ignored a 0 interface (wildcard)!!!!!!\n");
16            continue;
17          }

18          if(listener->event_type != FMIP6_IW_EVENT_QUAL_THRESHOLD
19             && listener->event_type != 0)
20          continue;

            //get signal level:
21          sig_level = fmip6_mn_iw_get_sig_level(listener->if_index);
22          if(!sig_level){
23            FMDBG("Failed to retrieve signal level for iface %d\n",
                     listener->if_index);
24            continue;
25          }

            //Verify if the threshold has been crossed in the desired direction and
            //trigger an event if so
26          if((listener->direction == FMIP6_IW_EVENT_QUAL_THRESHOLD_DOWN
                && sig_level < listener->threshold
                && listener->last_measured_sig_level >= listener->threshold)
                || (listener->direction == FMIP6_IW_EVENT_QUAL_THRESHOLD_UP
                && sig_level > listener->threshold
                && listener->last_measured_sig_level <= listener->threshold))
27             {
28              iw_link_event evt;

29              FMDBG2("Threshold(%d) for iface(%d)crossed! Notifying listener.\n",
                        listener->threshold, listener->if_index);
```

```
            //init the event
31          evt.src_if_index = listener->if_index;
32          evt.event_type = FMIP6_IW_EVENT_QUAL_THRESHOLD;
33          evt.threshold = listener->threshold;
34          evt.callback_args = listener->callback_args;

            //dispatch
35          (listener->callback)(&evt);
            //remove if this is not a peristent listener
36          if(!listener->persistent)
37             fmip6_mn_iw_remove_link_listener(listener);
38          }else{
39             FMDBG3("WON'T notify:sig_lvl=%d, thresh=%d, prev_thres=%d, dir=%d\n",
                       sig_level, listener->threshold,
                       listener->last_measured_sig_level, listener->direction);
40          }

            //keep that so that we don't repeat the notification for that same event.
41          listener->last_measured_sig_level = sig_level;

42 }

43       _pthread_rwlock_unlock(&fmip6_iw_listener_list_lock);
44       _pthread_rwlock_unlock(&fmip6_iw_scan_lock);
}
```

The code is largely self-explanatory, so we skip describing it any further. Readers may note that the code needs to run with exclusive access, hence the use of lock and unlock constructs for the thread. At some point, a callback happens, eventually bringing the control to the following function module:

```
void fmip6_mn_handle_link_qual_event(iw_link_event * event){
car_descriptor * target_nar;
ap_descriptor * target_ap;

1        FMDBG4("Entry\n");

2        if(!list_size(&available_cars_list)
            || !list_size(&last_detected_ap_list))
3        {
4         FMDBG3("Aborting handover due to lack of necessary predictive info\n");
5         FMDBG3("(%d available_cars and %d usable APs) info\n",
                     list_size(&available_cars_list),
                     list_size(&last_detected_ap_list));
6         return;
7         }

8         FMDBG("RECEIVED A LINK QUALITY EVENT ON IFACE %d for threshold %d!\n",
                 event->src_if_index, event->threshold);

         //XXX maybe we should scan here before going on with the handover ...
         // especially if we have a wifi interface dedicated to scanning
         //XXX well let's assume that we are in the case when the event means
         //that the link is really going away - let's try and do a predictive
         //handover (XXX what we really should be doing is getting such an event
         //on numerous occasions and in some cases only refresh the car list for
         //example, or only start a scan if we have a secondary scanning
         //interface ....). We will get the first AP in the list of APs as it is
         //sorted upon quality and then handover to the router that's behind it.

9        target_ap = fmip6_ap_list_get_first_entry(&last_detected_ap_list);

10       if (target_ap == NULL){
11          FMDBG("Failed to do a handover! AP list was empty (Strange!!)\n");
12          return;
13       }
```

```
14        target_nar = fmip6_car_list_find_nap_lla(
                     &available_cars_list, &target_ap->l2_addr);

15        if (target_nar == NULL){
16            FMDBG("Failed to do a handover! CAR list was empty!\n");
17            return;
18        }
19        fmip6_mn_do_hover(event->src_if_index, target_nar);
20        FMDBG4("Exit\n");
}
```

Lines 1 - 8 do a sanity check and declare that an event has happened. Observe the comments following line 8. It is not particularly desirable to scan during a handover since it will add considerable latency. Implementations should consider *selective scanning* on different link events before handover *and* keep the neighborhood access router list refreshed. In line 9, the AP at the head of the queue is extracted, and in line 14, the corresponding target access router is resolved. The handover function is called in line 19, which we review next.

```
int fmip6_mn_do_hover( int if_index, car_descriptor * nar)
{
1         struct in6_addr* coa = NULL;
2         struct in6_addr* par = NULL;

3         FMDBG4("Entry\n");

          //make sure our quality listener is unregistered cause it'll go nuts during
          //the handover

4         if(fmip6_cfg_monitor_quality_triggers()){
             fmip6_mn_iw_unregister_link_listener(link_quality_listener);
5         }

6         coa = md_get_current_coa(if_index, 1);
7         par = md_get_default_rtr_addr(if_index, 1 );

8         FMDBG3("Our current coa seems to be: %s\n", inet6_ntoa(coa) );
9         FMDBG3("Our current access router is: %s\n", inet6_ntoa(par) );
10        FMDBG3("Preparing to switch to AR:\n");

11        fmip6_car_list_print_element(nar);

12        if(memcmp(&nar->mn_ncoa, &in6addr_any, sizeof(struct in6_addr)) == 0){
13           if( !ipv6_autoconf_addr_gen(if_index, &nar->nar_prefix,
14                  nar->nar_prefix_len, &nar->mn_ncoa)){
15                     FMDBG("Failed to generate an RFC 2462 compliant prefix.");
16                     return -1;
17                }

18           FMDBG3("Generated an autoconf compliant ncoa=%s\n",
                    inet6_ntoa(&nar->mn_ncoa));
}
```

Line 4 is important, since responding to any link quality event when handover has already been decided can cause unstable behavior. The code in lines 12 - 17 formulates a new care-of address for the mobile node using IPv6 stateless address autoconfiguration.

```
        //in case we're moving before receving a BACK for a previous handover -
        //clear all tunnels still active for the current link (if any)

19      homng_remove_tunnel_addr(coa);

20     //make sure that the we know about the target AP.

21     if(fmip6_ap_list_find_lla(&last_detected_ap_list,&nar->nap_lla) == NULL){
22        FMDBG("Failed to perform a handover towards NAR: \n");
23        fmip6_car_list_print_element(nar);
24        FMDBG("Failed to find an AP corresponding to the specified L2 address!\n");
25        return -1;
26     }

27      memcpy(&nar->mn_pcoa, coa, sizeof(struct in6_addr));
28      memcpy(&nar->mn_par, par, sizeof(struct in6_addr));

        //Register a listener so that we get notified whenever the corresponding
        //fback is received

29      predictive_fback_listener = homng_register_evt_listener(
          0, // all interfaces (could be vertical handover)
          HOMNG_FBACK_RECEIVED, // only receive NEW_COA events
          predictive_fback_received, // the function that will handle mvmt evts
          nar,//record the nar so that the evt handler could go on with the hover.
          0);//not persistent - we only want to be notified for this fback

        //Register a listener so that we get notified if no fback is received

30      predictive_fbu_failure_listener = homng_register_evt_listener(
          0, // all interfaces (could be vertical handover)
          HOMNG_FBU_TRAN_FAILED, // only receive NEW_COA events
          predictive_fbu_failed, // the function that will handle mvmt evts
          nar,//record the nar so that the evt handler could go on with the hover.
          0);//not persistent - we only want to be notified for this fback

        //SEND AN FBU TO THE PAR.

31     if( coa == NULL
            || par == NULL
            || homng_send_fbu(if_index, coa, par, &nar->mn_pcoa, &nar->mn_ncoa)<0)
32     {
            //Sending FBU failed. Never mind. Log, go on with hover and and resend
            //after we arrive at NAR
            FMDBG("Failed to send an fbu! Proceding with handover anyway!\n");
            //XXX schedule an FBU for after we arrive at NAR.
33     }

        //the rest of the predictive handover will be completed once the fback has been
        //received or the fbu transaction has failed.

34      return 0;
}
```

The above code is also largely self-explanatory. The `homng_send_fbu()` function will eventually call the `send_fbu_tran()`, where the FBU packet is constructed and sent.

```
int send_fbu_tran(msg_transaction* tran)
{
1       int if_index = tran->if_index;
2       struct in6_addr *src = &tran->src;
3       struct in6_addr *dst = &tran->dst;
4       struct in6_addr *pcoa = &((fbu_descriptor*)(tran->msg_descriptor))->pcoa;
5       struct in6_addr *ncoa = &((fbu_descriptor*)(tran->msg_descriptor))->ncoa;
6       struct ip6_mh_fast_binding_update *fbu;
7       struct iovec iov[4];//hdr +  alt_coa + home_add + lla
8       int iov_ind = 0;
9       int ret = -ENOMEM;
10      struct in6_addr_quad quad;
11      ll_address lla;

12      FMDBG4("Entry\n");

13      memset(iov, 0, 4*sizeof(struct iovec));
14      fbu = (struct ip6_mh_fast_binding_update *)
              mh_create(&iov[iov_ind++], IP6_MH_TYPE_FBU);
15      if (!fbu){
16         FMDBG("mh_create failed!\n");
17         return -ENOMEM;
18      }
```

The base packet of type Mobility Header is created.

```
        /*The following two flags are a must ... XXX add support for L and K flags*/
19      FMDBG2("Transaction ID is %d\n", tran->transaction_id);
20      fbu->ip6mhfbu_seqno = htons(tran->transaction_id);
21      fbu->ip6mhfbu_flags_reserved = IP6_MH_FBU_ACK | IP6_MH_FBU_HOME;
22      fbu->ip6mhfbu_lifetime = htons(1);//FIXME this should come from the a config
```

The Sequence Number, Lifetime and the flags are filled in.

```
        /*Create the alternative coa using NCoA if available*/
23      if(ncoa != NULL){
24         if( mh_create_opt_altcoa(&iov[iov_ind], ncoa)) {
25             FMDBG("mh_create_opt_altcoa failed!\n");
26             free_iov_data(iov, iov_ind);
27             return -ENOMEM;
28         }
29         iov_ind++;
30      }
```

The Alternate CoA option is created and filled with new CoA.

```
        /* Get L2 address of the sending interface */
31      if ((lla.length = get_l2addr(if_index, lla.address)) < 0)
32      {
33         FMDBG("get_l2addr failed with %d\n", lla.length);
34         free_iov_data(iov, iov_ind);
35         return -EINVAL;
36      }

        /* Section 6.3.3 says MH-LLA is a MUST */
37      if (lla.length > 0
            && mh_create_opt_lla(&iov[iov_ind], &lla) < 0)
38      {
39          FMDBG("Failed to create an L2 Address option");
40          free_iov_data(iov, iov_ind);
41          return -ENOMEM;
42      }
43      iov_ind ++;
```

A Mobility Header LLA option in the FBU allows the PAR to readily obtain the LLA for using in the HI message.

```
44      quad.src = src;
45      quad.dst = dst;
46      quad.local_coa = pcoa;
47      quad.remote_coa = NULL;
48      quad.bind_coa = NULL;

49      _pthread_rwlock_wrlock(&fmip6_mn_mh_lock);

50      ret = mh_send(&quad, iov, iov_ind, NULL, if_index);
51      if (ret <= 0)
52         FMDBG("mh_send failed  ret: %d\n", ret);
53      else{
54         FMDBG3("Sent a Fast Binding Update\n");
55      }

56      _pthread_rwlock_unlock(&fmip6_mn_mh_lock);

57      free_iov_data(iov, iov_ind);

58      return ret;
```

The addresses are filled in lines 44-48. The FBU packet is sent, under exclusive access in lines 49-56. The memory associated with the message is freed and control returns to the calling function in lines 57, 58.

15.2.2 Fast Binding Acknowledgment

The previous router sends a Fast Binding Acknowledgment (FBack) message after processing an FBU message. The previous router may be configured to send the FBack message only after it receives a HAck message to ensure that the proposed new care-of address is acceptable to the new router. In some scenarios where the risk of wrongful redirection of traffic is considered small, the FBack may be sent immediately.

The format of the FBack message is shown in Figure 15.6. It is nearly identical to the Mobile IPv6 Binding Acknowledgment message, but with a different Type number.

The header fields are described below.

- IP fields:
  - Source address: The IP address of the Previous Access Router
  - Destination address: The New CoA; optionally, previous CoA as well.
- Status:
  8-bit unsigned integer indicating the result of the Fast Binding Update. Values less than 128 indicate that the FBU was accepted. The following such Status values are currently defined:

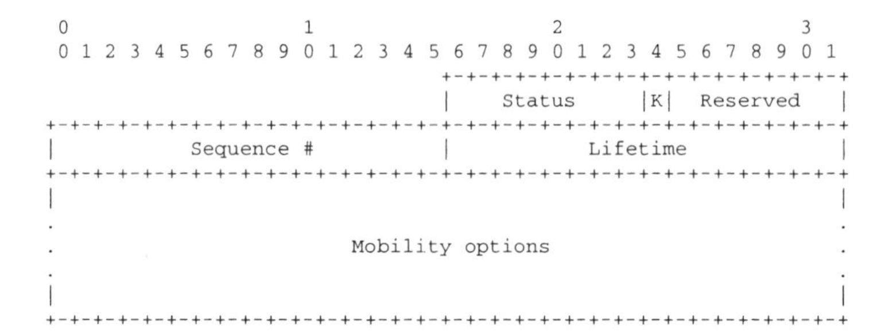

Fig. 15.6 Fast Binding Acknowledgment (FBack) Message

0 FBU accepted
1 FBU accepted but new CoA is invalid. Use new CoA supplied in "alternate" CoA

Values of the Status field greater than or equal to 128 indicate that the FBU was rejected by the receiving node. The following such Status values are currently defined:

128 Reason unspecified
129 Administratively prohibited
130 Insufficient resources
131 Incorrect Interface Identifier length

- 'K' flag: See [2].
- Reserved: MUST be set to zero.
- Sequence Number: Copied from FBU message for use by the mobile node in matching this acknowledgment with an outstanding FBU.
- Lifetime: The granted lifetime in seconds for which the sender of this message will retain a binding for traffic redirection.
- Mobility Options: MUST contain "alternate" CoA if Status is 1.

A mobile node may receive an FBack on its previous link itself, but this is not necessarily assured. A mobile node does not have to wait for FBack on its previous link. For this reason, FBack is always sent to the new care-of address. It is optionally also sent to the previous care-of address. Note that sending to previous care-of address needs to be done prior to the tunnel setup; otherwise, such a message will also be tunneled to the new care-of address.

The protocol recommends that the routers exchange handover messages before a tunnel is set up. As described previously, the Handover Initiate message provides signaling for the new router so that it can differentiate incoming packets (due to handover) from any other traffic. Recall that in the absence of such an indication, the new router would simply perform Neighbor Discovery operations which can lead to discarded packets if the mobile node has still not attached to the new router and, in the worst-case, wrongful redirection of traffic to an unsuspecting node. When a Handover Acknowledge message is received, the previous router can safely assume that the traffic redirection would not cause unintended consequences. When the proposed new care-of address in the FBU is not acceptable to the new router, it should provide an alternate address to use. For instance, in some scenarios, the router itself may provide addresses to use, perhaps acting as a DHCP Proxy. In such cases, the FBack contains the alternate care-of address to use. In a special case, the alternate care-of address provided could be the previous care-of address itself.

When a mobile node leaves its previous link without having received an FBack, the mobile node always sends an FBU from the new link. The FBU could use the same sequence number space as the Mobile IPv6 Binding Update. Sequence number matching allows a mobile node to associate its FBU with an FBack. The behavior is identical to base Mobile IPv6 operation.

Both the FBU and FBack messages can reuse the Mobile IPv6 software with different Type assignments.

```
 0                   1                   2                   3
 0 1 2 3 4 5 6 7 8 9 0 1 2 3 4 5 6 7 8 9 0 1 2 3 4 5 6 7 8 9 0 1
+-+-+-+-+-+-+-+-+-+-+-+-+-+-+-+-+-+-+-+-+-+-+-+-+-+-+-+-+-+-+-+-+
|     Type      |     Code      |           Checksum            |
+-+-+-+-+-+-+-+-+-+-+-+-+-+-+-+-+-+-+-+-+-+-+-+-+-+-+-+-+-+-+-+-+
|   Subtype     |S|U| Reserved  |           Identifier          |
+-+-+-+-+-+-+-+-+-+-+-+-+-+-+-+-+-+-+-+-+-+-+-+-+-+-+-+-+-+-+-+-+
|   Options ...
+-+-+-+-+-+-+-+-+-+-+-+-
```

Fig. 15.7 Handover Initiate (HI) Message

15.3 INTERACCESS ROUTER COMMUNICATION

The Handover Initiate message serves as a means of informing the new router about an impending handover whenever the previous router is able to process an FBU message before the mobile node attaches to the new router. This is desirable for at least two reasons: First, obtain permission from the new router to use new care-of address so that the mobile node is not able to redirect traffic to some existing node, either inadvertently or intentionally. Second, facilitate transfer of any network-resident contexts in conjunction with HI signaling. Even when the previous router processes an FBU message sent from the mobile node's new link, there are advantages to exchanging the HI message with the new router; the contexts can still be useful for the new router. Also, the new router may seek to verify the authenticity of a mobile node by querying the previous router with which it shares a trust relationship. The previous router sends an HI message after processing an FBU message. The format of the HI message is shown in Figure 15.7.

- IP fields:
  - Source Address: The IP address of the Previous Access Router
  - Destination Address: The IP address of the New Access Router
  - Hop Limit: Set to 255, according to RFC 2461.
- ICMP Fields:
  - Type: The Experimental Mobility Protocol Type. This is specified in [3].
  - Code: 0 or 1.
  - Checksum: The ICMPv6 checksum.
- Subtype: 4.
- 'S' flag: When set, this flag requests a new CoA from the receiver. May be set when Code = 0, MUST be 0 when Code = 1.
- 'U' flag: When set, requests the receiver to buffer packets meant for the node identified in the options. Used when Code = 0, SHOULD be set to 0 when Code = 1.
- Reserved: Set to zero by the sender and ignored by the receiver.
- Identifier: Set by the sender so replies can be matched to this message.

The options in the HI message include the mobile node's LLA, new care-of address and previous care-of address. The previous router supplies the mobile node's LLA so that the new router can recognize the mobile node once it attaches itself. The FBU itself may not contain the mobile node's LLA, mainly for compatibility with Mobile IPv6 packet formats. However, including the LLA option in the FBU can readily enable the previous router to include it in the HI message. It is especially important during intertechnology handovers; the mobile node will most likely be required to use different LLAs.

The new router has to verify if the new care-of address is conflict-free. The only sure way is to perform a DAD probe, wait for over a second and then ascertain address uniqueness. DAD probe delay will almost always be undesirable. Given the small likelihood of collision, it is tempting to ignore the DAD operation all together. However, there is room for "non-random" collision, where a mobile node effects collision either accidentally or intentionally. Where this threat is a concern, the new router may look up its database of previously authenticated nodes to observe for collisions. Such a database can be assumed to be present on routers where the threat needs to be addressed. If such a database does not exist or if the threat is not a concern, a router should still consult its forwarding tables to see if there are any collisions. This would ensure verification against at least those nodes for which the router has forwarded packets recently.

When the new care-of address is acceptable, the new router starts defending it, typically by sending Proxy Neighbor Advertisements when any node solicits the address in question. The new router may also buffer any incoming packets for the new care-of address. The size of such a buffer is implementation-dependent, and the new router may choose to wait until an FNA is received before attempting to resolve forwarding to the new care-of address. However, since an FNA may be lost, performing Neighbor Solicitation after a period of time corresponding to the average handover delay might actually be useful. This is left as an exercise to the reader.

In some instances, the new care-of address may be assigned by the new router. In a special scenario, the assigned address is previous care-of address itself. If previous care-of address is supported, for instance in an IPv4-only network, the new router must create a host route entry for previous care-of address, as well as a reverse tunnel for the previous router. Supporting previous care-of address on the new router's subnet link is generally discouraged in IPv6 since there is no scarcity of addresses for the foreseeable future. So, such a support should be used in exceptional cases only for IPv6, for instance a particular deployment chooses to support traffic for previous care-of address without the mobile node changing its address.

The new router responds to HI with a Handover Acknowledge HAck message shown in Figure 15.8.

The only option present in HAck is the IP address option, providing a new care-of address when assigned addressing is used. However, there are multiple Code values providing the status of disposition of the HI message.

```
 0                   1                   2                   3
 0 1 2 3 4 5 6 7 8 9 0 1 2 3 4 5 6 7 8 9 0 1 2 3 4 5 6 7 8 9 0 1
+-+-+-+-+-+-+-+-+-+-+-+-+-+-+-+-+-+-+-+-+-+-+-+-+-+-+-+-+-+-+-+-+
|     Type      |     Code      |           Checksum            |
+-+-+-+-+-+-+-+-+-+-+-+-+-+-+-+-+-+-+-+-+-+-+-+-+-+-+-+-+-+-+-+-+
|   Subtype     |   Reserved    |          Identifier           |
+-+-+-+-+-+-+-+-+-+-+-+-+-+-+-+-+-+-+-+-+-+-+-+-+-+-+-+-+-+-+-+-+
|   Options ...
+-+-+-+-+-+-+-+-+-+-+-+-+-
```

Fig. 15.8 Handover Acknowledge (HAck) Message

The HI and HAck messages can specify options for carrying network-resident contexts. This would allow context transfer to take place in conjunction with fast handovers. We address this in the next chapter.

15.4 ANNOUNCING ATTACHMENT

As soon as a mobile node detects that it has moved to a new subnet, it announces itself to the new router. This is done using the FNA message in [4]. In the event that the mobile node has not received an FBack, it can also encapsulate the FBU within the FNA. The current header type of FNA is Mobility Header, and it contains the LLA option formatted for the Mobility Header with the source IP address set to new care-of address. When a router receives an FNA, it must verify that an entry for new care-of address with a different link-layer address other than the router's own link-layer address is not present to ensure that new care-of address is unique. It must then forward any buffered packets for new care-of address, and stop defending the address for the mobile node. Packets might be present if an FBU has already been processed. If an encapsulated FBU is present, it must forward it like any other packet even if it has processed a HI message. This will allow the previous router to respond with a recent FBack message that matches the sequence number in the most recent FBU message.

If the new router detects a collision for new care-of address, it drops the inner FBU packet. This prevents binding the previous care-of address to an unintended address at the previous router. Subsequently, the new router sends a Router Advertisement with a Neighbor Advertisement ACK (NAACK) option. This message is sent to the new care-of address but using the LLA present in the FNA. The NAACK option may be followed by an IP address option (present after the Reserved field) in which a new care-of address is supplied. See Figure 15.9 for the format of the NAACK option.

The purpose served by the FNA message can also be accomplished by sending an unsolicited Neighbor Advertisement message in the ND protocol. This will not allow encapsulation however, since the unsolicited NA is of type ICMP. Hence, the router will not be able to drop the FBU even if it is able to detect a collision. Hence, only a recovery operation is possible. For this, the NAACK option is sent immediately in a Router Advertisement forcing the mobile node to send another FBU immediately.

```
 0                   1                   2                   3
 0 1 2 3 4 5 6 7 8 9 0 1 2 3 4 5 6 7 8 9 0 1 2 3 4 5 6 7 8 9 0 1
+-+-+-+-+-+-+-+-+-+-+-+-+-+-+-+-+-+-+-+-+-+-+-+-+-+-+-+-+-+-+-+-+
|     Type      |    Length     | Option-Code   |    Status     |
+-+-+-+-+-+-+-+-+-+-+-+-+-+-+-+-+-+-+-+-+-+-+-+-+-+-+-+-+-+-+-+-+
|                           Reserved                            |
+-+-+-+-+-+-+-+-+-+-+-+-+-+-+-+-+-+-+-+-+-+-+-+-+-+-+-+-+-+-+-+-+
```

Fig. 15.9 Neighbor Advertisement Acknowledgment Option

15.5 SUMMARY

In this chapter, we have discussed salient features of the fast handover protocol specification. Just as in the Mobile IPv6 protocol, the (Fast) Binding Update is a crucial message. It can be sent before the actual handover itself however. This allows the protocol to maximize the overlap of link switching delay, which we studied in Chapter 14, with the IP protocol operations. In order to send FBU predictively, the mobile node needs knowledge of the target router IP address, link layer address and the prefix it can use to formulate a new care-of address. This information is provided by the Proxy Router Advertisement message as a part of neighborhood discovery, which we also studied in Chapter 14.

In addition to the protocol message formats, we have illustrated how the messages are implemented in software. We ask readers, especially those interested in further implementation details, to obtain the software and experiment with it themselves.

15.6 CHAPTER NOTES

Emil Ivov and Martin Andre created fmipv6.org and allowed us to use their source code for illustration here. Our sincere to thanks to them. The FreeBSD implementation is available in nautilus6.org. Some of the option formats discussed here may undergo revision or will likely be adapted according to the specific wireless technology needs.

Exercises

15.1 Review the code for sending FBack on the previous router. What are the pros and cons of waiting until HI and HAck exchange before responding with an FBack message?

15.2 Review the code for HI and HAck message exchange. Define protocol extensions to supply the network-resident context to the new router in an HI message. For instance, define a "QoS Context" to transfer from the previous router in the HI message.

15.3 In some scenarios, the mobile node may not be able to obtain any information regarding the target network before it attaches to it. In such cases, the latencies stemming from movement detection, new IP configuration and location update can be substantial. The DNA working group in the IETF is considering protocols to help improve the movement detection latency, and the Optimistic DAD proposal in the IPv6 group is considering the modifications to Neighbor Discovery when using a tentative address. Review both DNA and Optimistic DAD protocols and investigate how they could be used in conjunction with fast handovers.

REFERENCES

1. S. Bradner. "Key Words for Use in RFCs to Indicate Requirement Levels," BCP 14, RFC 2119, March 1997.

2. D. Johnson, C. Perkins, and J. Arkko. "Mobility Support in IPv6," RFC 3775, Internet Engineering Task Force, 2004.

3. J. Kempf. "Instructions for Seamoby and Experimental Mobility Protocol IANA Allocations," RFC 4065, Internet Engineering Task Force, July 2005.

4. R. Koodli (Editor). "Fast Handovers for Mobile IPv6," RFC 4068, July 2005, Internet Engineering Task Force.

5. T. Narten, E. Nordmark, and W. Simpson. "Neighbor Discovery for IP Version 6 (IPv6)," RFC 2461, Internet Engineering Task Force, December 1998.

16

Context Transfers

What's sort of interesting about the whole public relations disaster that is the Net, in some ways, is that the fundamentals are really good. – Meg Whitman

16.1 INTRODUCTION

In the previous chapters, we have studied fast handovers in detail. We have studied the basic concepts and the protocol details. With careful system design, the basic protocol can be used to minimize latency and reduce or eliminate packet loss during a subnet handover. In this chapter, we will investigate a companion protocol for achieving smooth handovers. Fast handovers are primarily concerned with routing operations such that packet forwarding can take place as soon as new link establishment occurs. Without such support, basic packet forwarding would not take place in a timely fashion.

In addition to expedited routing operations, however, additional *transport-level* operations may be necessary to further improve the handover experience. For instance, assume a VoIP packet stream running on a mobile node which is afforded certain QoS on an access link. At the IP layer, this typically involves appropriately classifying the packet stream as a VoIP stream and providing the desired queuing support to ensure low latency and near-zero packet loss. In order to also ensure that the stream follows an agreed-upon traffic contract, the stream is also metered and shaped (or policed) if necessary. These QoS operations introduce a state for each stream on a router (and on the host). This state is created as a result of either explicit signaling or by other means, such as prior configuration. Now if the mobile node undergoes handover to a new link, the state needs to be reestablished on the new router in order to provide the same QoS to the packet stream. A straightforward solution is to require the mobile node to reengage in signaling that would enable the new router to establish the necessary state again. Until the state is created, the VoIP packet stream will be treated like another best-effort stream. This can result in poor voice quality due to additional delay and possibly packet loss, depending on the traffic load on the router.

An alternative to signaling is to transfer the state in conjunction with handovers so that the packet stream is given the desired QoS as soon as packet transmission begins on the new link. The mechanisms to detect handover at the transport level and subsequent signaling are not required on the mobile node. In this chapter, we will investigate the topic of state or *context* reestablishment.

16.2 CONTEXT CREATION ON AN ACCESS ROUTER

In addition to QoS, there are other reasons why contexts are created on an access router. When a mobile node obtains connectivity at an access router, access control is performed in most provisioned networks [5, 12]. The mobile node is required to present its credentials such as its Network Access Identifier (NAI), which is used to determine its home network for authentication purposes. A suitable protocol is used to authenticate the mobile node, which has the effect of establishing a security context on the access router. Similarly, when a client of an IEEE 802.11 access point performs authentication using IEEE 802.1X [13], a certain state is created on the access point. Another example is header compression, in which the IP and the transport header state are stored on both the network and the terminal so that headers can be compressed and compressed headers can be decompressed correctly. The header compression state is built using several packets with normal IP headers using either an acknowledgement-based approach or an optimistic approach (where transmission of a predefined number of packets is deemed enough to assume a stable reference state). In summary, there is context creation due to access control, QoS and efficiency reasons. Such contexts are created using methods specific to each *feature* in question; the access control state is created through the protocol used for authentication, the QoS state via configuration or signaling, and the header compression state by in-band operation of the process itself.

An investigation of available options in establishing this context during mobility is worthwhile. The default is to allow each feature to reestablish the soft state. Before each feature can do this, an indication to initiate context establishment is necessary. Such an indication can come from continued absence of progress for each packet stream. For instance, if a stream does not get the QoS necessary subsequent to handover, it may be able to detect it based on the performance of the packet stream. The transport protocol acknowledgements may be useful in such a detection. However, such an approach may not always be reliable; a packet stream either may not notice the degradation or may only notice it after several round trips. A more explicit approach would involve the access router informing a mobile node about lack of any context for streams arriving with certain QoS codepoints in Diffserv [2], for instance. Such a notification would immediately inform the mobile node to reestablish contexts. Such notifications do not exist, in part because there has not been a need for them. Such notifications also need careful scrutiny, for example, to distinguish them from rogue requests. Another example of an explicit notification is an internal trigger within the mobile node's IP stack itself, which notifies each feature that the mobile node has undergone handover and that new context establishment is necessary. Perhaps a larger question involves the architecture. Signaling for "IP Services" is usually a top-down approach tied to applications requesting certain QoS. Signaling as a function of mobility requires bottom-up support. Further, such a support should be generic enough to support a variety of functions. Such *cross-layer* design for mobility purposes is beginning to receive attention; see [9] for an overview. However, we are far from conclusive results and many important results await research.

As an alternative to explicit signaling for reestablishing contexts, context relocation or transfer is a worthwhile design. The essential idea is to relocate *well-defined* contexts at an appropriate time so that packet streams undergoing handover do not perceive a degradation in performance. There are other benefits too. There is no need for per-feature signaling to reestablish contexts. This is important because such signaling would be time-sensitive, since packet streams are already in progress (as opposed to the very first time, when the contexts are established). Such signaling would also be susceptible to *mobility load*, where the control traffic due to handover could be significant and unreliable. When the need for explicit per-feature signaling can be avoided, so can the need for host internal communication between the various layers to initiate signaling subsequently. This should simplify host software design. Finally, there is ongoing debate about the role of mobility triggers on transport protocol performance [1].

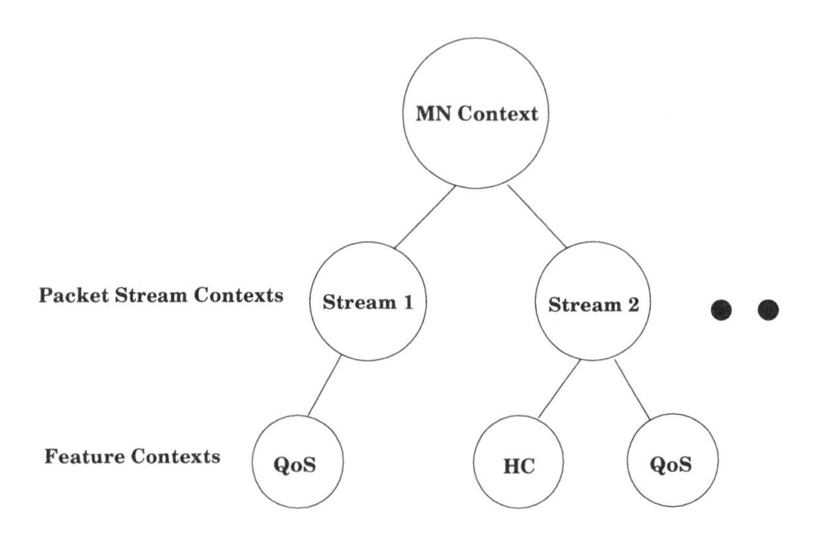

Fig. 16.1 Context Hierarchy

A design where each transport protocol or, even worse, each application is made to react in its own way to handovers would create a situation similar to one where each transport protocol decides to do routing in its own way. A handover event is a temporary change in routing, and as such should be handled by the IP layer so that all transport protocols can continue to function smoothly.

An example of performance benefits achievable through context transfer is outlined in [10]. The paper describes the cost of context reinitialization when header compression is used in narrowband cellular systems. We report two observations here: First, there is a linear relationship between the cost of context initialization and channel reliability. So, unreliable channel conditions, especially surrounding the "fringe handover areas," could require more bytes over the air to reliably establish contexts. Second, perhaps more importantly, the burstiness of control traffic can make the data traffic move slowly, thus causing a throttling effect. For instance, when Full IP Headers need to be transmitted over a channel designed primarily for carrying voice payload, the Full Headers can block the much smaller voice payloads behind them. This example illustrates the effects of signaling during an inopportune time for a single packet stream. The situation is even worse with multimedia.

Having motivated the need for context creation on the network and for relocating the contexts during handover, let us now consider the essential design elements.

16.3 CONTEXT TRANSFER DESIGN

Context transfer can be considered as a framework consisting of multiple design elements. First and foremost, a context object definition has to be uniform and standardized so that the object can be processed unambiguously from both the sender and the receiver. Such objects also need containers for transporting them from one network node to another. The transport itself has to provide adequate level of reliability and security. Finally, the transport of objects has to be closely synchronized with handover signaling to achieve maximum performance benefit. In summary, a context transfer framework should address the following:

1. Data structure representation of contexts for interoperability

2. Packet formats for context data structures for transfer and processing, and

3. Synchronization with handover signaling for effecting context transfers

We discuss each of these in the following sections.

16.3.1 Data Structure Representation

The need for a suitable data structure representation arises from the diversity of feature contexts as well as the way they are typically realized in practice. Typically, there are multiple features associated with each unidirectional packet stream, including QoS, header compression, security, and so on. Each of these features may be supported using different protocols and mechanisms. For example, an access router may support IPv6/UDP/RTP and IPv4/TCP header compression and Intserv [11] and Diffserv QoS. One or more feature contexts belong to a packet stream context [1], and one or more packet stream contexts belong to a mobile node's context. This hierarchy is shown in Figure 16.1.

As we can observe from Figure 16.1, a feature context is the basic unit for context transfer. Hence, it is useful to define a *Feature Profile Type* (FPT) as an object that uniquely identifies the structure of the data related to a feature context. Each instance of a FPT clearly defines the state variables associated with the particular feature context. For example, a QoS Profile Type (QPT) for Diffserv defines the packet classification, packet metering and packet marking control variables, whereas a Compression Profile Type (CPT) for IPv6 defines all the static and changing fields in the IPv6 header. Given an appropriate FPT, each feature context itself then consists of a tuple of the form [FPT, associated state parameters].

[1] Sometimes, a packet stream may also be referred to as a *microflow*.

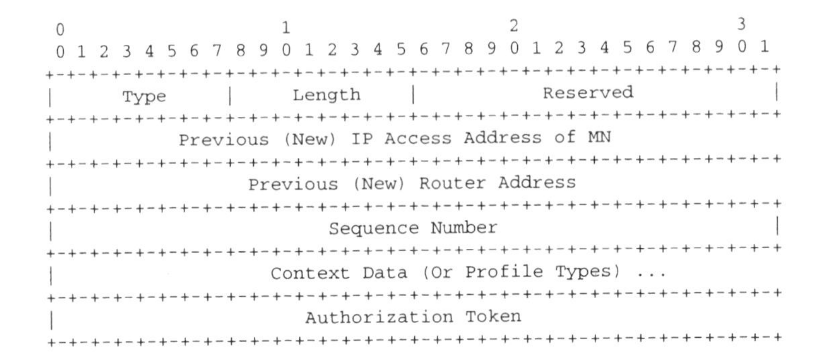

Fig. 16.2 Generic Context Transfer Message Format

An FPT serves multiple purposes. First, it provides a definition of the necessary and sufficient state parameters associated with a particular feature context. This definition includes the way a feature context is realized by protocols and mechanisms in order to provide interoperability. Without an FPT, a receiver will not know how to interpret the data and recreate the context for further use. An FPT can also be viewed as a *handle* to request specific feature contexts or to initialize or even authorize specific feature contexts in signaling messages. Finally, an FPT object can be used to verify if a target router supports the feature context(s) in question. Such information can be used for target router selection purposes.

16.3.2 Context Data Structure Format

With a suitable data structure representation, the feature contexts need to have appropriate packet formats. The messaging for context transfer involves mobile nodes and access routers. The mobile nodes communicate with access routers providing FPTs, for instance. The access routers exchange messages requesting and providing the feature context data. The appropriate messages used between a mobile node and its access router, and between the access routers, are defined in [7]. Here we will only discuss some important parameters that need to be carried in the messages. We illustrate such parameters using Figure 16.2

The Type and Length fields identify the type of the context transfer option (e.g., request, response) and its length, respectively. The Previous IP address of the mobile node is needed when the New Access Router (NAR) needs to fetch the context from the Previous Access Router (PAR). The New IP address of the mobile node is needed when the NAR has to associate the received contexts (corresponding to previous IP address) to the mobile node's new IP address. Furthermore, the mobile node may supply the PAR's address when context transfers take place subsequent to handover. Following the IP addresses, various context transfer requests or the contexts themselves are enumerated in the [type, length] format.

Finally, the authorization token is included to protect against malicious or bogus nodes that may attempt to steal a mobile node's contexts by prematurely providing a false indication that the mobile node arrived at the new access router. The authorization data includes all the context data protected using an appropriate security association between the mobile node and the access router.

The authorization token can be computed as follows:

$$First(X,\ HMAC\_SHA1(Key,\ (MN\ IP\ Address \\ \mid Sequence\ Number \mid Context\ Data))) \tag{16.1}$$

where X is the number of acceptable bits chosen for security, $SequenceNumber$ is a unique number used for each computation of the token for replay protection, and $ContextData$ is the feature context profiles whose transfer is requested. A default profile includes all the feature contexts so that enumerating each feature context profile is not necessary. The parameter Key is the shared secret between the mobile node and the access router where contexts are present. Such a shared secret can be established when obtaining access to the network, and is not specific to the context transfer protocol itself.

16.3.3 Using Context Transfer Options with Handover Signaling

The third and perhaps a crucial part of the design is coordinating the context transfer with handover, since that's when context transfer is most useful. Going one step further, the usefulness of insufficiently synchronized context transfer is moot considering the strong requirement for various features to be available immediately upon handover. This requires that context transfer and fast handover [8] be designed to work together. Figure 16.3 provides an illustration.

The Fast Binding Update message in fast handovers is used for establishing the route table entry at the previous router in order to be able to reroute packets to a new IP address. This message can be sent either before a mobile node leaves its link or immediately after it regains a new link. So, this message is ideally suited for initiating context transfer as well. However, the message must include the target router's IP address and a sequence number, as well as the authorization token. The default Context Transfer (CT) extension for carrying in an FBU message is presented in Figure 16.4.

The CT extension serves the same purpose as the Context Transfer Activate Request (CTAR) message sent to an access router in [7]. When carried in an FBU, the extension is sent to the previous router, and includes the new router's IP Address. The extension can also be carried in the FNA message, in which case the access router's IP address is that of the previous router. In any case, the mobile node is identified using its previous IP address. We explain this further below.

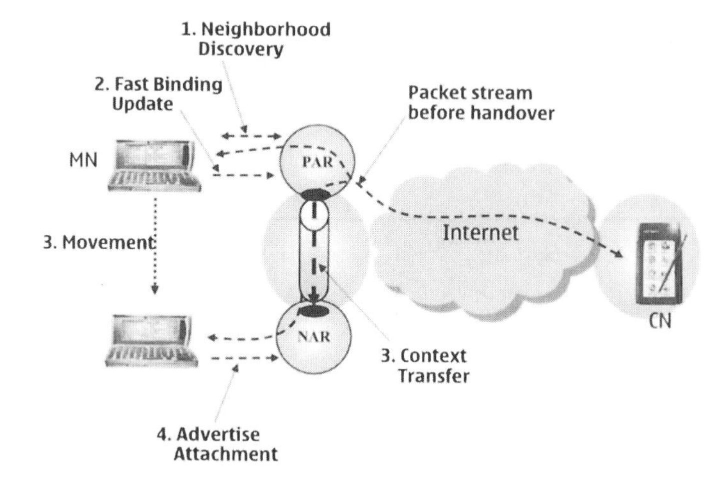

Fig. 16.3 Context Transfer with Fast Handover Signaling

```
 0                   1                   2                   3
 0 1 2 3 4 5 6 7 8 9 0 1 2 3 4 5 6 7 8 9 0 1 2 3 4 5 6 7 8 9 0 1
                                +-+-+-+-+-+-+-+-+-+-+-+-+-+-+-+-+
                                |  Type = CT    |    Length     |
+-+-+-+-+-+-+-+-+-+-+-+-+-+-+-+-+-+-+-+-+-+-+-+-+-+-+-+-+-+-+-+-+
| Option-Code   |    Pad0=0     |  Default Context Profile      |
+-+-+-+-+-+-+-+-+-+-+-+-+-+-+-+-+-+-+-+-+-+-+-+-+-+-+-+-+-+-+-+-+
~                  MN's Previous IP Address                     ~
+-+-+-+-+-+-+-+-+-+-+-+-+-+-+-+-+-+-+-+-+-+-+-+-+-+-+-+-+-+-+-+-+
~                   New (Previous) Router's IP Address          ~
+-+-+-+-+-+-+-+-+-+-+-+-+-+-+-+-+-+-+-+-+-+-+-+-+-+-+-+-+-+-+-+-+
|                       Sequence Number                         |
+-+-+-+-+-+-+-+-+-+-+-+-+-+-+-+-+-+-+-+-+-+-+-+-+-+-+-+-+-+-+-+-+
|                     CT Authorization Token                    |
+-+-+-+-+-+-+-+-+-+-+-+-+-+-+-+-+-+-+-+-+-+-+-+-+-+-+-+-+-+-+-+-+
```

Fig. 16.4 Context Transfer Extension Format

Assume that a mobile node sends the CT extension in an FBU message prior to leaving its current link. As a response, the previous router can send the feature contexts to the new router in the HI message. The Context Transfer Data (CTD) message in [7] can be used as an option in the HI message to do such a transfer. When context transfer happens in such a predictive fashion, the new router must ensure that the received contexts are made available to the legitimate mobile node.

Presumably, different designs for ensuring that only the legitimate mobile node is able to make use of the contexts are feasible. One such design supplies the new router with the security credentials in context transfer itself. The scope and implications of security context transfer are still being worked on [3, 4]. One of the debatable issues is transfer of key material from one router to another. Another design point could involve preauthenticating with the new router even before attaching to it. Such a design must ensure that the pre-authentication procedures do not happen during handover, which would affect handover performance. In either case, the mobile node itself must be made to present proof of ownership of contexts to the new router. This calls for the inclusion of the CT extension in a message such as FNA. Using the security parameters sent in the HI message and those present in the FNA CT extension, the new router can compute its own authorization token and match it against the token present in the FNA CT extension. When the tokens match, the new router makes the contexts immediately available.

It is possible that the new router has no contexts present when it receives the FNA CT extension. This can happen when the FBU message itself is lost. It also means the mobile node has not received an FBack message. In such a case, the contexts are retrieved reactively when the FBU is processed by the previous router. After verifying the token, the previous router sends the contexts in the HI message. It might seem suboptimal to perform reactive context transfer. However, it can still provide performance benefits, as we shall see in Chapter 22.

The response (i.e., the CTDR message in [7]) to context transfer can be piggybacked on the HAck message. In this way, the entire context transfer process can be synchronized with fast handovers in order to achieve maximum benefit.

16.4 SUMMARY

Context Transfer is a mechanism to smoothen the effects of handover on transport protocol and hence the application performance. Although fast handover expeditiously establishes routing and enables a mobile node to quickly regain IP connectivity, it does not (rightfully) address the problem of recreating a state that is needed to forward packets. The simplest of such state is that of access control, without which packets can be dropped even if a mobile node has quickly regained its IP connectivity. The other contexts of interest include QoS and header compression. One could always argue that such states can be reestablished by a mobile node. As we saw, however, this argument requires explicit understanding of handover at multiple protocol layers, including IP (which needs to trigger a handover event to various transport protocols), transport (which needs to initiate signaling to re-create the state), and applications (which may need to adjust their behavior until an appropriate state is established). While cross-layer interaction might be an interesting research topic in itself, perhaps it is not needed for establishing access router contexts. In a truly seamless handover experience, an application is completely unaware of a handover. Context transfer offers a mechanism to achieve this objective. At the same time, this is a relatively new area which can benefit from further research and implementation.

Context transfer should be synchronized carefully with fast handovers to achieve maximum performance benefit. In a following chapter, we will review an experimental study using fast handovers and context transfers together.

Exercises

16.1 Given access control, header compression and QoS contexts, in what order must context transfer be managed, and why?

16.2 Is reliable transfer of contexts a concern for context transfer protocol? If so, what is the effect of congestion control on the timeliness requirement of context transfer? Which contexts need to be reliable, if any, and which do not?

16.3 Firewall state is one of the critical contexts for successful operation of a protocol such as TCP. However, a firewall is typically not the same as an access router, and as such is not aware of handovers. Assume that a mobility pattern encounters a change of firewalls. How can existing sessions continue smoothly? How could one do firewall context transfer?

16.4 Design an experiment to study the effect of handover on TCP. Then investigate TCP throughput with buffering support on the access router. What enhancements are still needed? Now consider the TCP Performance Enhancement Proxy state and outline how that could be managed during a handover which requires change of proxies.

REFERENCES

1. B. Aboba (Editor). "Architectural Implications of Link Indications" (work in progress), draft-iab-link-indications, Internet Engineering Task Force, December 2005.

2. S. Blake, et al. "An Architecture for Differentiated Services," RFC 2475, Internet Engineering Task Force, December 1998.

3. D. Forsberg (Editor) et al. "PANA Mobility Optimizations" (work in progress), draft-ietf-pana-mobopts, Internet Engineering Task Force,, October 2005.

4. J. Bournelle (Editor) et al. "Use of Context Transfer Protocol (CXTP) for PANA" (work in progress), draft-ietf-pana-cxtp, Internet Engineering Task Force,, March 2006.

5. B. Lloyd and W. Simpson. "PPP Authentication Protocols," RFC 1334, Internet Engineering Task Force,, October 1992

6. J. Kempf (Editor). "Problem Description: Reasons For Performing Context Transfers Between Nodes in an IP Access Network," RFC 3374, Internet Engineering Task Force, September 2002.

7. J. Loughney, M. Nakhajiri, C. E. Perkins and R. Koodli. "Context Transfer Protocol," RFC 4067, Internet Engineering Task Force, July 2005.

8. R. Koodli (Editor). "Fast Handovers for Mobile IPv6," RFC 4068, Internet Engineering Task Force, July 2005.

9. R. Koodli. "Mobility and Cross-layer Design", Panel presentation at the First IEEE/ACM Workshop on Mobility in the Evolving Internet Architecture (MobiArch), San Francisco, CA USA, December 2006.

10. C. Westphal and R. Koodli. "IP Header Compression: A Study of Context Establishment," Proceedings of the IEEE WCNC, 2003.

11. J. Wroclawski. "The Use of RSVP with IETF Integrated Services," RFC 2210, Internet Engineering Task Force.

12. "Medium Access Control Security Enhancements," IEEE standard 802.11i, http://standards.ieee.org/getieee802/download/802.11i-2004.pdf

13. "IEEE Standard for Local and Metropolitan Area Networks: Port-Based Network Access Control," IEEE Standard 802.1X, 2004.

17

Hierarchical Mobility Management

Engineers like to solve problems. If there are no problems handily available, they will create their own problems. Normal people don't understand this concept; they believe that if it ain't broke, don't fix it. Engineers believe that if ain't broke, it doesn't have enough features yet. – Scott Adams

—

17.1 INTRODUCTION

Although the fast handovers protocol is primarily concerned with reducing latency and packet loss during handovers, it is, in a sense, independent of the actual mobility protocol used over the Internet. The protocol does not affect the Mobile IP protocol itself; it ensures that the latencies in the Mobile IP protocol are addressed. In other words, a mobile node would still perform the Mobile IPv6 operations including Binding Update to its home agent, Return Routability with its correspondents, and subsequent updates to their Binding Cache entries. These Mobile IPv6 operations can be seen as introducing additional signaling overhead in certain deployments. The Hierarchical Mobile IPv6 (HMIPv6) protocol is designed primarily to address this overhead. However, it turns out that the protocol can also be useful in a limited sense for *location privacy* in terms of the IP addresses revealed to the correspondents. We will discuss this topic in Chapter 23.

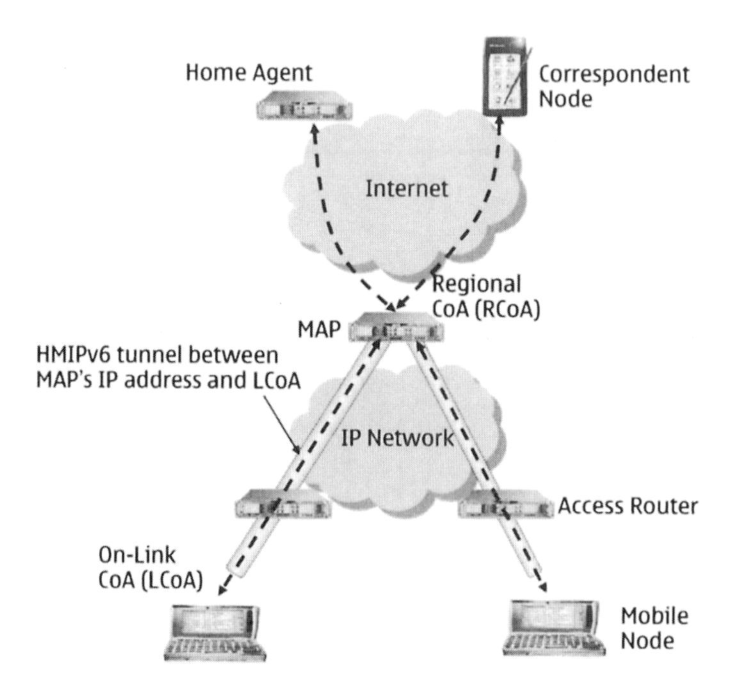

Fig. 17.1 Hierarchical Mobility Reference Model

Since Hierarchical Mobile IP is concerned with minimizing Mobile IPv6 signaling, one of the design goals is to make subnet mobility transparent to correspondent nodes and the home agent so that IP address changes are hidden from them. This calls for the introduction of a new functional entity which can proxy an address that remains fixed over a substantially large IP network. The mobile node can freely roam within the jurisdiction of such an entity, changing its subnet IP address and yet its peers be unaware of its mobility. This functional entity is called the *Mobility Anchor Point* (MAP), whose tasks include assigning and managing a Regional Care-of Address (RCoA) for each mobile node, binding the RCoA to the subnet On-Link CoA (LCoA), and managing the traffic between the RCoA and LCoA. Often, these functions can be realized using the home agent functionality. For this reason, the protocol operation is fairly straightforward.

17.2 HIERARCHICAL MOBILITY MODEL

Before going further into the protocol details, let us look at the basic terms and the reference model. See Figure 17.1.

- A MAP is a router on the visited network which advertises its IP address and a prefix for the mobile node to use. The higher the topological position of MAP in the access network topology, the lower the Mobile IPv6 signaling overhead since the mobility can now span a larger IP network.

- RCoA is an additional CoA that a mobile node configures when it uses HMIPv6. This address is valid on a subnet attached to MAP, and is seen by the home agent and the mobile node's correspondents. A mobile node typically auto-configures this address using IPv6 stateless address autoconfiguration [2].

- LCoA is the CoA that a mobile node configures on the subnet link it is currently attached to. This is the CoA that mobile node would have used for Mobile IPv6 if it were not using HMIPv6. In other words, the mobile node must not use LCoA for Mobile IPv6 purposes with its home agent and the correspondents.

- Local Binding Update (LBU) is a message that a mobile node using HMIPv6 sends to the MAP in order to bind RCoA and LCoA together. An HMIPv6-aware mobile node must be able to process a new option present in Router Advertisements.

17.3 PROTOCOL OPERATION

The HMIPv6 protocol involves the following phases: MAP discovery, MAP registration, and packet forwarding.

17.3.1 Mobility Anchor Point Discovery

A mobile node that wishes to use HMIPv6 first needs to discover a MAP. This discovery is accomplished using a new MAP option in Router Advertisements. See Figure 17.2. When a mobile node processes this option, it obtains the MAP's IP address, Lifetime of the prefix for RCoA, and other information such as the "Distance" of MAP from the current access router and a "preference" value for a MAP. Using this information, a mobile node can establish a Security Association (SA) with a MAP and subsequently bind its RCoA to its LCoA.

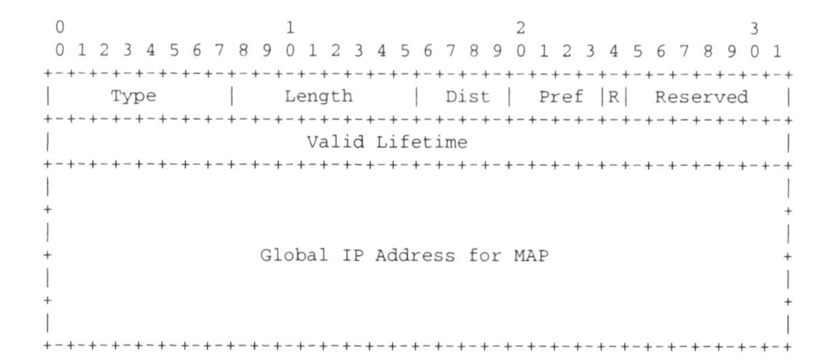

Fig. 17.2 MAP Option for Neighbor Discovery

The MAP option itself needs to be available to all access routers in order for them to be able to advertise it. HMIPv6 offers two choices: static configuration on each access router or dynamic MAP discovery. With static configuration, a network administrator manually enters the MAP option on each access router's database of the interface configuration. Dynamic discovery involves routers learning of a MAP by listening on adjacent Router Advertisements containing the MAP option. A network administrator configures a node to act as a MAP on at least two interfaces (one towards the Internet whose prefix is used for RCoA computation and the other towards the visiting mobile node for tunneling purposes), allowing it to send Router Advertisements to routers downstream in the network. The routers that receive such advertisements must also be configured to resend the MAP option in their advertisements on some *selected* interfaces, after incrementing the Distance ("Dist" in Figure 17.2) field in the MAP option. When exchanging the MAP option between routers, HMIPv6 also relies on mutual authentication in order to securely propagate the MAP option.

When an HMIPv6-aware mobile node receives a Router Advertisement with a MAP option, it verifies if the Lifetime is zero for failure indication of an existing MAP. However, no particular MAP failure detection algorithm is mandated. Instead, router implementations are suggested to use ICMP Echo Request to ascertain MAP liveness. We leave more sophisticated MAP failure detection and correction algorithms as an exercise for the reader. [1] If the Lifetime is non-zero, then the mobile node has to choose the MAP with the highest "preference" value. The higher this value, the higher is the assumed availability of the MAP service.

[1]We note that a MAP can be viewed as a single-point of failure. So, failure detection and recovery mechanisms are important for reliability.

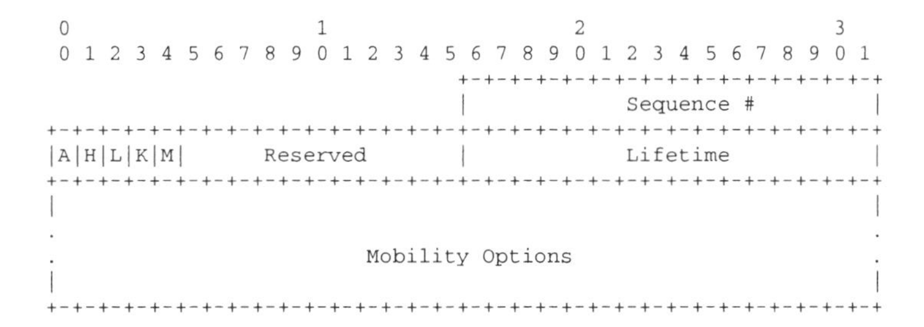

Fig. 17.3 Local Binding Update Format

17.3.2 Mobility Anchor Point Registration

Once it has discovered and selected a MAP, the mobile node has to configure an RCoA using the prefix advertised in the MAP option and create a binding between the RCoA and LCoA at the MAP (assuming that it has already configured a valid LCoA). Hence, the mobile node has to send an LBU to the newly discovered MAP. However, this LBU must be protected using a shared SA between the mobile node and the MAP. The HMIPv6 protocol currently does not specify exactly how this SA is set up. It does, however, describe how IKE (which we studied in Chapter 3) can be used to set up the SA. See Section 12.1 in [1]. It is important that the SA is set up for the RCoA using LCoA as the source IP address in the IKE negotiations. It is only after this stage that the mobile node can send an LBU, whose format is shown in Figure 17.3.

The LBU message only adds an 'M' flag to a Mobile IPv6 Binding Update message to indicate that it is a MAP registration. The source IP address is LCoA and the home address is RCoA. The MAP performs DAD for RCoA and then sends a Binding Acknowledgment (BAck) which is identical to the Mobile IPv6 BAck. If the BAck does not contain a Type 2 Routing Header with RCoA, the mobile node silently discards the message. Prior to discarding the message, however, the mobile node should inspect the error code in the BAck to see if a new RCoA needs to be computed and a fresh LBU sent subsequently.

If a successful binding cannot be created at the MAP, the mobile node falls back to using base Mobile IPv6. In such a case, it sends a Binding Update message to the home agent using LCoA as the CoA. The mobile node uses the LCoA as the CoA in the Return Routability tests with its correspondents. If a binding can be established at the MAP, then the mobile node uses RCoA as the CoA with the home agent and the correspondents. In order to do so, the mobile node uses RCoA as the source IP address in all its packets, but tunnels them to MAP using LCoA as the source IP address for the tunnel.

When a binding exists between LCoA and RCoA, the MAP forwards the inner packet towards the home agent and the correspondents. In any case, the mobile node must wait for the resolution of LBU before it sends a Binding Update to its home agent or before it engages in Return Routability with its correspondents. Furthermore, the Lifetime of the Binding Update sent to the home agent and the correspondents must not be greater than the Lifetime in the BAck received from the MAP.

17.3.3 Packet Forwarding

The packet forwarding is straightforward. A MAP acts as a home agent. And, the mobile node acts as if it is using reverse tunneling with its home agent. A mobile node establishes a tunnel interface for all its packets once it receives a successful BAck from the MAP. This tunnel interface uses LCoA as the source IP address and MAP's IP address as the destination address for the tunnel for out-going packets. For incoming packets, the tunnel interface must check for the same addresses with the source and destination address fields swapped. The IP address of the inner tunnel header is RCoA. However, the mobile node may also use LCoA in packets that do not have to traverse the MAP.

The MAP, acting as a home agent, establishes a binding cache entry for RCoA upon successful MAP registration. This allows it to tunnel all the incoming packets to the mobile node's LCoA. In the reverse direction, a binding entry (between the source IP addresses of the tunnel and the inner packet) ensures that it can forward the inner packet to the home agent and the correspondents.

The HMIPv6 protocol also describes inter-MAP handovers in which a mobile node can traverse from one MAP domain to another. The protocol defines algorithms for movement detection recommending *eager switching* to a new MAP as soon as it is discovered but maintaining its binding at the old MAP until it is no longer accessible. The protocol also defines a default method for selecting a MAP from among multiple ones should they exist. In addition, the protocol includes operations involving multiple simultaneous MAP bindings. These flexible modes of operation, however, need further validation in terms of experimentation and deployment. We refer the readers to [1] for details of such operations.

17.4 SUMMARY

Hierarchical mobility is primarily concerned with reducing the signaling overhead associated with subnet mobility. It relies on a service offered by a visited network entity called Mobility Anchor Point in order to make handovers within the jurisdiction of MAP transparent to the mobile node's home agent and correspondents. A MAP behaves just like a home agent, hence the operation of the protocol itself is quite straightforward. A mobile node has to discover and choose an appropriate MAP however. It also needs to establish a security association in order to bind the on-link and regional care-of addresses. These operations differ in practice from those involving a typical home agent since the assumed trust model and provisioning of credentials is not the same. Hierarchical mobility can hide the movement of a mobile node within a network of operation of the MAP. Only the regional care-of address is exposed to the mobile node's peers. This might be useful for some mobile nodes, although the regional care-of address itself can provide sufficient information about the location of a mobile node.

Exercises

17.1 Return Routability is based on a (loose) notion of address ownership. A mobile node is required to prove that it owns one address (HoA) and is reachable at another (CoA). When using HMIPv6, the mobile node always uses the RCoA with its correspondents. Within the MAP domain, however, it uses the LCoA. What are the trust and address ownership implications of using RCoA, which is *not* the on-link address, with Return Routability?

17.2 A home agent is not required to verify if a mobile node is actually present at a particular CoA when it processes the Binding Update to create a binding. The situation is similar for MAP. What happens when a node claims an RCoA and binds it to an LCoA that does not belong to it? What are the implications for Return Routability?

17.3 Assume that an HMIPv6-aware mobile node is using RCoA in its route-optimized communication with one or more correspondents. Identify the steps it needs to take if it receives a Router Advertisement with MAP option that contains a Lifetime of zero.

REFERENCES

1. H. Soliman, C. Casteluccia, K.El Malki, and L. Bellier. "Hierarchical Mobile IPv6 Mobility Management (HMIPv6)," RFC 4140, Internet Engineering Task Force, August 2005.

2. S. Thomson and T. Narten. " IPv6 Stateless Address Autoconfiguration," RFC 2462, Internet Engineering Task Force, December 1998.

Part IV

Applying IP Mobility

Any new technology has its share of challenges in deployment. First, technology development and deployment do not always go hand in hand. This is especially the case when the technology is sufficiently large enough for deployment. Second, new technologies always need to find their position in an existing ecosystem of networks of routers, firewalls and network address translators. Some argue that this has become so complicated that an overhaul, rather than co-existence, is needed. This is particularly the case for IPv6. Finally, applications need to be available to make use of the new technology.

These challenges are not specific to Mobile IPv6, which in fact has inherited at least some of these challenges from IPv6. There is widespread acceptance that the current pool of IP addresses will be exhausted, beginning especially in those parts of the world where mobile technology use is exploding. Furthermore, deployment can be incremental covering the access networks and popular applications, followed by more widespread deployment (which need not be uniform across the entire Internet). Finally, Mobile IPv6 does not require applications to be rewritten; applications do not even know about it. Since Application Programming Interfaces have been standardized for IPv6, applications written for IPv6 work with Mobile IPv6. In fact, they work better, since they have mobility support!

In previous chapters we have studied numerous concepts, reviewed the Mobile IPv6 protocol in detail and studied advanced mobility protocols. This part of the book addresses some applications of the technology discussed earlier. The emphasis here is on how a particular network or system is using Mobile IPv6 technology to support mobility on the Internet. Mobile IPv6 is currently specified for use in CDMA packet systems and is being specified for the WiMax systems. We study the use of Mobile IPv6 in CDMA systems. We then look at how enterprises are adopting mobility in general; we discuss how security and mobility come into play, for instance, in Virtual Private Networks. The focus is on Mobile IP concepts in general rather than on Mobile IPv6 itself. We then consider mobility support over WLAN for VoIP. We consider WLAN-specific bottlenecks and review IP delays described in the previous part of the book. It is instructive to study how link-layer and IP operations need to be carefully orchestrated to achieve the desired level of performance.

In the chapters that follow, you will see figures with multiple sequences of messages being exchanged. Essentially, they illustrate protocol messages between different network nodes as a function of time, with time increasing from top to bottom. So, a message sequence that appears at the top of the figure takes place before a sequence that takes place underneath it. Such diagrams are sometimes, unsurprisingly, referred to as "message sequence diagrams".

18

Mobile IPv6 in CDMA Packet Data Networks

It's important to remember that the relationship between different media tends to be complementary. When new media arrive they don't necessarily replace or eradicate previous types. Though we should perhaps observe a half second silence for the eight-track. There that's done. What usually happens is that older media have to shuffle about a bit to make space for the new one and its particular advantages. Radio did not kill books and television did not kill radio or movies - what television did kill was cinema newsreel. TV does it much better because it can deliver it instantly. Who wants last week's news? – Douglas Adams

—

18.1 INTRODUCTION

The CDMA cellular networking system is one of the most widely deployed mobile networking technologies in use today. This system has a specific architecture for supporting packet data services beginning with the so-called *1XRTT* systems. In this model, a Mobile Station (MS) connects to a Packet Data Serving Node PDSN through the wireless base station, obtains an IP address and engages in packet data communication. In the following, we review the elements of this architecture and the procedures for establishing IP connectivity. We subsequently discuss how IP mobility is supported.

Before jumping into the functional description of the CDMA packet data architecture, we need to define some terms. Figure 18.1 is a useful reference as we do this, perhaps also for the entire chapter itself.

A detailed description of the wireless network architecture based on the Internet protocols is provided in [2].

- An *Access Provider Network* is a cellular network that provides access to a mobile user. The CDMA packet data network is meant to conform with the IMT-2000 network architecture [20].

- *AAA* is a generic set of functional modules used for authentication, authorization and accounting. A Home AAA refers to a server that resides in a mobile user's *Home IP Network*, which is similar to the Mobile IP home network. However, a Home IP Network is the home to the user's Network Access Identifier (NAI) [3]. In other words, any AAA entity can reach the mobile user's Home IP Network and verify the user's association with the Home IP Network using the NAI. The Home IP Network is assumed to provide IP-based data services to the user. The counterpart of the Home AAA is a Visited AAA, which resides in the network that a mobile user is currently visiting.

- *Packet Data Service* refers to the ability of the network to offer IP-based communication and services. A packet data session is a continuous exchange of IP packets by the user's Mobile Station with its correspondents. During a single session, the user may handover from one access point to another or from one access router (PDSN in Figure 18.1) to another. Depending on which particular packet data service is used (see below), a session may or may not continue during a handover.

- *Simple IP* is a type of packet data service in which the user's MS is assigned a dynamic IP address which is valid over some geographic area. When the user roams beyond that area, the IP address is no longer valid and any sessions in progress need to be terminated (and possibly restarted).

- *Mobile IP* is the other type of packet data service in which an MS is assigned either a dynamic or a static IP address from the user's Home IP Network. This IP address remains valid no matter where the user roams, including an IP network that is not IMT-2000 compliant.

- *PPP Session.* PPP [16] is used as a data link layer. The PPP session is maintained between the MS and the PDSN. This session state is maintained when the MS is dormant, and when the MS undergoes handover between access points but remains attached to the same PDSN.

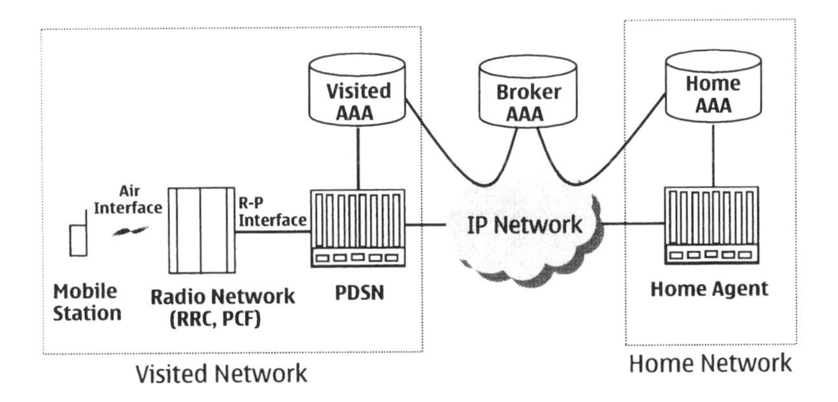

Fig. 18.1 CDMA Packet Data System Elements

- *R-P Session* refers to the session maintained between the Radio Network (RN) and the PDSN for a particular PPP session. When the MS changes an RN, the R-P session is relocated from the old RN to the new one.

Let us now consider the functional elements of the architecture illustrated in Figure 18.1. The Radio Network consists of a CDMA Radio Resource Control (RRC) function and a CDMA Packet Control Function (PCF). The RRC is in charge of establishing, maintaining and terminating radio resources for packets sent between an MS and a PCF. In addition, the RRC manages the resource states, i.e., whether a session is active or dormant. The RRC may also perform authentication before granting radio access support for encryption over the air interface.

The PCF provides a bridge between the radio and the PDSN. The PCF mediates resource allocation for packet streams with the RRC and relays packets to and from the PDSN. It also maintains knowledge of radio resource status and buffers packets arriving from the PDSN when the resources are not sufficient for forwarding packets to the MS. When the MS undergoes handover to another RRC, the PCF relocates state information so that the receiving PCF can reestablish a packet data session with the PDSN.

A PDSN is the MS's default access router. It is also the PPP termination point. The PDSN performs both PPP authentication (including LCP, IPCP, PAP or CHAP [see below]) and AAA authentication. The PDSN receives user profile parameters as part of the authentication, provides IP addressing support, performs ingress filtering, provides Differentiated Services QoS functions (such as packet classification, metering and marking), records usage of packet data, and relays the statistics to the AAA server for accounting. The PDSN interacts with the PCF to maintain PPP sessions during handovers.

The home agent works with the Home AAA server to provide an authenticated packet data service with session continuity for the MS. The AAA server in the visited network relays the authentication requests from a PDSN to the Home IP Network and relays the authorization responses from the Home IP Network back to the PDSN. The visited AAA server also provides the user profile including the QoS information to the PDSN. The Home AAA server authenticates the MS based on the requests from the visited AAA. The Home AAA server interacts with the home agent to enable dynamic home agent and home address assignment which we will discuss at length later.

With this brief overview of the CDMA packet data architecture, we proceed to explain how Mobile IPv6 works over CDMA networks.

18.2 MOBILE IPV6 OPERATION

There are multiple distinct phases before a MS becomes Mobile IPv6 capable. The first phase is link establishment.

18.2.1 Data Link Layer Establishment

The PPP protocol is used as the data link layer. See Figure 18.2. The PPP establishment involves three phases: the Link Control Protocol, Authentication Protocol, and Network Control Protocol. Once a MS establishes connectivity with the RN, the PDSN receives an R-P connection establishment message from the PCF. At this point, the PDSN sends a Link Control Protocol (LCP) *Configure-Request* message to the MS. The LCP is part of the PPP protocol and is used for establishing and configuring a link (with parameters that can be used by the network layer to transmit and receive data), as well as for link maintenance and link termination.

The LCP also includes an option for carrying the type of authentication to use before a link is successfully established. The two authentication types supported are the Challenge Handshake Authentication Protocol (CHAP) [17] and the Password Authentication Protocol (PAP) [12]. The MS replies either with *Configure-Ack* if the proposed authentication protocol is acceptable or with *Configure-Nak* if the proposed authentication protocol is not acceptable. The MS is usually configured to support CHAP and may also support PAP. So, the Configure-Nak message is usually sent when the MS would like to use PAP, whereas the PDSN has proposed CHAP. If the MS is configured with neither authentication protocol, it sends a *Configure-Reject* message. In this event, if the PDSN is configured to accept an MS without CHAP or PAP, it will respond with a fresh LCP Configure-Request without an authentication protocol option.

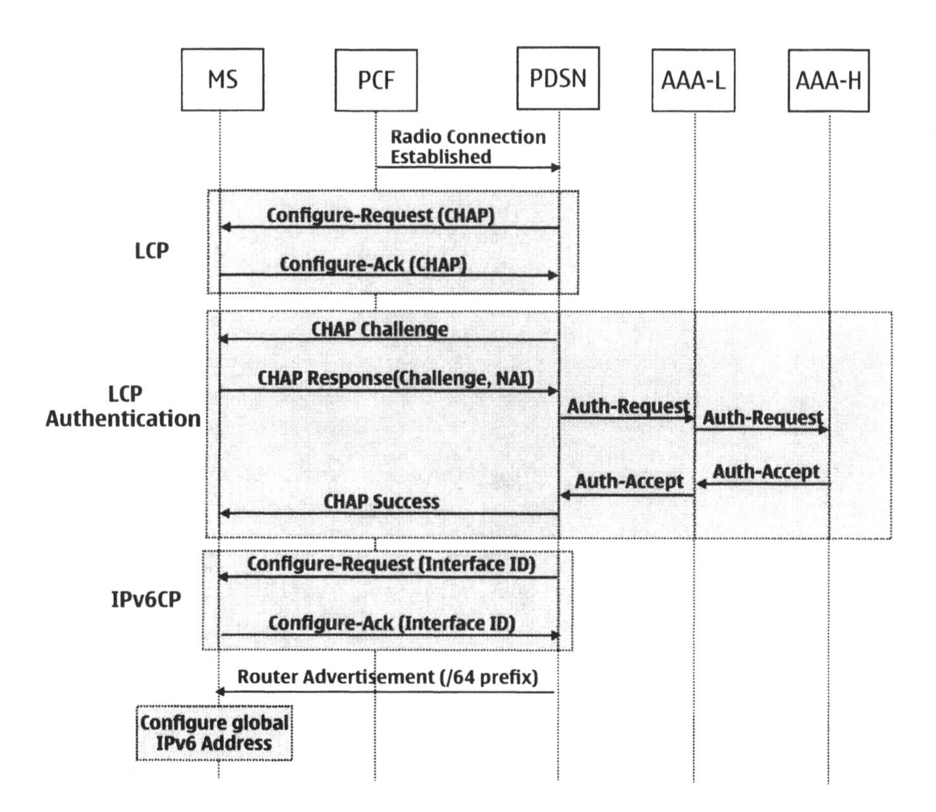

Fig. 18.2 Data Link Layer (PPP) Establishment

The authentication takes place in accordance with the protocol agreed upon during the LCP phase. We will discuss CHAP here. Immediately after the LCP phase is completed, the PDSN issues a Challenge to the MS. The MS responds to the Challenge with a CHAP-Password computed using the secret key it shares with its home network. The response also includes the NAI of the user. The PDSN then constructs an *Authentication Request* message containing the challenge, NAI and the response to the challenge and sends it to its local (i.e., visited) AAA server, which then routes the request to the Home AAA using the NAI. The Home AAA verifies the password and authenticates the MS.

The result of the authentication are sent back to the local AAA, which subsequently forwards the result to the PDSN. If the authentication is successful, PDSN indicates to the MS that the PPP link has been successfully established. This PPP link now needs to have the IPv6 address, which is obtained using the IPv6CP procedure [9]. The IPv6CP is the Network Control Protocol (NCP) phase of the PPP.

The IPv6CP involves negotiating a unique IID for the MS. The PDSN offers a unique non-zero Interface IID in a Configure-Request message to the MS which allows it to compute a link-local IPv6 address. The MS responds when the IID is acceptable using a Configure-Ack message that includes the IID. The other scenarios in the negotiation process are described in [9]. Since the MS and the PDSN are the only endpoints on this PPP link, no DAD needs to be performed.

Immediately after the IPv6CP phase, the PDSN sends a Router Advertisement in which it includes a globally unique 64 prefix. The MS then computes a global IPv6 address using stateless address autoconfiguration. Since the same IID is also used for the global IPv6 address, DAD is not necessary for the global IPv6 address.

18.2.2 Bootstrapping Home Network Parameters

An MS needs to bootstrap its home network parameters including its home agent, Home Link Prefix and home address configuration once it is IPv6-enabled. This bootstrapping takes place using the stateless Dynamic Host Configuration Protocol (DHCP) [7] between the MS and the PDSN. The PDSN obtains the home network information during the PPP authentication phase. Recall from Section 18.2.1 that the CHAP response from the MS is sent to the Home AAA for authentication in the Access Request message. The same message also carries the user's NAI. The Home AAA server, during the authentication of the user, detects that the roaming user has a Mobile IPv6 subscription. The Home AAA server can assign a home agent the Home Link Prefix as well as a home address itself, and supply the information in the Access-Accept message as a *Vendor-Specific Attribute*.

After a successful access authentication, the MS bootstraps its home network parameters with the PDSN using the stateless DHCP protocol. See Figure 18.3. The MS sends an Information-Request message containing an Option Request Option [1] (Section 22.7, [6]) specifying the information that it wishes to obtain from the PDSN. The Information-Request message also contains the Client Identifier Option (Section 22.2, [6]) in which the MS includes its NAI, and sets the DHCP Unique Identifier (DUID) to Enterprise Number. [2] The PDSN provides options containing the home network parameters based on the Client Identifier (and possibly DUID) in the Reply stateless DHCP message. The options include either the Home Agent option with Home Link Prefix Length or the Home Link Prefix.

[1] Yes, it is called Option Request Option!

[2] DHCP uses DUID to identify clients and servers. See Section 9 in [6].

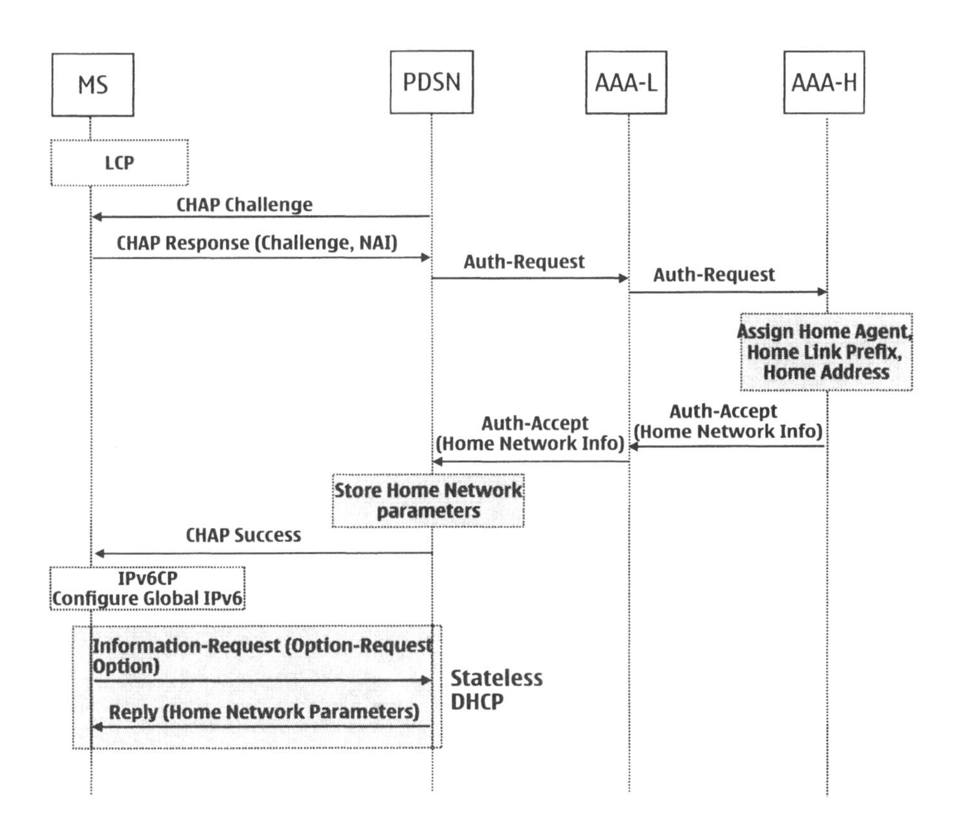

Fig. 18.3 Bootstrapping Mobile IPv6 Parameters

With the Home Agent option and the Home Link Prefix Length, the MS can compute a home address using stateless address autoconfiguration. When only Home Link Prefix is supplied, the MS can compute the home address again using stateless autoconfiguration, but it needs to perform dynamic home agent discovery. The PDSN may also supply a home address (together with the home address Lifetime) itself if the Home AAA server supplied it during the initial PPP authentication. The Home Agent option, the Home Link Prefix option, the assigned Home Address option and the assigned Home Address Lifetime option have Vendor-option codes 1, 2, 3 and 4 respectively.

18.2.3 Mobile IPv6 Home Registration

The MS needs either the Home Agent option with the Home Link Prefix Length or the Home Link Prefix option when it bootstraps the home network parameters with the PDSN. With both the options, the MS can configure a home address. Alternatively, the home address may be assigned to the MS by the Home AAA server. With the Home Link Prefix option, however, the MS also has to perform Dynamic Home Agent Address Discovery, as specified in Section 11.4.1 of the Mobile IPv6 specification [10]. Essentially, the MS sends an ICMP Home Agent Address Request message to the Mobile IPv6 Home Agent's Anycast Address. The home agent which receives this request responds with an ICMP Home Agent Address Discovery Reply message in which a list of home agents operating on the link is provided. The MS registers with one of the home agents in the list beginning with the first address and selectively proceeding to the following ones if it is unable to register with any of the previous addresses.

When the MS knows both its home address and its home agent's address, it can send a Binding Update to its home agent registering its newly configured IPv6 address. This Binding Update message is typically secured by IPsec as we discussed in Chapter 9. The CDMA packet data system, however, makes use of the AAA infrastructure which was already established for IPv4 for security credentials provisioning. In addition, this system relies on dynamic home agent and home address assignments, which already use AAA infrastructure for authentication and key generation. Furthermore, the need to support mobile nodes which do not support the IKE protocol for bootstrapping secret keys and the additional round trips of latency incurred during the operation of the IKE protocol are apparently unacceptable to the CDMA packet data system designers and operators. For these reasons, a new design is needed for bootstrapping the secret keys for securing the Binding Update and the Binding Acknowledgment messages. In the following, we describe the interaction between Mobile IPv6 and AAA in order to first establish a shared secret between the mobile node and the home agent and then to secure the Binding Update itself.

```
 0                   1                   2                   3
 0 1 2 3 4 5 6 7 8 9 0 1 2 3 4 5 6 7 8 9 0 1 2 3 4 5 6 7 8 9 0 1
                +-+-+-+-+-+-+-+-+-+-+-+-+-+-+-+-+-+-+-+-+-+-+-+-+
                |  Option Type  | Option Length |   Subtype     |
+-+-+-+-+-+-+-+-+-+-+-+-+-+-+-+-+-+-+-+-+-+-+-+-+-+-+-+-+-+-+-+-+
|                         Mobility SPI                          |
+-+-+-+-+-+-+-+-+-+-+-+-+-+-+-+-+-+-+-+-+-+-+-+-+-+-+-+-+-+-+-+-+
|                    Authentication Data...
+-+-+-+-+-+-+-+-+-+-+-+-+-+-+-+-+-+-+-+-+-+-+-+-+-+-+-+-+-+-+-+-+
```

Fig. 18.4 Message Format for the MS-AAA, MS-HA Authentication Option

A MS sends a Binding Update message with the following new Mobility Header options:

1. *MS - AAA Authentication Option.* In this option, the MS includes an SPI field to reference a particular type of authentication to use. It includes the result of a hash function computed using the shared secret between the MS and the AAA over a MAC of the Mobility Data comprising of the new IPv6 care-of address, home address and the contents of the Mobility Header, not exceeding the SPI field of this option. See Figure 18.4, which is specified in [13]. If the SPI field is set to CHAP_SPI, the home agent receiving this Binding Update acts as a AAA client and contacts the AAA server for mobile node authentication.

2. *NAI Option*: This allows the home AAA server to identify the user.

3. *Message Identifier Option*: This is used to provide replay protection for the Binding Update messages. The Mobile IPv6 protocol uses a 16-bit Sequence Number which must be monotonically increasing. When home agents are dynamically selected or when an existing home agent powers up, the previous Sequence Number is lost. In the CDMA system, a Timestamp of the physical time is used as the Message Identifier. As long as the perceived time values on the MS and the home agent are within an acceptable window, there is no need to store this value anywhere. In fact, the Timestamp value can be used directly in the Sequence Number field.

When the home agent processes this Binding Update, it realizes from the Authentication Option's Subtype value that the Home AAA needs to authenticate the user and provide a shared secret to secure the Binding Update. It also knows that CHAP is to be used with the AAA server when Mobility SPI is set to CHAP_SPI. The home agent constructs an Access-Request message in which it includes the NAI, MS-AAA Authentication Data (whose computation is described below), the care-of address and the home address from the Binding Update message. It also computes a MAC of the Mobility Header contents up to the SPI field in the MS-AAA option using its own shared secret with the AAA server, and sends the Access-Request message. [3]

If the authentication succeeds, the Home AAA server includes an Integrity Key in the Access-Accept message. The Integrity Key is computed using the MS-AAA shared secret over a set of parameters including the home agent's IP Address and the home address. The authentication may succeed, but the home address may already be in use in the AAA server's database, in which case the Integrity Key is supplied anyway to allow the home agent to include the MS-HA Authentication option in the BAck message.

If the MS-AAA authentication fails, the AAA server sends an Access-Reject message, and no Integrity Key is provided. The Home Agent silently discards the Binding Update since it would not have the shared secret to authorize the BAck message.

[3]The home agent also includes a "User-Password" derived from the shared secret between the home agent and the AAA server in the Access-Request message.

After receiving the Access-Accept message, the home agent checks for the presence of home address authorization failure indication. It then verifies the freshness of the Binding Update using the Message Identifier present in the Binding Update. Finally, the home agent sends a BAck message to the MS. In this message, it includes a MS-HA Authentication Option computed using the Integrity Key supplied by the AAA server over the contents of the Mobility Header. Since the MS can compute its own Integrity Key using the shared secret with the AAA server, it can authenticate the BAck message. If its own computation matches, the MS can start using the Integrity Key in subsequent Binding Updates to the home agent. Indeed, the MS immediately sends a Binding Update to the home agent in which it includes the MS-HA Authentication Data computed using the newly derived Integrity Key serving as the shared secret between the MS and the home agent. At this point, the Binding Update is processed as in [10], including apparently performing the DAD operation for the home address.

The home agent can support multiple home registrations for different home addresses as long as the NAI remains the same. Each registration, however, needs authorization from the Home AAA server. These multiple registrations can be for the same care-of address. The Binding Cache Entry (BCE) is indexed by NAI in addition to the home address. If the home agent receives a Binding Update with an MS-HA Authentication option when no BCE is present for the NAI and the home address tuple, the home agent must have the shared secret to verify the MS-HA option and create a BCE. When the MS has previously authenticated itself to the Home AAA server, this shared secret will be the Integrity Key. If such an authentication has not taken place, the Home Agent may still use a shared secret provisioned by other means; however, this behavior is not currently specified. The home agent also supports the Return Routability messages but it does not support IPsec encryption of those messages.

The various types of Authentication Data are calculated as follows:

$$\begin{aligned} MS - AAA\ Authentication\ Data = \\ HMAC\_SHA1(MS - AAA\ Shared\ Secret, \\ MAC(MobilityData)), \end{aligned} \quad (18.1)$$

where

$$\begin{aligned} MAC(MobilityData) = \\ SHA1(Care\ of\ Address \mid Home\ Address \mid \\ Mobility\ Header\ Contents) \end{aligned} \quad (18.2)$$

where, Mobility Header Contents consists of data up to and including the SPI field in the MS-AAA Authentication Option.

$$MS - HA\ Authentication\ Data = \\ First(96, HMAC\_SHA1\ (MS - HA\ Shared\ Secret, MobilityData) \quad (18.3)$$

where

$$MobilityData = \\ (Care - of\ Address|\ Home\ Address|\ Mobility\ Header\ Contents), \quad (18.4)$$

where, Mobility Header contents consists of data up to and including the SPI field in the MS-HA Authentication Option.

$$IK = SHA\_1(MS - AAA\ Shared\ secret \mid \text{“}IntegrityKey''\\ |Message\ Identifier \mid Home\ Agent\ Address \mid Home\ Address) \quad (18.5)$$

So, we have seen in this and and the previous sections how a mobile node is able to perform Mobile IPv6 operations in a CDMA packet data network. Specifically, we have studied how it first establishes link and IP connectivity as it authenticates itself (Section 18.2.1, followed by bootstrapping of Mobile IPv6 parameters (Section 18.2.2, concluding with Home Registration (Section 18.2.3. As we can see, an access network such as CDMA brings its own set of deployment considerations which need to be taken into account when using Mobile IPv6 in such systems. In the rest of this chapter, we focus on using Mobile IPv6 in the so-called IMS. Before that, we briefly explain what an IP Reachability Service is.

18.3 IP REACHABILITY SERVICE

An IP Reachability Service is defined as the ability of the home network to update the Domain Name System server with the current IP address. Some nodes may wish to be reachable to their peers by means of hostnames in the DNS. Since the home address may be assigned dynamically, the entries in the DNS system could get inconsistent every time a new home address is assigned. Hence, for nodes without statically assigned home addresses, the home network can update the entries in the DNS every time a new home address is assigned to a MS.

When a home network node, either the home agent or the AAA server, updates the DNS with a Resource Record [11] for an IP address, it makes the Time To Live (TTL) in the update zero. This has the effect of forcing all queries for the hostname to be made to the authoritative server for the user in question. No DNS caching takes place. This avoids the possibility that the cached entries in the correspondent node are stale when the MS has acquired a new home address.

18.4 MOBILE IPV6 AND IP MULTIMEDIA SUBSYSTEM (IMS)

The IP Multimedia Subsystem (IMS) is defined in the 3rd generation projects to specify how IP-based packet data services can be supported by the cellular systems [19]. In this framework, the IETF's *Session Initiation Protocol (SIP)* [14] is a crucial element for facilitating multimedia applications. It is used as a trigger to initiate authentication, resource authorization, accounting and charging for IMS purposes, although the protocol itself is used only to initiate sessions. In addition, SIP offers *personal mobility* for a user, so that the user can expect to establish communications from new domains according to roaming agreements with the user's home domain. Personal mobility requires user identification and the association of the user with a particular device. Mobile IPv6, on the other hand, offers device mobility for all applications and services. Device mobility requires device identification (often embodied in its IP address), and application transparency is highly desirable so that all existing applications can continue to work on the device as it moves from one point of connection to the next. For these reasons, the interplay between SIP and Mobile IPv6 is important. In the following sections, we investigate this in greater detail without delving into CDMA-specific access details.

We begin with a brief review of SIP.

18.4.1 Overview of SIP

SIP provides a way for initiating sessions between users using two or more Internet devices. Each user is uniquely identified by a SIP Universal Resource Indicator (URI), which is used when sending session initiation requests. The general format of a SIP URI defined in [14] is as follows:

sip:user:password@host:port;uri-parameters?headers

A brief description of the fields in the URI follows:

- *sip:* defines a scheme according to the guidelines in RFC 2396 [4] and follows a form similar to *mailto* found on www URLs. The scheme allows specification of SIP header fields and message body construction according to well-known standard interpretation.

- The *user* part refers to a resource at the specified host field which comes after the user field. The user can be a human name ("Alice"), a computer name ("spock"), a telephone number ("+1-414-555-1212") and so on. The *userinfo* of a URI includes the user field, the password field and the '@' sign.

- The syntax supports a password associated with the user, but it is *not* recommended for use since it exposes the confidential information to eavesdroppers.

- The *host* refers to a Fully-Qualified Domain Name (FQDN) or an IP address. Often the *hostpart* refers to the public domain with which a user is associated. For instance, the userinfo and hostpart are typically used as "Alice@talk.com", which is an example of a public SIP address. Note that the hostpart here does not necessarily identify the actual device "Alice" is currently using, but only the domain with which she is is associated. The SIP Proxies use a location service (see below) to determine the SIP endpoint device where "Alice" is reachable whenever SIP requests arrive.

- The port field is simply the transport protocol port number to which the messages are sent. SIP defines a default port number of 5060.

- The *uri-parameters* operate on the specified URI to provide more flexible processing. These parameters determine choice of a transport protocol, TTL for UDP multicast, selecting a Proxy directly by overriding the address in the host field etc. Readers are referred to [14] for the details.

- Finally, the headers field includes name and value pairs referring to contextual information associated with the URI. For instance, a header field could include "subject=project" and may include multiple such pairs separated by the ampersand sign.

SIP Proxies facilitate communication between SIP peers by performing such functions as user tracking and call routing. The Proxies are discovered using the methods described in [15]. A SIP user sends a *REGISTER* message to a local *Registrar*, which acts as a "front end" to a logical entity called *location service* that stores the current location (an FQDN or an IP address within the domain of the Registrar) of the user. The registration allows the Proxy to query the location service in order to retrieve the current location of the user. In practice, the Registrar, the location service and the Proxy are typically implemented in a single physical device.

A typical session or a dialog in SIP begins when a peer sends an *INVITE* message to another peer, mediated by Proxies associated with the peer and itself. The initiating peer's Proxy forwards the INVITE to the intended recipient's Proxy, which then forwards the INVITE to the actual recipient. The sender indicates the address to respond to the INVITE in the *Via* field, and the sender indicates an address to directly send future requests to in the *Contact* field. The recipient's response, in the form of a three-digit numeric code, is routed back through the Proxies, sometimes traversing several intermediate Proxies.

An INVITE message can carry a description of the session parameters of interest. For instance, QoS parameters and IP address endpoints for the media stream could be signaled using the *Session Description Protocol (SDP)* [8]. These parameters in an existing SIP dialog can be modified using the so-called *Re-INVITE* message.

If a user begins operation in a new access network, the user should register with a Proxy as soon as possible in order to be able to receive incoming dialog requests. In this way, through registration upon roaming, SIP provides personal mobility.

18.4.2 Personal Mobility and Device Mobility

SIP provides user tracking by means of registrations, which ultimately map to the device (a SIP Endpoint) the user is using. For mapping to the device to take place, the hostpart of the URI in the REGISTER message can be either a domain or a hostname. When it is a domain, the Proxy needs to perform a DNS SRV query to first obtain the hostname and then perform a DNS "A/AAAA/A6" record query to obtain the IP address of the device. When the hostpart is a hostname, the Proxy needs to perform a DNS A/AAAA/A6 record query directly to obtain the IP address. The DNS A records are specified in [11], AAAA records are specified in [18], and the A6 records are specified in [5].

Since Mobile IP defines two addresses, a crucial problem is which address to use for registration purposes. If the hostname is mapped to the care-of address in the DNS, that entry must be updated upon every change of subnet. Also, a mobile node must send REGISTER every time an IP address changes even if the Proxy with which it is associated does not change. Since a mobile node may not necessarily be trusted to update the DNS database, a Proxy has to perform this operation in addition to updating its own location service. Furthermore, a DNS lookup by the Proxy is needed for every request arriving for the user since a cached DNS entry could indeed turn out to be stale.

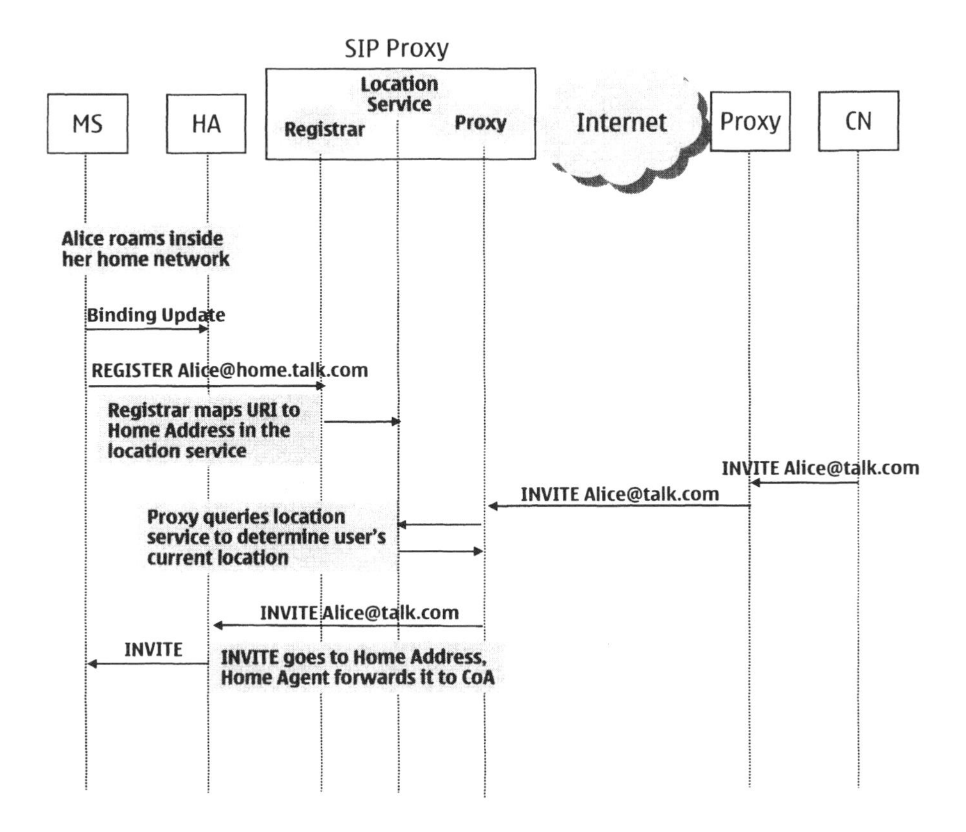

Fig. 18.5 SIP Registration of Mobile IP home address

Mapping the hostname in the URI to the home address does not have the drawbacks identified above. See Figure 18.5. The mapping in the DNS database remains fixed, regardless of the current location of the mobile node. A Proxy would always forward the arriving requests to the home address, and those requests would reach the care-of address by means of the Mobile IP operation. Also, a mobile node would simply register its URI, whose hostname maps to its home address in the DNS database, once, when it discovers a new Proxy. This choice of home address for URI mapping allows SIP, which is an application-layer protocol for establishing interactive communication on the Internet, to be relieved of subnet mobility. Hence, a clean separation of purposes can be achieved while fully exploiting the functionalities of both SIP and Mobile IP.

When we consider a roaming SIP user, things become more interesting. Assume a user of CDMA roaming to a visited network. Once the user's mobile node gains access to the CDMA network, it can discover the SIP Proxy. The Proxy is called *Call State Control Function* (CSCF) in the 3rd generation systems, the home proxy is referred to as *Serving CSCF* or *S-CSCF* and the visited proxy is called *Proxy-CSCF* or *P-CSCF*. So, once the MS discovers the P-CSCF, it needs to REGISTER with it. Before such registration can take place, IMS authentication must take place. This is done using the *Authentication and Key Agreement* or AKA protocol defined in [1]. The result of IMS authentication is a security association establishment between the MS and the P-CSCF so that the latter can process and forward SIP messages belonging to the MS.

The REGISTER message from the MS to the P-CSCF contains the URI that still maps to the home address. See Figure 18.6. However, the P-CSCF forwards the registration to the S-CSCF such that the current address for the user now points to the P-CSCF by adding its own URI in the *Via* field. In other words, the host part of the URI in the REGISTER message from the P-CSCF contains a URI that maps to the visited domain or the IP address of the P-CSCF itself. This allows the S-CSCF to update its location service database for the user to contain the P-CSCF address. No DNS-specific actions are necessary. When a new request for communication with the user arrives, the S-CSCF routes the INVITE to the P-CSCF, which then forwards the message to the Home Address of the MS.

Some readers may have noticed that the incoming INVITEs would have to traverse the home network since the Contact field of the MS always maps to the home address. These INVITEs are then intercepted up by the home agent like any other packets and forwarded to the MS. So, the basic operation, even if less optimal, ensures reliable packet forwarding. If route-optimized communication between the P-CSCF and the MS is desired, the MS can initiate the Return Routability procedure with the P-CSCF and subsequently establish a Binding Cache Entry at the P-CSCF. Since a P-CSCF is an IPv6 node that supports mobile nodes, it is reasonable to expect it to be a Mobile IPv6 correspondent node which implements route optimization. The Mobile IP Binding Update follows the P-CSCF discovery but must precede the SIP registration, and the SIP software only sees the home address in the source IP address field of the IP header of the SIP packet. The SIP implementations that make assumptions regarding the source IP address should be mindful of the presence of the home address in this field when Mobile IP route optimization is used.

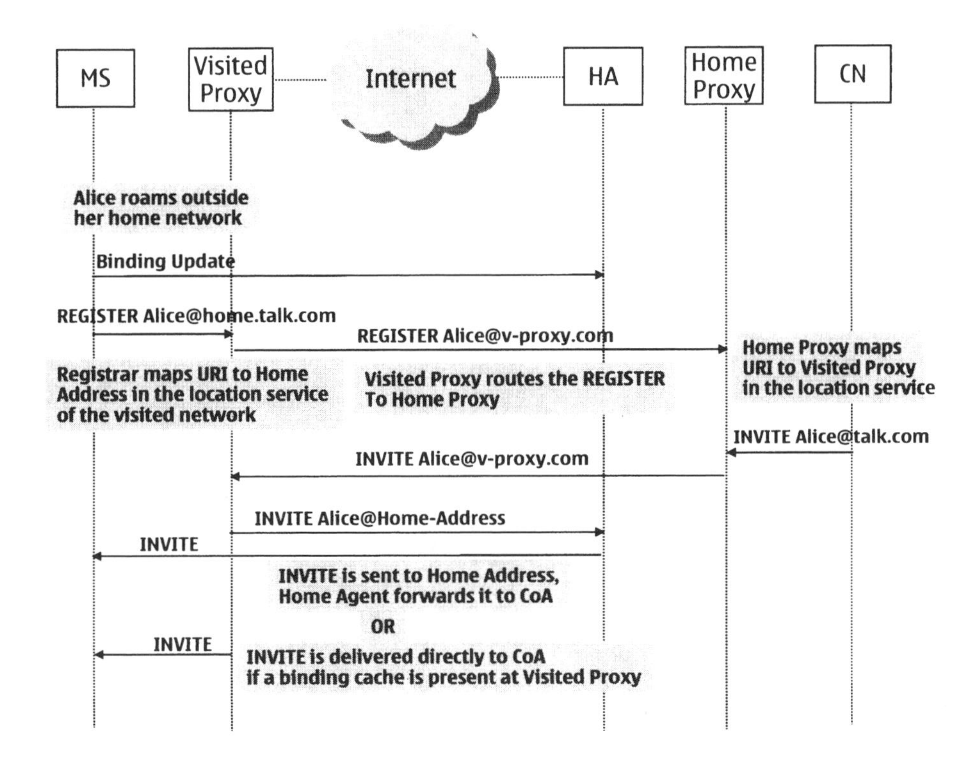

Fig. 18.6 SIP Registration of a Roaming User

Alternatives to Return Routability with the P-CSCF are worthy of investigation. The URI in the Contact field must be resolvable in order for the P-CSCF to forward incoming requests, and this resolution could be made possible via DNS lookups. Since the user was authenticated by the home domain before the user was allowed to register, the same home domain could be trusted to have securely updated the DNS entry for the user's URI to the home address. In other words, the P-CSCF can accept the home address provided by the MS by trust in the home domain's ability to correctly enter DNS records. When this trust model is acceptable, the P-CSCF can directly accept Binding Updates from the MS, provided they are properly secured. In order to secure the Binding Update, the MS must use the security association established with the P-CSCF during IMS authentication. The MS may use the Binding Authorization Data option to prove the authenticity of the Binding Update. We leave the details of message formats as an exercise for the reader.

Perhaps an even more interesting (and controversial) optimization involves using SIP signaling to install Binding Cache entries at a correspondent node! Recall that Mobile IPv6 requires address tests to be conducted in order to verify the ownership of home and care-of addresses before a correspondent node can bind those addresses together for route-optimized communication. Since no URIs are used in Mobile IP, no confirmation of addresses can be done using the DNS. With SIP, however, a mobile node can supply its home address directly in the Contact field, and a correspondent can verify it using the URI when the URI is detailed enough to map to the user's device address directly and not to the user's Proxy IP address in the DNS. This depends on whether the user is willing to expose the device hostname or domain details in the INVITE message; typically, only registrations with the Proxy contain the end-point details. However, when providing hostname details is acceptable to a user, the correspondent node can verify the Contact address by resolving the URI in the DNS. This lookup by itself does not guarantee that the initiator actually owns the address. The initiator may have provided a valid URI belonging to some other user. In fact, this issue is not specific to our topic. Any SIP message can have arbitrary addresses specified in the Via and Contact fields. The Proxies are entrusted with the task of ensuring that those fields contain values which are currently registered in the location service database. So, in our case, a correspondent has to trust the chain of Proxies to ensure that the value specified in the Contact field (which maps to the home address) is indeed verified by the initiator's Proxy, with which the correspondent's Proxy has some kind of trust relationship. With this assumption of trust and with the acceptable trust in the DNS system, the correspondent can believe that the initiator also owns another address with which it can communicate. When this is feasible, it is possible to define mobility extensions that allow a correspondent to install a Mobile IP Binding Cache Entry immediately after processing the SIP messages, but activating it only after the mobile node sends a Binding Update (which follows the exchange of SIP signaling).

As we observed at the beginning of the previous paragraph, any approach that attempts a closer integration of SIP and Mobile IP has to carefully consider various security implications, which we have not covered extensively here. However, it is a promising area to consider, interesting at least. As protocols proliferate, often with purpose-built security architectures, there is a risk of the burden of security handling being passed on to end users. This is not always desirable, and *single sign-on* approaches are worthy of investigation.

18.5 SUMMARY

In this chapter, we have discussed how Mobile IPv6 is used in CDMA packet data networks. Beginning with link establishment, we have considered authentication, IP address acquisition, Mobile IPv6 parameter bootstrapping, and finally the Binding Update procedure. As we saw, a protocol such as Mobile IPv6 is subject to the constraints imposed by a particular deployment in terms of access authentication, shared secret establishment and so on. It is important for this reason alone to design protocols that do not require significant changes to the existing deployments. In practice this is not always achievable however. In the end, extensions to existing protocols and mechanisms are typically defined in order to accommodate the newer ones.

We also discussed personal mobility and device mobility. We saw that SIP is well-positioned to provide the former, whereas Mobile IP is best-suited for device mobility. We discussed how the two could co-exist and provide the maximum benefit by having an appropriate mapping of parameters. We also presented some early ideas on how SIP signaling could be used to achieve better results in route optimization.

Exercises

18.1 Compare the authorization assumptions behind supporting IP address update using SIP Re-INVITE and Mobile IPv6 Binding Update (As we mentioned earlier, a SIP endpoint *can* update its peer with a change of IP address by sending a Re-INVITE message. The Mobile IPv6 Return Routability was designed to address a similar address authorization problem.)

18.2 Single sign-on has been a holy grail of access authentication. Users are often confronted with multiple authentication mechanisms arising out of the need from multiple layers in the protocol stack (e.g., MAC, IP and application layers). In principle, a single sign-on should solve this problem, relieving users of the idiosyncrasies associated with disparate access control methods. Investigate the feasibility of using SIP authentication to derive a necessary SA for Mobile IPv6. What are the security implications? What happens when the domains of applicability of an authentication mechanism change?

18.3 One of the challenges in providing services to a roaming user is where to locate the service control. In the current deployment models, it is always at the home network; the visited network provides access and requested QoS but not the service logic itself. As an unfortunate extension of this model, most cellular packet networks today rely on tunneling from the home network even for data packets. With route optimization in Mobile IPv6, peers can communicate directly however. How could one devise mechanism(s) to allow a service provider to maintain service control while allowing users to communicate most efficiently? What policy extensions are needed to achieve the technical solution?

REFERENCES

1. 3rd Generation Partnership Project; Technical Specification Group Services and System Aspects; 3G Security; Security Architecture (Release 5), 2002.
2. Wireless IP Architecture Based on IETF Protocols, 3GPP2 P.R0001, Version 1.0.0, July 2000.
3. B. Aboba and M. Beadles. "The Network Access Identifier," RFC 2486, Internet Engineering Task Force, January 1999.
4. T. Berners-Lee, R. Fielding, and L. Masinter. "Uniform Resource Identifiers (URI): Generic Syntax," RFC 2396, Internet Engineering Task Force, August 1998.
5. M. Crawford and C. Huitema. "DNS Extensions to Support IPv6 Address Aggregation Renumbering," RFC 2874, Internet Engineering Task Force, July 2000.
6. R. Droms (Editor). "Dynamic Host Configuration Protocol for IPv6 (DHCPv6)," RFC 3315, Internet Engineering Task Force, July 2003.
7. R. Droms. "Stateless Dynamic Host Configuration Protocol (DHCP) Service for IPv6," RFC 3736, Internet Engineering Task Force, April 2004.
8. M. Handley and V. Jacobson. "SDP: Session Description Protocol," RFC 2327, Internet Engineering Task Force, April 1998.
9. D. Haskin and E. Allen. "IP Version 6 over PPP," RFC 2472, Internet Engineering Task Force, December 1998.
10. D. Johnson, C. Perkins, and J. Arkko. "Mobility Support in IPv6," RFC 3775, Internet Engineering Task Force,June 2004.
11. P. Mockapetris. "Domain Names - Implementation and Specification," RFC 1035, Internet Engineering Task Force, November 1987.
12. B. Lloyd and W. Simpson. "PPP Authentication Protocols," RFC 1334, Internet Engineering Task Force, October 1992.
13. A. Patel et al. "Authentication Protocol for Mobile IPv6,", RFC 4285, Internet Engineering Task Force, January 2006.
14. J. Rosenberg et al. "SIP: Session Initiation Protocol," RFC 3261, Internet Engineering Task Force, June 2002.
15. J. Rosenberg, and H. Schulzrinne. "Session Initiation Protocol (SIP): Locating SIP Servers," RFC 3263, Internet Engineering Task Force, June 2002.
16. W. Simpson (Editor). "The Point-to-Point Protocol (PPP)," RFC 1661, Internet Engineering Task Force, July 1994.
17. W. Simpson. "PPP Challenge Handshake Authentication Protocol (CHAP)," RFC 1994, Internet Engineering Task Force, August 1996.
18. S. Thomson et al. "DNS Extensions to Support IP Version 6,", RFC 3596, Internet Engineering Task Force, October 2003.
19. "IMS Network Architecture", 3GPP Technical Specification TS 23.002.
20. "IMT-2000 Network Aspects", http://www.itu.int/ITU-T/imt-2000/network.html

19

Enterprise IP Mobility

Computers in the future may weigh no more than 1.5 tons. –Popular Mechanics, forecasting the relentless march of science, 1949

—

Blackberry 5810, weighing 4.7 ounces

19.1 INTRODUCTION

Remote access to enterprise network resources has always been a popular application. Even before the dawn of WLAN, roaming employees used the Plain Old Telephone Service (POTS) for dial-up access to corporate data. With WLAN, mobility across enterprise campus networks has become feasible. Roaming users inside an enterprise WLAN network can maintain their application sessions in spite of mobility. As a next step, such enterprise users can also move from WLAN to Wireless WAN and still maintain their sessions. In this chapter, we will discuss specific characteristics of enterprise networks and how mobility solutions could be designed in such environments.

The enterprise networks present some unique challenges to mobile networking. First, security is of paramount importance for fairly obvious reasons. This implies that any access must be authorized and subsequently kept confidential. When mobility is involved, this criterion must continue to be met. Second, a huge majority of these networks are IPv4 networks. So, mobility solutions must work with IPv4. Third, the emerging peer-peer applications such as VoIP need to work in an environment protected by firewalls which permit mostly TCP traffic. Finally, with enterprise data and voice networks converging towards a single network for cost and efficiency reasons, performance of VoIP over WLAN must be comparable to the standards of enterprise users.

Broadly, we can classify the solutions for the challenges into three categories: secure mobility, middle-box management and real-time handovers. Secure mobility means having continued access to corporate data and resources in a secure manner even in the presence of handovers. VPNs have been used to provide secure remote access. Making VPNs work without having to reestablish them because of mobility is what we mean by secure mobility. Middlebox management refers to making perimeter security devices such as firewalls permeable for real-time applications such as voice, and maintaining a session despite movement across multiple interior firewalls or even multiple provider firewalls. It also refers to managing wireless access points to provide the secure access and protocol machinery necessary for improving latencies during handovers. We have already considered real-time handovers before: providing support for applications such as VoIP during handovers. We emphasize that all of these are emerging areas with no default deployed solutions. Our discussion will focus on protocols that provide the required functionality for a secure mobile enterprise.

19.2 VPN MOBILITY

IPsec has been used as a protocol for secure remote access. Typically, each enterprise device, such as a laptop, is configured with security credentials so that it can initiate a secure channel established from anywhere on the Internet as long as the device has a routable IP address. The user is typically presented with a companion solution, such as *SecurID* card, which produces a one-time password from a shared secret between the user and the security gateway. When the password matches, an encrypted tunnel is established between the user's laptop and the security gateway.

The Internet Key Exchange (IKE) is the protocol used for exchanging key parameters between two endpoints. When IKE is successful, the two endpoints establish a security association (SA) which is based on the IP addresses. When packets are to be sent or when they are received, it is important that they be classified correctly using the selectors maintained as part of the SA in order to apply security transforms (i.e., encryption or decryption using an agreed-upon algorithm and key). Since IP addresses are used as indices for all traffic classification, when an IP address changes the security transforms cannot be applied as agreed upon before; hence an existing VPN session "breaks." The user needs to reestablish the VPN session using a peripheral such as SecurID. So, the problem for secure remote access mobility is to keep an existing VPN session alive even when an IP address changes.

Even though VPNs are primarily used with remote access scenarios, a WLAN campus network of an enterprise can also be seen as an instance of remote access. The VPNs are used even when the WLAN uses link-layer authentication such as IEEE 802.1X. This is partly because an enterprise may not necessarily trust any wireless access point which has provided link access. Also if a user roams from enterprise WLAN to a WWAN, VPN access is needed anyway. Within a campus WLAN network, VPN mobility takes place when a mobile node crosses subnet boundaries. When such a node crosses the provider boundary, the IP address changes again.

Another instance of VPN mobility involves a user in a typical remote access network changing subnet boundaries. This could happen in a large enough WLAN campus or when the user switches from WLAN to GPRS or CDMA packet data networks. In such scenarios, protocol enhancements are needed to maintain an existing VPN session.

19.2.1 IKEv2 Mobility and Multihoming (Mobike)

The Mobike protocol provides extensions to the IKE version 2 (IKEv2) protocol [3] in order to securely update a VPN termination point (such as a VPN Gateway) that an address change has occurred. An IP address change can occur because of physical mobility or because a mobile node with multiple interfaces switches its interface for ongoing traffic. The protocol itself is recently standardized in IETF [2]. We will discuss the main design points the protocol is addressing.

First, a peer provides an indication that it is Mobike-capable during the initial IKEv2 exchange (called IKE_AUTH request), which itself does not involve any address updates. This exchange may include multiple IP addresses that the peer would like to use in anticipation of possible change in its IP connectivity. Providing a peerlist of addresses, the argument goes, allows the peer (which is either mobile or multihomed or both) to start using an alternate address in the list when its current address is no longer preferred even without having to inform the VPN Gateway explicitly. In other words, the protocol attempts to explicitly support direct and indirect indications of change of an IP address. This allowance however, appears to be restricted by the need for explicit signaling in the final protocol specification [2].

After the initial indication of Mobike capability and peerlist IP address exchange, at some point an address change for the existing VPN session may be necessary. At this time, the mobile node sends an IKEv2 *INFORMATIONAL* message with Mobike extension (UPDATE_SA_ADDRESSES notification) to indicate a change of address. When notification is used, the VPN Gateway can ensure the validity of the new address by performing Return Routability; it can verify that the new address provided by the mobile node indeed belongs to it in the routing sense. This involves the VPN Gateway providing a cookie meant for the new address, and the mobile node responding back with the cookie just provided. Without Return Routability, a node with Mobike capability can redirect traffic to another node either inadvertently or maliciously. Hence, even if a peerlist is provided, the VPN Gateway should enforce proof of address ownership before it updates the IKE SA. Furthermore, since the Mobike exchange will be protected by an existing IPsec SA, the return routability messages are protected from eavesdropping, unlike those in Mobile IPv6. An exception when return routability may not be used is if the new IP address in question is already part of a certificate communicated during the IKEv2 exchange.

A VPN Gateway can also switch the address for the mobile node proactively if it determines that the current address is not working, e.g., if it receives an ICMP Host Unreachable message. The IKEv2 protocol specifies a mechanism called *Dead Peer Detection*, in which an IKE endpoint probes the liveness of its counterpart by sending an empty Informational message to which it expects an acknowledgment. If no acknowledgment is received within a timeout followed by a few retransmissions and their timeouts, the peer is deemed dead. Subsequently, the IKE SA is deleted. For Mobike, however, the IKE endpoint (VPN Gateway) can test the liveness of other available addresses before deleting the IKE SA. If any of the addresses in the peerlist is reachable, the VPN Gateway can start sending traffic to it.

The IKEv2 notifications with Return Routability provide the basic support necessary to securely update an address for the IPsec protected tunnel. Since the address change is only for the outer IP tunnel between the client and the VPN Gateway, the inner transport address remains intact, thus allowing ongoing sessions to continue.

Since an IPsec SA is a child of the IKE SA, the extent of IPsec SA update per Mobike update is also a moot design point. A simple approach is to update all the IPsec SAs created using the IKE SA in question. Providing the flexibility of selective IPsec SA updating requires the protocol to carry more options, but it may be useful in moving SAs based on such factors as connection speed and pricing. The Mobike protocol has also considered a suspend and resume mode of operation in which a peer notifies that it is going dormant but would like the SAs kept alive so that it can resume traffic at a later point in time. Again, the exact details of such an operation are unlikely to be specified in the base document.

It is worthwhile to ask "How do Mobike and Mobile IP relate to each other?" There are similarities. Mobike manages the outer tunnel so that VPN applications using the IP address present in the inner packet can survive movement across subnets. Mobile IP, manages the care-of address present on the outer tunnel so that applications using the home address present in the inner packet can survive subnet movements. Mobike is a security protocol that tries to support session persistence during mobility. Mobile IP is a mobility protocol that provides session persistence and reachability while ensuring that protocol operations are adequately secure. Given their origins, the designs emphasize different aspects of functionality. In any case, it is informative to investigate various interworking scenarios. In the following, we present one possibility. Readers are encouraged to construct others.

Many enterprises already provide remote access via VPNs.[1] So, the firewalls allow traffic destined for VPN Gateways easily. This suggests using Mobike when the mobile node is roaming outside of a corporate enterprise and encounters subnet movements. For instance, imagine a roaming enterprise user connecting to the corporate Intranet via a WLAN hotspot. Such a user may roam freely between subnets spanning the WLAN or connect to a WAN such as GPRS [7] or EV-DO [8]. The mobile node can run the Mobike protocol to keep the VPN session alive.

On the other hand, if the user is within a protected enterprise perimeter, such as a company campus, there may not be a need to traverse the remote access VPN Gateway when link-layer access control and encryption are enforced. For instance, when IEEE 802.1X [9] is deployed, the mobile node can encrypt its traffic over the air using the keys derived during authentication. [2] In such a "VPN-blind" environment, a mobile node may still traverse different subnets. In order to maintain session persistence, and possibly reachability, a fixed transport address is necessary. This motivates the need for using a Mobile IP home address.

So, the scenario of interest for us is mobility with remote access as well as mobility within an enterprise WLAN without VPNs. Furthermore, a user must be able to roam from remote access to direct enterprise access without having to restart any of the applications. For instance, a user with an EV-DO data access card and a WLAN access may switch from cellular (remote access) to direct enterprise access via WLAN when roaming from home to enterprise campus. With these requirements in mind, let us consider the following solution.

[1]For our purposes here, we will consider IPsec-based VPNs.

[2]Of course, deployments could still use interior VPN Gateways

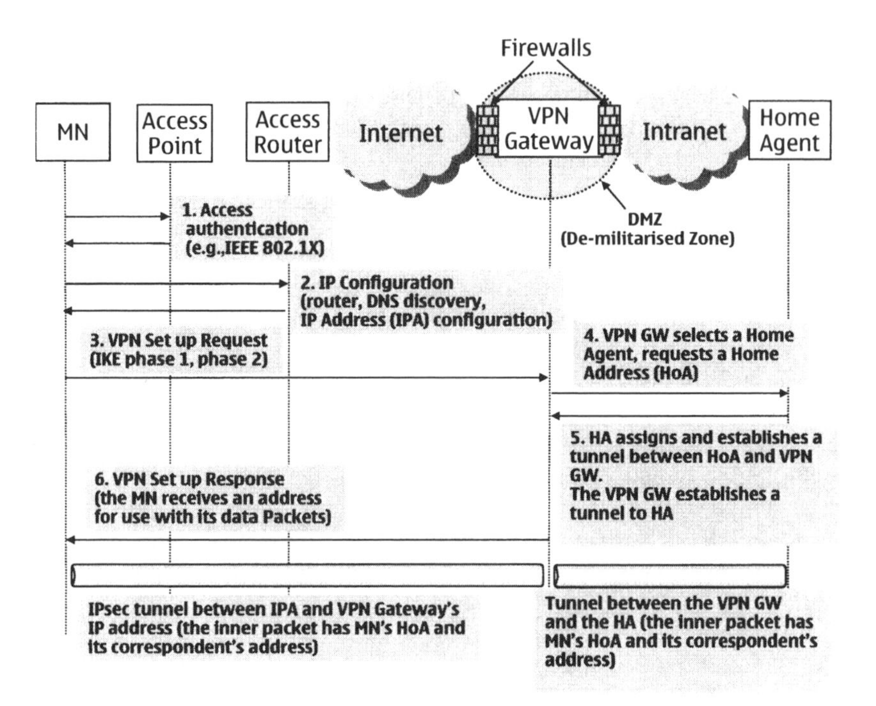

Fig. 19.1 Mobile Node Outside the Enterprise Network

Assume that the mobile node is outside the enterprise network and hence runs the VPN application to connect to the corporate network. See Figure 19.1. The IPsec VPN client runs the IKE protocol in order to establish an IPsec tunnel. As a part of client credentials, the mobile node's home agent address and possibly a home address are stored in the company security database. The VPN Gateway sends a new *Home Address Request* message to the home agent providing the mobile node's NAI. The home agent itself may assign a home address dynamically or else use a statically assigned address. If the home agent can provide a home address, it creates a Binding Cache Entry, mapping the home address to the VPN Gateway's IP address. This establishes a tunnel between the home agent and the VPN Gateway for all packets arriving for the home address. The home agent sends a *Home Address Reply* message to the VPN Gateway, which then establishes a tunnel to the home agent for packets arriving from the mobile node. In essence, the new pair of messages between the VPN Gateway and the home agent result in allocating the home address as the transport address for the VPN session and establish a bidirectional tunnel between the VPN Gateway and the home agent.

So, when the mobile node sends packets, the inner packet has the allocated home address as the source IP address and the correspondent's IP address in the destination IP address. The outer packet contains the mobile node's IP address (valid on its point of attachment) and the VPN Gateway's address in the source and destination IP addresses, respectively. After it strips off the outer tunnel packet and inspects the source IP address of the inner packet, the VPN Gateway encapsulates the packet again and sends it to the home agent which then forwards the packet to the correspondent. In the reverse direction, the home agent tunnels the packets arriving for the home address to the VPN Gateway, which inspects the destination IP address of the inner packet and then tunnels the inner packet to the mobile node.

Now, assume that the mobile node roams into the enterprise network. See Figure 19.2. The mobile node first needs to detect that it has roamed to its enterprise network. Reliably detecting this depends on the deployment of WLAN, including configuring SSIDs, authentication mechanisms used before granting access and so on. We do not go into those details here. Subsequently, the mobile node obtains a new IP address from its enterprise network. It can then send a Mobike message, assuming that the firewall is configured to accept such messages and subsequently data traffic from hosts inside the enterprise, to the VPN Gateway again and continue its sessions as before. However, this would unnecessarily introduce traffic on the VPN Gateway from hosts inside the enterprise. It also makes the VPN Gateway the physical bottleneck for all roaming WLAN traffic.

Since the transport IP address is the home address, the mobile node can use Mobile IP. So, the mobile node uses the outer address as the care-of address and tunnels all its traffic to the home agent after updating the Home Agent with its care-of address. The home agent then tunnels all the traffic to the mobile node directly and tears down the tunnel to the VPN Gateway. The mobile node can either allow the VPN session timeout at the VPN Gateway or send an explicit message to disconnect the VPN session.

In the foregoing discussion, we have considered an interworking scenario between a security protocol with mobility extensions and a mobility protocol that needs security for its operation. It should be clear that there are other ways one could construct these scenarios. For instance, Mobile IP can be used even when roaming outside the enterprise network inside the VPN tunnel. Different trade-offs exist in terms of efficiency and complexity of operation. We encourage the readers to investigate different scenarios.

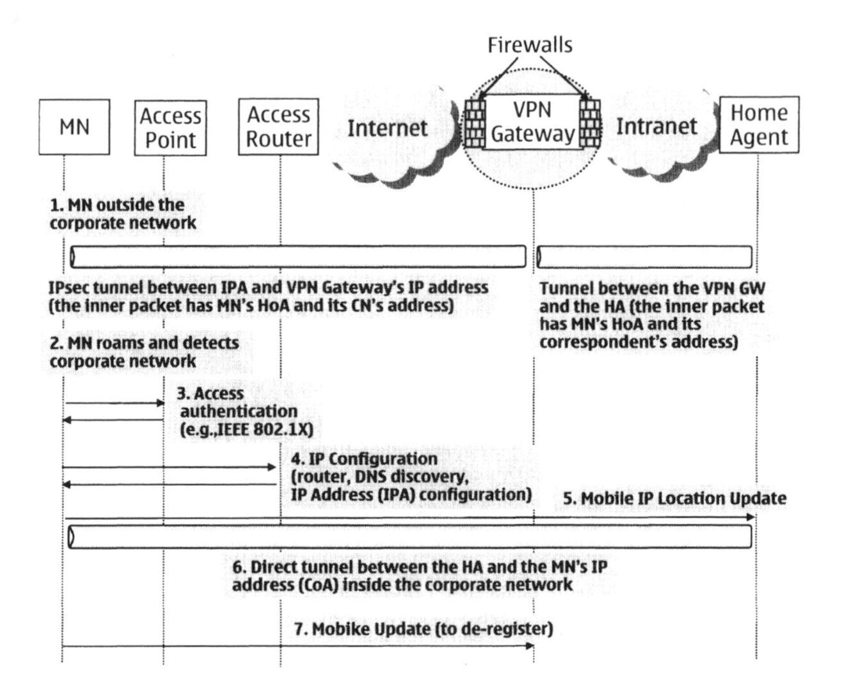

Fig. 19.2 Mobile Node Inside the Enterprise Network

19.3 MIDDLEBOX MANAGEMENT

Consider again our scenario of roaming between remote access and direct enterprise access in the previous section. Assume that a user begins wireless access within the enterprise network and then roams to a remote access network. See Figure 19.3. So, the user initiates a VPN session from the remote access network after handover. If the mobile node used a home address, as we described in our interworking scenario, then the existing sessions *can* continue if the VPN Gateway allocates the same home address as the transport address for the VPN session (as we described using Figure 19.1). However, the interior firewall which inspects the packets that have no VPN tunnel header would have no TCP state information for the on-going sessions, since the initial TCP exchange did not traverse this firewall. So, even though the existing sessions can be made to work by carefully orchestrating interworking between Mobike and Mobile IP, the firewalls may not allow packet flows to continue because no state exists for those flows.

Consider another version of the same scenario. Assume that the user roams from one remote access provider to another after establishing a VPN session. Also assume that firewalls protect the remote access networks from unauthorized access. So, when a user establishes a VPN session from within the remote access network, the firewall establishes relevant TCP or UDP state in order to allow traffic to the user's device. The exterior firewall at the enterprise network, on the other hand, allows traffic only for the VPN Gateway, and the interior firewall maintains the transport layer state (TCP connection status, port numbers, etc.).

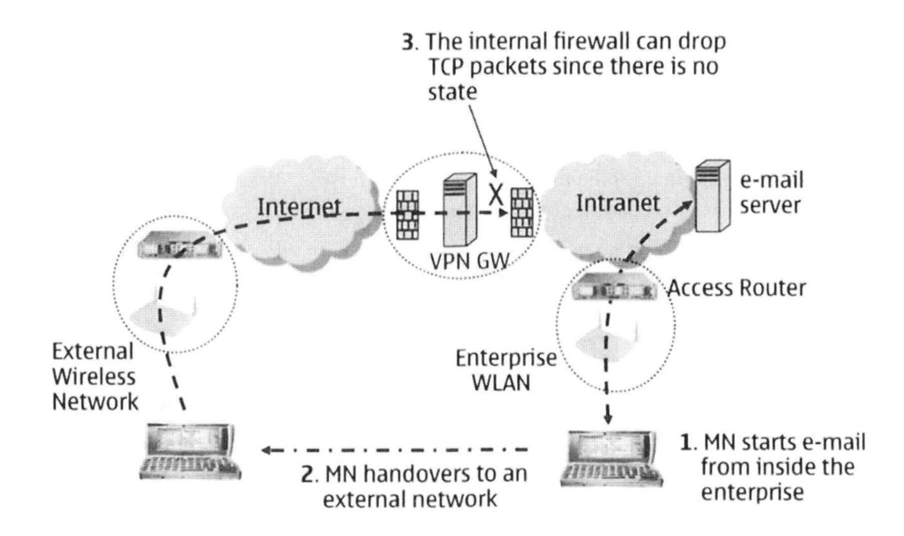

Fig. 19.3 Mobility and Firewall Interaction

When the user roams from one remote access network to another, the firewall at the enterprise network may only see VPN update (Mobike) messages, and subsequently the VPN tunnel packets between the mobile node and the VPN Gateway. However, the firewall at the new remote access network would not have the TCP state to allow data traffic inside the VPN tunnel to *continue*. This is because the firewall was not involved in inspecting the initial TCP exchange. [3]

In these scenarios, firewall state management is critical for ensuring session continuity. However, no well-accepted, if not standard, techniques exist for performing firewall state management during mobility. Context transfer is a promising technology, but firewalls are not access routers. Synchronizing transfer of contexts resident on network elements other than the access routers has not been specified and no deployment experience exists.

[3]Of course, if the firewall merely inspects the outer tunnel contents, it would have the necessary state for the outer tunnel since the Mobike exchanges take place from the new remote access network.

19.4 ADAPTING IPV6 FAST HANDOVERS FOR IPV4 NETWORKS

The need for fast handovers in enterprise WLAN networks involves primarily VoIP over WLAN. In the near future, enterprise collaborative applications including real-time video-conferencing and streaming could also impose fast, smooth handover requirements. So, it is crucial to support optimized handovers in an enterprise network which requires strong performance guarantees.

In the deployment of the first generation of enterprise WLAN networks, support for fast handovers is considered to be primarily between access points within the same subnet. The technology components of interest include minimizing *scan latency* and *re-authentication latency* during inter-access-point handovers (See Chapter 20. However, many vendors are beginning to realize that handovers limited to a single subnet are restrictive, and are providing proprietary solutions for intersubnet handovers. As this market grows, so will the scalability requirement. And, the network administrators will perhaps want the benefit of managing the WLAN just like other Ethernet subnets. In any case, scenarios will emerge where providing performance in a standardized fashion when crossing subnet boundaries will be necessary. With this background, we describe the fast handover protocol for IPv4-only enterprise networks.

The design principles of an IPv4 fast handover protocol are the same as those of the IPv6 fast handover protocol: minimize movement detection and IP configuration latency while disengaging Mobile IP messages from the critical path. We describe the protocol operation in detail in the following sections. The protocol itself can run without Mobile IP. However, the packet formats presented in Mobile IP [6] are used so that existing software can be re-used. The Mobile IPv4 Working Group in IETF is specifying the protocol for Fast Mobile IPv4 operation described below [5].

19.4.1 Protocol Operation

In order to eliminate movement detection latency, Proxy Router Solicitation and Advertisement messages are used. These messages are useful for new router discovery as well. However, these messages must be in ICMPv4 format [5]. IP address configuration, on the other hand, is not as easy due to the limited address space in IPv4. We consider three scenarios. First, the access routers are not involved in IP address management for the mobile node. The mobile nodes acquire a new IP address upon attaching to a new subnet link as they normally do, using DHCP for instance. The access routers do not participate in Mobile IP (i.e., act as foreign agents). Second, an access router acts as a foreign agent, using the same IP address used by a multitude of mobile nodes. In this scenario, an access router provides its own IP address for the mobile node to use upon connecting to the new link. Third, an access router may allocate an IP address to a visiting mobile node by some means, for instance from a pool of IPv4 addresses for temporary use by visiting mobile nodes. In all the scenarios, the tunnel endpoints are the access routers.

19.4.1.1 Scenario 1: Routers Not Involved in Address Assignment When the mobile node sends the FBU in a "predictive" mode from the previous link, it uses the following values in the message:

- Home Address field is set to PCoA
- Home Agent field is set to PAR's IP address
- Care-of Address is set to NAR's IP address (discovered using the PrRtAdv message)
- Destination IP Address is that of PAR's address, and
- Source IP Address is mobile node's address on PAR's link. This could be the home address or the co-located care-of address when the mobile node is using Mobile IP

The care-of address in Mobile IP can either be a co-located care-of address or a Foreign Agent care-of address. When a Foreign Agent care-of address is used, the PCoA field is set to the home address, and not the Foreign Agent care-of address. This is because the fast handover support is for the address the mobile node is using on its interface (which in this case happens to be the home address). When co-located care-of address is used, the PCoA is the co-located care-of address itself.

The Care-of Address field is always set to the new router's address, unlike in the case of Mobile IPv6 fast handovers. This is done because a mobile node cannot formulate an IPv4 address just as easily as it can with IPv6.

FBU message format is described in [5].

The FBU is sent from the new link regardless of whether an FBack has been received previously. In other words, the FBU serves as a notification of attachment to NAR. Such an FBU, sent in a "reactive" mode, uses the following fields:

- Home Address field is set to PCoA
- Home Agent field is set to PAR's IP address
- Care-of Address is set to NAR's IP address (discovered using the PrRtAdv message)
- Destination IP Address is that of NAR's address, and
- Source IP Address is mobile node's address on PAR's link. This could be the home address or the co-located care-of address when the mobile node is using Mobile IP

When the NAR receives the FBU, it may already have a host route entry established for the PCoA. In that case, NAR can immediately start forwarding the packets including the FBack. In any case, NAR forwards the FBU message to PAR, replacing the source and destination IP addresses in the FBU message above with its own IP address and PAR's IP address respectively.

As in [4], the HI and HAck messages are used to set up the tunnel between the routers. The tunnel itself is established as a response to the FBU message. The PAR sends an HI message with Code = 0 when it receives the FBU with source IP address set to PCoA. The PAR sends an HI with Code = 1 when it receives the FBU with source IP address not set to PCoA (i.e., when received from NAR). The different Code values allow the NAR to disambiguate processing when an HI needs to be sent as a response to predictive and reactive modes of operation.

The previous router, as a result of HI and HAck exchange, creates a route table entry to tunnel packets arriving for the mobile node's IP address to the NAR's IP address. The new router creates a tunnel for packets arriving from mobile node's PCoA to the PAR. The mobile node may possibly run DHCP again after attaching to a new subnet. Since those operations are decoupled from handover messages, they do not affect the ongoing applications.

The mobile node may use its DHCP address as the co-located care-of address if it wishes to use Mobile IP even if the access router does not support it. The details of protocol operation are left as an exercise.

19.4.1.2 Scenario 2: The Access Routers as Foreign Agents In this scenario, Mobile IP is used with the foreign agent's IP address as the care-of address. The mobile node sends packets using its home address as the source IP address, and the foreign agent encapsulates packets and tunnels them to the home agent. In the reverse direction, the foreign agent decapsulates the tunneled packets from the home agent and forwards the inner packet (addressed to the mobile node's home address) to the mobile node.

As a response to FBU, PAR establishes an entry for home address in its routing table so that arriving packets are tunneled to the NAR's address. The arriving packets from the home agent are first decapsulated before the route lookup for the inner home address takes place. The new router also must decapsulate the tunneled packet from the PAR and must use a host route entry to forward the inner packet. The tunnel between the routers is established using the HI and HAck exchange, as earlier.

19.4.1.3 Scenario 3: Assigned Addressing Support In this mode, the mobile node obtains an NCoA from NAR in the FBack message, either on the previous link or on the new link. Even so, the mobile node still binds the PCoA to NAR's IP address in the FBU message since it is unaware of the actual address that NAR would return. The mobile node may use NCoA to register with its home agent. This scenario could reduce the time it takes to perform Mobile IP registration, but otherwise identical to previous scenario.

19.5 SUMMARY

In this chapter, we discussed problems specific to enterprise mobility. In particular, we looked at how VPN mobility can be supported across different roaming scenarios. We saw that address allocation for transport protocols is a crucial problem in this context. We then investigated the problems that arise due to firewall traversal in the event of mobility. Finally, we saw how VoIP could be supported for IPv4 by adapting the fast handovers for IPv6 protocol. In Chapter 20, we investigate how fast handovers operate over WLAN networks.

REFERENCES

1. S. Deering. "ICMP Router Discovery Messages," RFC 1256, Internet Engineering Task Force, September 1991.
2. P. Eronen (Editor). "IKEv2 Mobility and Multihoming Protocol (MOBIKE),", RFC 4555, Internet Engineering Task Force, June 2006.
3. C. Kaufman (Editor). "Internet key Exchange (IKEv2) Protocol," RFC 4306, Internet Engineering Task Force, December 2005.
4. R. Koodli (Editor). "Fast Handovers for Mobile IPv6," RFC 4068, Internet Engineering Task Force, July 2005.
5. R. Koodli and C. E. Perkins, "Mobile IPv4 Fast Handovers" draft-ietf-mip4-fmipv4-03.txt (work in progress), Internet Engineering Task Force, February 2007.
6. C. Perkins (Editor). "IP Mobility Support for IPv4," RFC 3344, Internet Engineering Task Force, August 2003.
7. 3rd Generation Partnership Project. "General Packet Radio Service (GPRS)," 3GPP TS 23.060, Service description, Stage 2, Release 6 http://www.3gpp.org
8. IS-856 Rev A - cdma2000 High Rate Packet Data Air Interface Specification (1xEVDO)
9. "IEEE Standard for Local and Metropolitan Area Networks: Port-Based Network Access Control," IEEE Standard 802.1X, 2004.

20

Fast Handover in a Wireless LAN

It's hard to imagine being mobile without assuming wireless. - (somewhat obvious) self revelation

—

20.1 INTRODUCTION

WLAN has quickly become a technology choice for homes, enterprises, metro mesh networks, rural mesh networks and so on. Due to low cost, higher bandwidth and ease of use, it has found large scale acceptance. Enterprises have found it appealing especially for the "mobile VoIP" potential. It is not only feasible to replace a fixed telephone with a device equipped with WLAN, but also enhance the user experience available with VoIP. Increasingly, such "devices" are becoming mobile phones which offer roaming to cellular networks when a user is out of coverage of WLAN. So, enterprises can trade the cost of their fixed telephony for mobile telephony while improving the productivity. Given the recent advances in mobile computing, mobile rich-media communication is not too far behind.

We saw in Chapter 14 the actions performed by a mobile node on a typical day:

while (true) do
{

1. Establish link connectivity
2. Establish IP connectivity
3. Request network resources
4. Run application(s)
5. Undergo handover
6. Go back to step 1

}

We return to fast handover in this chapter, focusing on implementation of the protocol over a wireless medium. We use WLAN as the wireless access medium and investigate the radio-specific problems as well as the mechanisms for optimally integrating the IP protocol operations with the WLAN system. In other words, we are interested in the details of Step 1 and Step 2 as they affect handovers (Step 5). This is important to support mobile VoIP and mobile rich-media communication.

In the following, we will study the procedures affecting connectivity and context management in a WLAN system. We begin with an overview of the link establishment in a WLAN system.

20.2 ESTABLISHING CONNECTIVITY IN WLAN

There are three distinct entities involved in connectivity establishment. A wireless client, referred to as a *STAtion* in 802.11 jargon, establishes radio link connectivity with an access point. Prior to this, the client, the access point and an authentication server (AS) may be involved in the 802.1X [22] authentication procedure. The access point often takes the role of an *Authenticator* facilitating the exchange between the client and the AS. The entire authentication framework is specified in [25].

A wireless client gains connectivity with a specific access point by *associating* with it. The client *scans* for available access points and chooses a particular access point based on the Server Set IDentifier (SSID) and other Information Elements [23] including the security suite selectors present in the responses. An SSID is an ASCII string that identifies a particular access network, such as "nokiaresearch" or "ietf64". The security suite selectors specify the available security mechanisms for link access. Instead of actively scanning, a client can also obtain these parameters via beacon receptions. This *discovery* phase involves two parts; probe request and response, during which the security suites are exchanged, and association request and response, during which the security suite selectors are agreed upon by both entities. See Figure 20.1.

Subsequent to discovery phase, there may be *authentication* involved before an access point provides wireless access. Specifically, IEEE 802.1X port-based authentication can be enforced. The IEEE 802.1X itself defines a transport for carrying the *Extensible Authentication Protocol* (EAP) [1] over IEEE 802 LANs. And, EAP itself does not specify a particular authentication mechanism, but it specifies a transport protocol for authentication between a client and an AS. A mechanism such as EAP-TLS [2] has to be supported both by the client and the AS for the authentication to take place.

The result of successful authentication is session *key generation*. A Master Key (MK) derived by both the client and the AS is used to derive Pairwise Master Key (PMK), which an AS distributes to an access point. Both the MK and the PMK are session-specific; if the client establishes a new session, it needs to generate them fresh. In addition, a collection of Pairwise Transient Keys (PTKs) is derived for proving possession of PMK, securing data traffic and so on.

A feature of particular interest during handovers is *preauthentication*. Once the client completes initial authentication with the access point with which it is currently associated, it can initiate pre-authentication with any target access point which advertises the preauthentication capability. The same 802.1X authentication takes place between the client and the target access point, and the messages are routed by the access point with which the client is currently associated. The result of a successful pre-authentication is generation of a PMK security association. The client can then use the PMK to reassociate with the target access point during a handover without having to go through the authentication process from scratch. The IEEE 802.11 Task Group *r* or IEEE 802.11r is investigating various mechanisms to enable fast roaming across access points [26].

When 802 authentication is used, no specific state is stored on the access router; the access points maintain the SA state. However, there is no requirement that the Authenticator be located on the access point itself. Instead, the authentication procedure may be centralized in an access point controller, which may also act as a default router.

In some deployments, no link-layer access control may be used. Instead, IP or higher-layer VPNs are used to secure the communication. Such *open-AP* models merely allow radio access, but no useful communication can take place without setting up a VPN using IPsec [11] or SSL [5]. In some scenarios, the physical boundary may be large enough to disallow even radio connectivity. Since IP or higher-layer VPNs are needed anyway (for roaming users), open-AP models simplify deployment in some scenarios. There are no link-layer authentication latencies to address in this case.

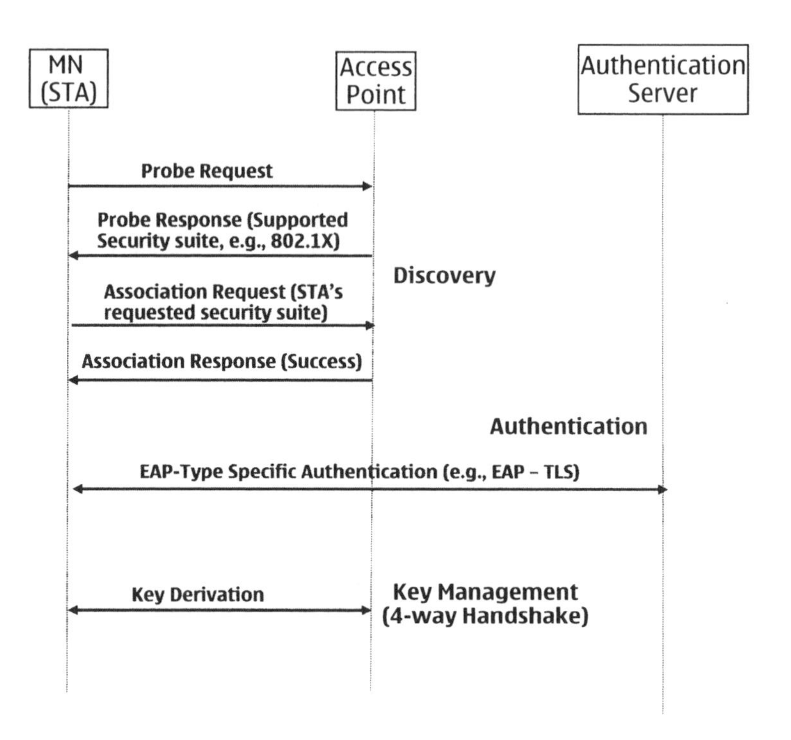

Fig. 20.1 IEEE 802.11i Security Model

20.3 HANDOVER BOTTLENECKS

20.3.1 WLAN Considerations

The decision to undergo handover often includes vendor-specific logic, but almost all clients consider the degrading signal strength and signal-to-noise ratio of the current access point. For instance, a client transmits packets without receiving acknowledgments, and then may reduce its transmission rate. If the client still does not receive any acknowledgments it may consider association with a new access point. This method of detection based on failed transmission is especially long, and most hardware now support physical layer logic for handover based on degrading signal strength [19]. In the following, we describe the steps that need to take place for WLAN handover.

1. Neighboring Access Point discovery during handover: A mobile node's wireless client scans for available access points prior to attempting to *join* one of them. The scan operation is especially delay-prone. For each channel, the client sends a probe, starts a probe timer and waits for probe responses. The probe includes the SSID, and the broadcast BSSID. [1] The client maintains two channel variables during the scan operation. By the time the probe timer reaches *MinChannelTime*, if the medium is idle, it scans the next channel. If the medium is busy, it waits until the probe timer reaches *MaxChannelTime*, collects all the received responses and moves to the next channel. See Figure 20.2.

[1]A BSSID is the MAC address of the access point. See Glossary. A broadcast BSSID is a BSSID with a value of all 1s.

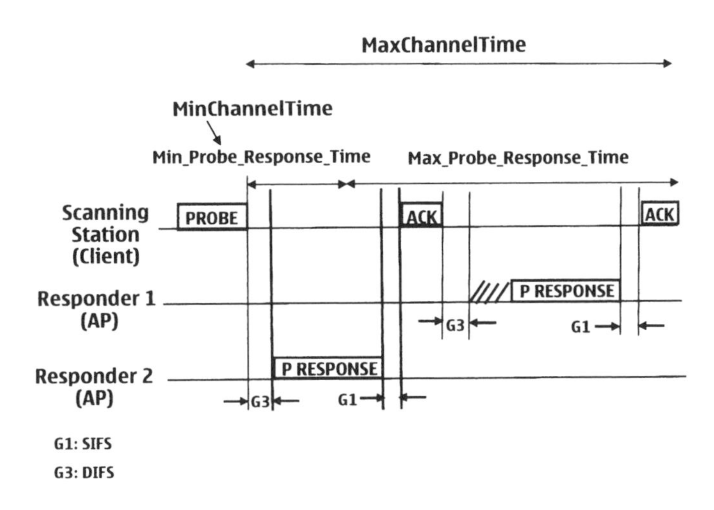

Fig. 20.2 IEEE 802.11 Active Scan Operation

The scan operation has been designed keeping the following in mind. First, the client has to obtain the list of all possible access points in the neighborhood. This is visible in the MaxChannelTime variable. Instead of choosing the first available access point heard during the MinChannelTime, the client waits until it hears from all the possible access points operating in that channel. Second, the client has no knowledge of the neighborhood. This is again visible in the expansive operation to collect all the access points in the neighborhood. Finally, the client wishes every access point on a channel to listen to its probe request; hence it broadcasts the probe and waits for response from each access point. Each PROBE in Figure 20.2 itself has to follow the channel access procedure defined in the 802.11 standard. These procedures define the *Inter-Frame Spaces* (IFS) depending on the state of the channel. The *Short IFS* (SIFS) is used during the ACK phase because the channel is already available for the Initiator (i.e., the client) to send the probe. The *Distributed Coordination Function IFS* (DIFS) is used by the Responder (AP) since it is not sure who else might be trying to send a Probe RESPONSE (P RESPONSE). The channel access requirement states that if the channel is seen as idle during carrier sensing for at least DIFS, a node can transmit frames. In Figure 20.2, Responder 1 waits for DIFS duration after it detects the last symbol of the ACK frame being sent, and backs off for a random period of time (shown by the slanted lines) before sending its own Probe RESPONSE.

Clearly, the entire scan operation depends on the number of channels to be scanned, whether each channel is busy or idle during MinChannelTime, and the duration of MaxChannelTime. The empirical studies report substantial variability [15]. According to [4], the best-case latency is around 40 milliseconds, while the worst-case latency is as high as 300 milliseconds.

2. Association or Reassociation with a new access point: The Discovery phase is a single roundtrip, with the latencies depending on the carrier access delays. If this phase is followed by Null authentication, as in an open-AP model, the resulting overall delay is typically only few milliseconds.

3. Reauthentication latency when IEEE 802.1X is used: If no preauthentication is used, the mobile node has to undergo authentication again. When the Discovery phase indicates that authentication is necessary, the mobile node has to perform the four-way authentication and subsequently derive keys before it can resume data transmission and reception, as illustrated in Figure 20.1.

The foregoing steps indicate that scan and reauthentication latencies dominate the delay due to WLAN operations alone. In Section 20.4, we will identify some mechanisms to adequately handle this delay to make it acceptable for real-time applications such as voice.

20.3.2 IP Considerations

As we have seen in Chapter 14, the primary sources of delay at the IP layer are from movement detection, IP configuration and location update. We briefly review each of these.

1. Movement Detection: The IP layer needs to be able to detect that the mobile node has moved to a new subnet as a result of its wireless client associating to a new access point. As we saw in Chapter 11, the existing solution in Mobile IP [10] is based on the mobile node receiving Router Advertisements, which are sent at preset intervals to all nodes on a link. The Router Advertisements are also sent unicast to a solicitation. The interval between unsolicited Router Advertisements is 3 seconds according to the Neighbor Discovery specification [17], and Mobile IPv6 specification redefines this to 30 ms. The solicited Router Advertisements, on the other hand, are typically not sent immediately following a request. They are staggered until additional requests are received or until the next unsolicited Router Advertisement is scheduled [17]. Both modes of receiving the router advertisement introduce tens of milliseconds of delay. Moreover, every change of access point introduces router solicitations even though subnet handover may not have occurred since the mobile node has no way of realizing which particular access point handover also means subnet handover.

2. IP Configuration: Once a mobile node determines that it has moved to a new subnet, it needs to configure a new IP address consistent with the new subnet prefix. Since movement detection is based on Router Advertisements, no router discovery itself is necessary unless the mobile node wishes to explore all available routers and configure a default router based on other preferences. In order to configure a new IP address, there are two options. In the stateless address auto-configuration method [18], a node configures its address without the need for a server by combining the advertised prefix in the Router Advertisement with its Interface Identifier. The IID may be a temporarily assigned pseudorandom number [16], which can be used to thwart traffic profiling of an address to a limited extent.

The other method of acquiring an address is through stateful assignment. The default stateful method is DHCP [8], in which a node contacts a DHCP server for address assignment.

As we have seen in Chapter 2, and Chapter 14, the DAD delay exceeds a second before an address is confirmed to be unique. In the stateful configuration, the address configuration delay depends on the server discovery and the performance of the server. The studies in [3, 7] report latencies of several tens of seconds for DHCP operation, which is clearly unacceptable for supporting real-time applications.

3. Location Update: The location update delay involves establishing a new binding at the home agent and then performing Return Routability (followed by Binding Update) with each of the correspondents. Depending on the correspondent's location, the delay could reach a couple of hundred milliseconds.

It is worthwhile to observe that delay imposed by the Mobile IP operations are applicable to any mechanism that needs to securely update a node's IP address with an arbitrary correspondent. Using a mechanism such as CGA [6], it is possible to reduce the delay by half a round-trip time; it is feasible to have a longer lifetime for the home address ownership so that HoTI and HoT exchange can be avoided. However, the care-of address test (CoTI/CoT) and Binding Update are required for each subnet handover, which could still account for a delay in excess of a 100 ms.

The WLAN-specific events and operations, as well as the protocol operations necessary at the IP layer, contribute to delay that adversely affects the performance of real-time applications such as Voice over WLAN. The delay also results in packet loss, which affects applications using TCP as well as voice. So, a handover design must specifically reduce the delay and eliminate or reduce packet loss.

20.4 IMPROVING HANDOVER SUPPORT

The bottlenecks identified in the previous section provide a glimpse of the design approach necessary to provide handover support for delay and packet loss. From the WLAN operations viewpoint, it is necessary to reduce scanning latency during handover as much as possible. It is also necessary to reduce reauthentication latency. From the IP protocol standpoint, the movement detection and IP configuration latencies have to be reduced, and the location update latency must be disengaged from the critical path for the average-case handover. When a handover does not follow the expected set of events, the delay must be made manageable from the IP operations point of view. Unfortunately, failover mechanisms at the WLAN MAC layer could still introduce considerable latency.

The general logic remains the same for handover trigger (see Section 14.2.4.1). That is, a client may initiate handover due to a host-internal trigger, a network trigger or a trigger from the user. The important consideration here is to execute the operations in a proper sequence so that maximum benefit can be obtained in terms of latency and packet loss. The steps involved in interworking WLAN procedures and IP operations for real-time handovers are discussed below. Readers already familiar with them may still wish to go over them to get an overall system description.

1. Link Connectivity: When its WLAN interface is enabled, a mobile node scans for all available access points, perhaps looking for some known SSID [23]. The mobile node then establishes link connectivity with an access point of its choice. This link establishment may involve performing access control operations such as IEEE 802.1X, or there may be no authentication when open-AP model is used.

2. IP Connectivity: Once a link is available, the mobile node performs IP configuration. It constructs a link-local IPv6 address and typically performs DAD for the newly formulated address [18]. Concurrently, the mobile node performs router discovery. The mobile node learns whether it can use stateless address autoconfiguration in the Router Advertisement message by verifying if the ‘M’ bit is set. If the router does not require managed address support, meaning that a stateful procedure is not required, the mobile node formulates a globally routable IP address. If the router requires managed address support, the mobile node typically runs DHCP. At this point, the mobile node is an IP endpoint that can communicate with any arbitrary correspondent on the Internet. The mobile node performs Mobile IP operations, registering its new location with its home agent, as well as with its correspondents.

3. Neighborhood Discovery: Once it is IP-capable, the mobile node resolves a subset of neighborhood access points to discover their respective subnet information. The mobile node may short-list access points based on the information it can gather from each of the access points. In any case, resolving an access point involves obtaining subnet information for each BSSID the mobile node has cached as a result of scanning. The mobile node sends a Router Solicitation for Proxy Advertisement (RtSolPr) message with one or more BSSIDs. The access router responds with a Proxy Router Advertisement (PrRtAdv) message in which a tuple containing new router's IP address, the MAC address and the subnet prefix information is supplied for each BSSID requested. We saw the source code for these messages containing the access point resolution in Chapter 15. As specified in [12], if an access point being resolved is attached to the current interface of the current access router itself, then only an indication as such is provided in the PrRtAdv message. All possible combinations are specified in [12].

A mobile node may also provide a wild card in the RtSolPr message. This requests its router to provide neighborhood information regarding all access points attached to a different subnet.

With neighborhood discovery, a mobile node builds a map of its available access points. Each entry in the map includes an access point identifier (BSSID), the IP address and the MAC address of the subnet router the access point is attached to and the subnet prefix valid on the link. Additional information may be carried in the future. For instance, the beacon advertisements may point the wireless client to a new IEEE service that provides additional attributes such as availability of handover support to cellular networks. The entries in the map may be ordered based on criteria such as QoS support or on signal strength indication. The map could be maintained as an easily accessible kernel-resident data structure.

4. Application Execution: The mobile node subsequently runs one or more applications, such as VoIP, streaming video, or e-mail. When these applications are running, the mobile node may need to handover to a new access point for multiple reasons. First, its existing access point may see an increase in user traffic, and hence the mobile node may receive decreased bandwidth for its applications. One could address bandwidth sharing by exerting appropriate admission control and supporting a service such as IEEE 802.11e [24]; nevertheless it needs careful attention and perhaps additional research. Second, the user may physically move away such that the Received Signal Strength Indicator (RSSI) continues to degrade. This may require the mobile node's wireless client to initiate link switching. In another scenario, the network itself may provide a hint to the mobile node to switch its link to another access point. For instance, the mobile node's access router itself may initiate a gratuitous PrRtAdv message to the mobile node in order to balance load on its access points. Finally, it is the user herself may switch the wireless client to a more convenient access point based on a variety of considerations, including available services, pricing preference and so on.

All these events constitute *triggers* for a handover. A host trigger is generated by the wireless client based on the perceived link conditions. A network trigger is generated by a network node providing a hint for handover.[2] A user trigger is generated by the user of the wireless client. In the network trigger, the target access point is known. In the user trigger as well, the target access point can be assumed to be known. In the host trigger, the new access point may not be known until the wireless client scans all available access points and chooses one. It is in this scanning process that the greatest delay is likely to be incurred.

There are two considerations in handling scan latency. First, how to maintain an up-to-date neighborhood map. Such a map ensures that the mobile node is reliably able to connect to an access point should a handover become necessary in the future. This operation involves scanning the known SSIDs, as in the neighborhood discovery phase. As a result, new entries may be created in the map, and the existing entries may need to be reordered or dropped to reflect the current state of the system. If the scan operation takes place when an application is running, packet loss may occur, depending on whether there is data to be sent or received during the scan interval. Performing a scan operation in its present standard form increases the likelihood of packet loss since all available channels need to be scanned. This specification, coupled with the absence of a separate control channel, leads to potential data loss independent of handover whenever a scan operation is performed. Clever techniques for performing the scan operation are still nascent; for instance, scanning only selected channels reduces the likelihood of packet loss. We review some of the proposed techniques in Section 20.5.

When a new access point is discovered, the mobile node also resolves it to obtain the subnet information by exchanging the RtSolPr and PrRtAdv messages.

The other consideration is how to reduce the scan overhead when a handover is required. This issue arises from the fact that many wireless clients appear to scan all available channels before attempting to join an access point. With a neighborhood map, it should be feasible to eliminate this bottleneck for the average case. Instead of attempting to scan all available channels, the mobile node may scan one channel at a time and probe if a *suitable* access point is available at that channel. If so, it could immediately attempt reassociation with that access point. Taking the neighborhood optimism one step further, the mobile node may directly attempt reassociation (i.e., skipping scan) with the first entry in its ordered neighborhood map. If reassociation fails, the mobile node can repeat the scan procedure for the next channel. In the worst case, the mobile node ends up scanning all the channels. In the average case, the delay is approximately half of the worst-case delay. In a typical case however, we anticipate the delay to be even smaller; a well-provisioned deployment of access points should provide coverage and capacity to the vast majority of client devices, thus effectively eliminating the possibility that a device is unable to connect to an access point for capacity reasons alone.

[2] In WLAN, the wireless client ultimately has to act on the hint to switch the links.

A tool such as the *Host AP Driver* [21] can be extremely useful in controlling the scan operation as well as the operation of joining an access point. This tool controls the operation of many wireless cards using the *Prism2* chipset. An implementation can set the *host_roaming* variable to 0, 1, or 2, which sets the scan and join (or roam) to firmware, the host driver and to manual control, respectively. For instance, setting it to 2 would allow the client operating system software to perform scan and roam based on any suitable logic of its choice.

5. Routing Update: Prior to reassociation with a target access point, the mobile node verifies if the new link implies a new subnet. If so, it formulates a new care-of address for use on the new subnet and sends a Fast Binding Update packet to the Previous Access Router (i.e., its current access router). As we learned in Chapter 14, FBU processing at the PAR establishes a route table entry for the previous care-of address to point to the new care-of address. Subsequently, the PAR begins tunneling packets arriving for the previous care-of address towards new care-of address. The mobile node may continue to use the link to send packets. However, the mobile node is likely to benefit from buffering its outgoing packets once link termination is determined. Such buffered packets can be transmitted once the new link is established.

6. Packet Transmission from the New Link: The mobile node reassociates with a new access point. This process could use preauthentication if the 802.1X is used. Using its neighborhood data structure, the mobile node determines that the new access point implies a new access router whose addresses as well as the interface subnet prefixes are known. As a result, the mobile node is able to send packets immediately using a prospective new care-of address. However, a correspondent will drop packets sent using the new care-of address until a binding is created between the home address and the new care-of address using the Mobile IPv6 Route Optimization procedure. Also, there is a very small likelihood that the chosen new care-of address is already in use by another node on-link (or off-link). To ensure that the new care-of address is confirmed, the mobile node sends the FNA message to the NAR.

7. Address Confirmation: The mobile node announces its attachment by sending the FNA message with its new care-of address as the source IP address. The NAR verifies the uniqueness of the new care-of address, for example by using its forwarding table entries or a database of access-permitted nodes. If there is a conflict, the NAR sends a Router Advertisement with the Neighbor Advertisement ACK option, in which it may supply an alternate new care-of address to use.

As we studied in Chapter 14, if there is no address conflict, the NAR can immediately start forwarding packets, including the FBack if it is present. If the PAR has already set up a binding by the time the FNA is processed, it is also likely that the NAR will already have some packets waiting to be delivered. If the PAR does not have a binding for the previous care-of address, the FBU establishes it, after the exchange of HI and HAck with the NAR. There is no need to reply to the FNA if there is no address conflict. The packet forwarding ensues immediately.

8. Reverse Tunneling: The mobile node tunnels its packets to the PAR until the Mobile IP operations with its correspondents are completed. These Mobile IP operations do not determine the forwarding of packets already in progress. The mobile node need not wait for FBack before sending these packets. But if it does receive a NAACK message, it must reconfigure a new address, send an FBU and only then reverse tunnel its packets to the PAR.

The foregoing description outlines an implementation guideline. It is also worthwhile to review [13]. Freeware implementations [20, 28] are perhaps the best source for experimenting with the actual software.

20.5 RELATED WORK

As we discussed earlier, the dominant delays in WLAN handover are due to scanning and reauthentication. There are proposals on how to contain the effect of these operations so that the resulting latencies are manageable. We begin with the scan operation.

The scan operation begins when a client decides that it needs to undergo handover. In its full form, this involves sending a probe request, waiting for one or more probe responses, and repeating this sequence for all the channels. A simple approach to reduce the associated delay is to prescan all available channels and order the target access points based on the host-specific preferences.[3] At the time the decision to undergo handover is made, the client may scan only the channel where the access point of preference is located to ensure that the desired signal quality is still available. If the access point is no longer there, the client may attempt the next access point in its preference list and so on. A good time to do prescanning is immediately after the wireless interface is enabled since there is no data traffic. Subsequently, the client can try selective scanning of one channel at a time when there is no data to be sent. This way, the list of available access points could be kept up-to-date. Prescanning, followed by selective scanning provides a means of disengaging scan latency during handover and effectively avoiding association with a "stale" access point at the time of handover. Selective scanning of access points requires a change of the existing specification.

[3]access points may advertise higher- order information, such as available QoS, security and so on, in the probe responses. See IEEE 802.21 [27].

In [19], an analysis of the 802.11 handover process begins by observing that the entire process includes detection, scanning and execution. The detection phase involves determining the need for handover, which includes scanning and reassociation. The detection process could be very long since a client cannot easily differentiate frame loss stemming from an out-of-range access point from other effects such as collision and fading. So, [19] proposes that loss of three consecutive frames could be used as an indicator of an access point that is out of range. The study also presents simulation results of the time needed to respond to active probes as a function of the number of channels used at various access point loads. Finally, the study reports simulation results for total scan time as a function of the number of used channels at various access point loads. These results indicate that when multiple channels are in use, the scan latency increases. This is because the likelihood of the clients having to wait for MaxChannelTime for each channel also increases, thus increasing the overall scan latency. The study conclude that one way to make this latency manageable for real-time applications is to scan a subset of all available channels.

A breakdown of delays due to 802.11 operations as well as IP layer latencies is provided in [4]. This document also reviews various proposals submitted to IEEE. We look at a couple of methods below.

The proposal in [9] describes a fast active scanning mechanism which resembles the selective scanning referred to above. The proposal describes directed probes to known target access points instead of broadcasting probe requests and eliminating the need for ACKs for probe responses. Effectively, this could make scan latency extremely small, even though a client may still scan for *all* access points in its list. The proposal assumes that the list of target access points is collected off-line by different mechanisms.

In [14], the concept of access point Neighbor Graphs is used to restrict the number of channels that a client has to scan during handover. The client also maintains a local variable *optimal channel time* for its minimum and maximum channel probe times. With the help of its current access point, the client scans a select list of neighbors using its optimal channel wait times and then picks a candidate for handover.

It appears that there is almost unanimous agreement that the current standard implementations of wireless clients produce highly variable and high latencies during handover. Mechanisms are being proposed to limit the number of access points to scan by using neighborhood lists and by sending directed probes during scanning. It is possible that a standard solution could emerge in the near-future.

Regarding reauthentication during handovers, preauthentication is a very feasible approach. One could imagine a client short-listing target access points and performing preauthentication with them before actual handover. When handover takes place to any one of the access points, no back-end authentication will be necessary. When the open-AP model is used, reauthentication latency is effectively eliminated.

20.6 SUMMARY

In this chapter, we looked at how WLAN and IP protocol operations contribute to handover delay. Specifically, we studied the WLAN link establishment procedure, including access authentication and the handover scan and (re)associtation procedures. The scan and authentication procedures contribute most to WLAN link-layer latencies. However, techniques exist to mitigate the effects of these operations. At the IP protocol layer, it is important to integrate the WLAN events (i.e., triggers) with the fast handovers messages in order to obtain the best performance. An excellent integration of L2 abstractions with the fast handover protocol is conducted in the Tarzan testbed [28]. Interested readers can download the source code and investigate. The latencies at the IP protocol layer are from movement detection, IP configuration and Binding Update, which we have investigated in detail in Chapter 14.

REFERENCES

1. B. Aboba et al. "Extensible Authentication Protocol," RFC 3748, Internet Engineering Task Force, June 2004.
2. B. Aboba and D. Simon. "PPP EAP TLS Authentication Protocol," RFC 2716, Internet Engineering Task Force, October 1999.
3. N. Akhtar, M. Georgiades, C. Politis, and R. Tafazolli. "SIP-based End System Mobility Solution for All-IP Infrastructure," IST Mobile and Wireless Communications Summit, June 2003.
4. A. Alimian and B. Aboba. "Analysis of Roaming Techniques," contribution to the IEEE 802.11 group, March 2004.
5. T. Dierks and C. Allen. "The TLS Protocol Version 1.0," RFC 2246, Internet Engineering Task Force, January 1999.
6. T. Aura. "Cryptographically Generated Addresses (CGA)," RFC 3972, Internet Engineering Task Force, March 2005.
7. A. Dutta et al. "Implementing a Testbed for Mobile Multimedia," Proceedings of the IEEE Globecom, 2001.
8. R. Droms (Editor). "Dynamic Host Configuration Protocol for IPv6 (DHCPv6)," RFC 3315, Internet Engineering Task Force, July 2003.
9. M. R. Jeong, F. Watanabe, and T. Kawahara. "Fast Active Scan for Measurement and Handoff," contribution to the IEEE 802.11 group, May 2003.
10. D. Johnson, C. Perkins, and J. Arkko. "Mobility Support in IPv6," RFC 3775, Internet Engineering Task Force,June 2004.
11. S. Kent and K. Seo. "Security Architecture for the Internet Protocol," RFC 4301, Internet Engineering Task Force, December 2005
12. R. Koodli (Editor). "Fast Handovers for Mobile IPv6," RFC 4068, July 2005, Internet Engineering Task Force.
13. P. McCann. "Mobile IPv6 Fast Handovers for 802.11 Networks," RFC 4260, Internet Engineering Task Force, November 2005.
14. A. Mishra, M. Shin, and W. Arbaugh. "Improving the Latency of the Probe Phase during 802.11 Handoff," contribution to the IEEE 802.11 group, May 2003.

15. A. Mishra, M. Shin, and W. Arbaugh. "An Empirical Analysis of the IEEE 802.11 MAC Layer Handoff Process," ACM SIGCOMM Computer Communication Review, Volume 33, Issue 2, April 2003.

16. T. Narten and R. Draves. "Privacy Extensions for Stateless Address Autoconfiguration in IPv6," RFC 3041, Internet Engineering Task Force, January 2001.

17. T. Narten, E. Nordmark, and W. Simpson, "Neighbor Discovery for IP Version 6 (IPv6)," RFC 2461, Internet Engineering Task Force, December 1998.

18. S. Thomson and T. Narten, "IPv6 Stateless Address Autoconfiguration," RFC 2462, Internet Engineering Task Force, December 1998.

19. H. Velayos and G. Karlsson. "Techniques to Reduce IEEE 802.11b MAC Layer Handover Time," Technical Report, Laboratory for Communication Networks, Department of Microelectronics and Information Technology, KTH, Royal Institute of Technology, Stockholm, Sweden.

20. http://www.fmipv6.org

21. Host AP Driver for Intersil Prism2. http://hostap.epitest.fi

22. "IEEE Standard for Local and Metropolitan Area Networks: Port-Based Network Access Control," IEEE Standard 802.1X, 2004.

23. "IEEE Standard for Local and Metropolitan Area Networks: Wireless LAN," the IEEE standard 802.11.

24. "Local and Metropolitan Area Networks: MAC Enhancements for Quality of Service," the IEEE Task Group TGe.

25. "Medium Access Control Security Enhancements," IEEE standard 802.11i, http://standards.ieee.org/getieee802/download/802.11i-2004.pdf

26. "Fast Roaming and Fast BSS Transitions," http://grouper.ieee.org/groups/802/11/Reports/tgr_update.htm

27. The IEEE 802.21 group. http://www.ieee802.org/21

28. http://software.nautilus6.org/TARZAN/

Part V

Emerging Topics in IP Mobility

The development of Mobile IPv6 is by no means complete. It has already inspired many related research efforts and will inspire many more as time goes on. Like IPv6, Mobile IPv6 is engineered for global deployment, but this broad applicability in itself introduces numerous different requirements that arise from various governmental regulations, interactions with new technologies, and the continued evolution of the Internet itself. We expect that there will be a continual push towards better performance in order to support more high-speed applications. There are likely to be other enhancements to improve battery life and bandwidth utilization and to facilitate integration with many other preexisting application scenarios.

In this part of the book, we introduce some interesting emerging topics that are part of the continuing evolution of Mobile IPv6. While Mobile IPv6 does solve the mobility management problem at the network layer, there are many other layers involved, and the solutions required for a complete system are prompting the further evolution of Mobile IPv6 along interesting new lines. For instance, Mobile IPv6 essentially discloses the home IP address of the mobile node. In fact, IP tunneling itself discloses this address, so that is nothing new, and this has been accepted behavior in the Internet since the beginning. But now, VoIP and Mobile IPv6 are bringing the Internet into competition with the world of mobile telephony. Since Internet technology is seen to be much less expensive and perhaps much more feature-rich, especially with the proliferation of 802.11 wireless solutions, people are beginning to demand that the Internet solution have all the features of the solutions available with the mobile telephone system. One primary feature is *location privacy*. This comes into direct conflict with the longstanding behavior of IP, and yet the pressure is so strong that we will no doubt see some important evolution in the behavior of IP itself.

The wealth of multiple wireless technologies integrated in a commercial mobile terminal also creates a great deal of flexibility, and Mobile IPv6 makes it easy to maintain the same IP address no matter which network interface is currently active. But when there are several possibilities, how does a device (or user) select a specific wireless interface? What are the user mobility requirements and performance needs when accessing networks such as WLAN and CDMA EV-DO? The answers to such questions are quite subtle and, some might claim, even impossible. Nevertheless, solutions are demanded, and no doubt at least one solution will be found to satisfy the many application scenarios.

We also illustrate the use of fast handovers and context transfers using header compression as a specific feature of interest. Our objective is to provide an understanding of the problems involved in supporting both routing and transport enhancements together, and header compression serves as a very demanding benchmarking application for performance. Finally, we discuss route optimization for Mobile IPv4 using Mobile IPv6 route optimization. This provides route optimization for both IPv4 and IPv6, and also has interesting properties. The underlying technique is quite relevant especially in the dual-stack Mobile IP operation.

These are just a few of the topics in this field; many more are to be found and explored. These are exciting times.

21

Multiaccess and Mobility

Users will also begin using their mobile devices to control and manage other Internet-enabled appliances (kitchen equipment, entertainment equipments, etc.) – Vint Cerf

21.1 INTRODUCTION

As the reach of mobile communication expands, advances in radio communication technologies, including miniaturization and integration and the ability to support Internet Protocol applications, are rapidly giving rise to *multiaccess* devices. With such devices, a user can connect to the Internet by any of the available networks or by more than one network for a best-connected experience. This choice, although no doubt empowering, leads to some interesting technical problems. How would a user discover available networks and their characteristics? How would a user select a network from available networks? How would handover take place? These and many other problems will be addressed in the near future. In this chapter, we will consider mobility between WLAN and a WWAN, the CDMA *EV-DO* network, and investigate how well voice handovers work under the *existing* conditions. We will also identify a set of problems in this domain.

21.2 MOBILITY BETWEEN NETWORKS

Over the years, the mobile cellular networks have improved their ability to support voice calls even when a user is crossing cell boundaries between radio base stations. As we observed in previous chapters, efforts are underway in developing standards as well as in the industry, to provide similar support using IP in networks such as WLAN and WiMax. Mobility between different access networks brings additional challenges, and is naturally addressed by IP. As before, supporting voice serves as a benchmark for any IP mobility protocol.

With multiaccess devices, it is possible to use the radio receivers simultaneously. As a contrast, the CDMA technology supports a mechanism called *soft handoff* in which a CDMA receiver can tune into multiple base stations simultaneously to receive the best possible signal. The same mechanism allows a receiver to handover to a better primary base station without completely losing connectivity and the signal. In a multiaccess device, however, the radios are independent and multiple connectivity refers to attachment at the IP layer rather than at the radio and the MAC layer. The multiple interfaces may not be switched-on simultaneously in order to conserve power. Hence, designing soft handovers with multiple network interfaces is vastly different.

The general principles of performing soft handovers between disparate interfaces can be described as follows. First, establish link and IP connectivity with a desired network. (Subsequently, one or more applications are run.) Second, continuously monitor the network interface for a drop in "quality" against a preset threshold. *Network quality* refers to parameter(s) of interest such as signal strength, available bandwidth or even a higher-layer parameter such as price. Third, once a trigger event happens so that a handover to another network needs to take place, perform the actions necessary to awaken [1] the chosen interface. Fourth, perform IP protocol operations to establish IP connectivity. (Both link establishment and IP establishment may include authentication procedures.) Finally, perform IP mobility protocol operations in order to handover the existing IP sessions and to start sessions on the new network. One could expect a software agent to automate this entire process so that users themselves do not need to be involved.

In the following sections, we describe some of the experiments we have conducted to understand the challenges in performing "soft IP handovers" between different access technologies.

[1]This does not necessarily mean powering up the network interface itself. The interface could just be dormant as well.

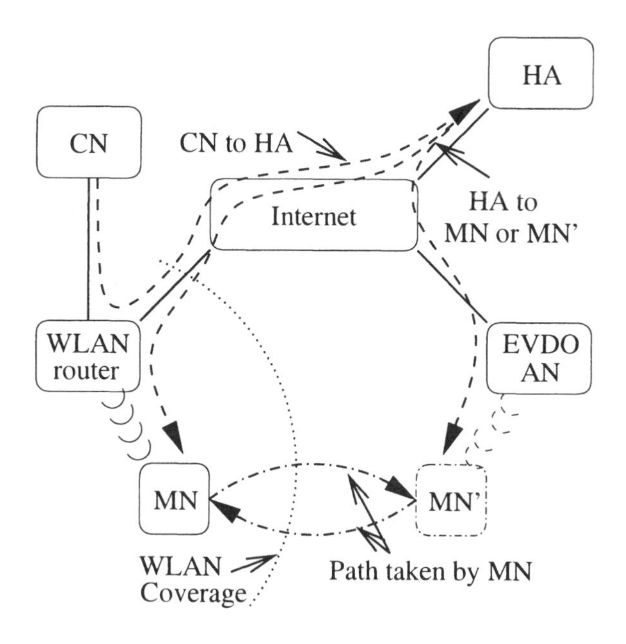

Fig. 21.1 Experimental Network Configuration

21.3 CASE STUDY: VOIP HANDOVERS

At a time when the usage of VoIP over WLAN is increasing, broadband wide-area wireless networks such as EV-DO are also being rolled out. The first phase of such broadband WANs is not meant to be used for VoIP; indeed, providers may offer such solutions in subsequent phases. However, an Internet application such as *Skype* could be used as a test tool to study the feasibility of a voice application even in existing networks. That's just what we intend to do here. Our objective is to study handovers between WLAN and a broadband WWAN using voice as an application with the intention of answering the following questions:

- Can a voice application sustain such handovers?

- What are the reasons for failed call handovers?

- What strategies are necessary for soft handovers?

We are also interested in analyzing various performance metrics such as protocol messaging delays, packet losses and inter-packet separation delays.

21.3.1 Experiment Framework

We illustrate the testbed using Figure 21.1, which consists of the following:

- A WLAN consisting of an IEEE 802.11b access point, a WLAN client embedded in an IBM T40 series laptop, and a fixed correspondent node that is also attached to the access point using Ethernet. The access point serves as a WLAN router providing a private address (i.e., Network Address Translation [NAT] functionality) to each of the hosts attached to the network.
- A residential broadband connection using the cable modem technology providing speeds of up to 4 Mbps towards the host (downlink speed) and 300 Kbps towards the cable modem network (uplink speed).
- A broadband WWAN using the EV-DO [1] technology, which can offer downlink speeds of up to 2 Mbps but typical sustained rates between 400 - 700 Kbps. The mobile node connects to the EV-DO network using a PCMCIA card. EV-DO is an air interface technology designed for packet data applications. The network architecture, connection (link and IP) establishment, and authentication procedures are the same as those discussed in Chapter 18.
- The popular application Skype [2], which is a peer-peer voice and interactive text application that does not rely on traditional infrastructure, such as call setup servers and switches, to provide its service. We provide an additional description of skype in the following.
- A Mobile IPv4 home agent located on the Internet, and a Mobile IPv4 mobile node functionality inside the host (IBM laptop) with WLAN and WWAN connectivity. The home agent address and the MN's home address are statically configured. We used the Birdstep (www.birdstep.com) Mobile IP client and the home agent software.

A Skype Client (SC) installed on a user device informs a Skype Super Node (SN) about its presence when the device has Internet connectivity. Users can maintain different profiles such as "I am Online," "I am Away," "Do not disturb" etc. The SN uses the status information of a user when establishing a call between users. Once a call or a session is established, the data typically flows between the user devices directly without involving the SN. However, we have observed a scenario in which the SN is involved in forwarding data because of a firewall's presence. This leads us to a discussion of what transport protocol is used by Skype.

Whereas typical Internet VoIP applications use RTP [4] for transport, Skype uses a proprietary protocol. So, the Skype fields in the UDP or even the TCP header cannot be parsed for any useful information such as a sequence number. The SC first attempts to set up a call over UDP. If that fails, for instance due to the presence of a firewall, the SC falls back on using TCP.

The generally acceptable audio quality in Skype could be at least partially attributed to the codecs it uses. The *iLBC* narrowband codec supports 20 ms and 30 ms frame lengths and is designed to conceal the effect of even a large number of packet losses. The *iSAC* wideband codec provides adaptive encoding, with rates ranging from 10 to 32 Kbps. It is also designed to operate in the presence of multiple packet losses. In addition, Skype uses the *NetEQ* jitter buffer manager in order to cushion the effect of variable arrival times of audio packets.

21.3.2 An Experiment

Each run includes the following steps:

1. The mobile node's WLAN interface associates with the WLAN network, and obtains an IP address. The mobile node configures the access point as its default gateway.

2. The mobile node registers its care-of address with the home agent on the Internet. Note that in this model, the mobile node is never at home since its home address is neither on the WLAN network nor on the EV-DO network.

3. The mobile node initiates a Skype call to the correspondent node. The Skype client on the correspondent node accepts the call.

4. The mobile node and the correspondent node begin a VoIP conversation.

5. The mobile node starts ambling [2] away from the WLAN Access Point. As the WLAN signal strength from the WLAN access point begins to drop below a pre-set threshold, the connection manager application on the mobile node attempts to establish the EV-DO connection. We used the "Excellent" threshold for monitoring the WLAN signal strength. The IBM T40 provides "Very Good," "Good," "Fair" and "Poor" qualitative grades for WLAN signal strength in addition to "Excellent".[3] Hence, our experiment uses a very conservative approach in which the client starts looking for an alternative radio at the slightest drop in quality, as reported along with above grades.

6. Once the EV-DO link is established, the mobile node obtains an IP address from the EV-DO network.

7. The mobile node sends a Registration Request [3] message over the EV-DO network to the home agent providing its new care-of address. The Registration Request message is the IPv4 analog of the Binding Update message to the home agent. Until this message is processed and a new route is established, the home agent continues to send packets from the correspondent node towards the WLAN network.

8. Once the home agent processes the Registration Request, a new tunnel is established, and the Skype packets arriving from the correspondent node are routed to the new care-of address over the EV-DO network. The home agent also sends the Registration Reply message to the mobile node. Until this message is received, the mobile node continues to use the WLAN for sending packets to the correspondent node.

[2]This results in handovers occurring at a comfortable human pace, as might be the case in a real-world usage scenario as well.

[3]The Excellent threshold corresponds to a signal strength between -90 to -83 dBm.

9. The mobile node and correspondent node continue the VoIP conversation over the EV-DO network.

10. The mobile node turns around begins moving back towards the WLAN network. As the WLAN signal strength increases beyond "Excellent," the mobile node associates with the WLAN and obtains an IP address.

11. The mobile node sends a Registration Request message over the WLAN network to the home agent providing its new care-of address. Until this message is processed and a new route is established, the home agent continues to send packets from the correspondent node towards the EV-DO network.

12. Once the home agent processes the Registration Request, a new tunnel is established, and the Skype packets arriving from the correspondent node are routed to the new care-of address over the WLAN network. The home agent also sends the Registration Reply message to the mobile node. Until this message is received, the mobile node continues to use the EV-DO for sending packets to the correspondent node.

13. The mobile node and correspondent node continue the VoIP conversation over the WLAN network.

14. The VoIP call is terminated after the handover takes place.

21.3.3 Analysis

We make certain observations regarding the network topology. First, the mobile node undergoes handover between two access networks: the EV-DO network and the ISP network providing Internet access for the WLAN network. Each of these access networks consists of a radio network and the wired network that connects them to the Internet. Second, the home agent is connected to the Internet by a wired broadband connection. We assume that the network paths between the home agent and the access network gateways (i.e., a PDSN and a cable modem headend router) exhibit consistent and even perhaps similar characteristics for delay and packet loss. Finally, the network path between the correspondent node and the home agent is fixed in all the experiments. These observations allow us to concentrate our analysis just on the two access networks. In other words, the results could be attributed primarily to the idiosyncrasies of the two access networks.

21.3.3.1 Link Establishment and MIP Delays: Skype calls in 60% of our experiments dropped due to various reasons. Recall that each handover involves link establishment and IP connectivity establishment on a radio interface. These procedures do not take the same amount of time on individual radio interfaces. Indeed, they are an order of magnitude longer on the EV-DO interface. Another observation is that the transition from WLAN to EV-DO could happen more swiftly because a mobile node is more likely to run out of WLAN coverage than the reverse. Because of this asymmetry, a relatively larger number of calls may be dropped when handing over from WLAN to EV-DO.

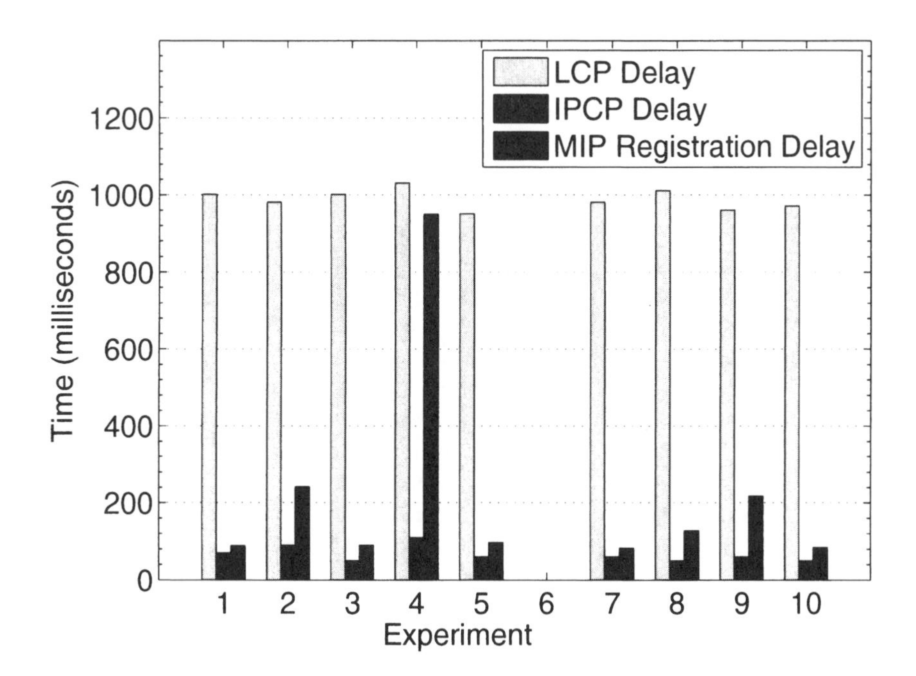

Fig. 21.2 PPP and MIP Registration Delays.

The EV-DO network connection establishment is especially delay-prone. So, even if the mobile node begins establishing the connection in advance, there may not be sufficient time before the call is switched over to the EV-DO network smoothly. We observed this in one of the experiments, where PPP connection establishment took considerably longer, forcing the mobile node to move out of WLAN coverage, resulting in a dropped call. This is illustrated in Figure 21.2 for Experiment 6, which experienced a 17-second delay before PPP was set up. Fortunately, all the other experiments exhibit a fairly constant delay of 100 and 1000 ms for LCP and IPCP, respectively. These delays, even though much greater than the time it takes to establish a WLAN connection, provide a good reference for handover design.

Once a wireless link is set up and an IP address is obtained, the Mobile IP client sends a Registration Request to the home agent in order to update the tunnel end-point. The home agent replies with a Registration Reply message. This sequence is especially longer over the EV-DO network. Again in Figure 21.2, we show this registration delay for handovers from WLAN to EV-DO. The figure shows that the delay is between 100 and 200 ms. This delay adds to the budget for handover planning. It is also on the critical path, meaning that only after the home agent is notified is the Skype call actually switched from one network to another.

21.3.3.2 Inter-Packet Delay: The packets on the EV-DO network can experience long delays. [4] The long delay sometimes leads Skype to abandon the call. This is captured in the cumulative distribution function of the Inter-Packet Delay (IPD) separation. IPD is the interval between successive packets sent at the correspondent node or received at the mobile node. For isochronous operation of VoIP, IPD needs to have smaller variation. Clearly, the network will introduce jitter. However, the larger the deviation, the harder it is to provide a smooth playout.

Figure 21.3 shows the IPD measured at the correspondent node, as well as at the mobile node when using different interfaces. As expected, the figure shows very little variation in IPD at the correspondent node; the delay range of 20 ms to 40 ms represents the packet sampling rate at the sender.

On the other hand, the CDF at the mobile node when using the EV-DO network shows a very large variation in IPD. Even though Skype uses codecs that use adaptive encoding as a function of network performance, the large delay variation appears to negatively affect the quality adaptation. We have seen IPDs the order of seconds contributing to complete silence. We believe this is due to the scheduling algorithm used by the EV-DO network in the downlink.

With WLAN, over 90% of the packets were received within 20 ms intervals. Larger intervals were presumably caused by jitter and packet loss along the network path.

Figure 21.4 shows the time series nature of the IPD during the three phases of the experiments. Phase 1 and phase 3 are when the mobile node is in WLAN, and phase 2 is when it is in the EV-DO network. Clearly, both WLAN and EV-DO are bursty. However, WLAN bursts are concentrated within 5 ms, and there are no long pauses on the order of seconds (which we cannot show in the figure). EV-DO, on the other hand, shows interesting periodicity in multiples of 1 ms for the IPD. We believe this is due to the scheduling policy used. In any case, there are times when the IPD is on the order of seconds. The EV-DO access network buffers packets and sends them in bursts (in which packets are separated by a 1 ms delay), but the bursts themselves are separated by large delays. It appears that this "silence" and "talk-spurt" nature of scheduling has a time series which is not fully compatible with the operation of even the best-effort VoIP applications such as Skype.

[4]This was expected since the network was not meant to support real-time applications. The subsequent version of the EV-DO network is expected to support VoIP.

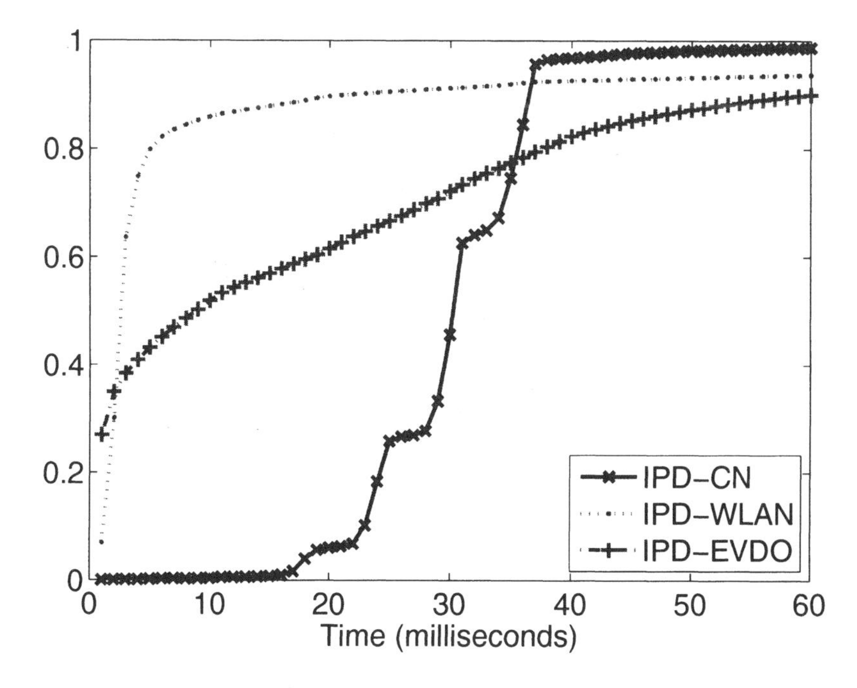

Fig. 21.3 CDF of the Inter-packet delays (IPD).

21.3.4 Summary

Our experiments were an effort to understand and evaluate the performance of real-time applications as a function of mobility between a WLAN and a WWAN. As we saw in the previous sections, there are several deficiencies. First, the transport capability of wireless link itself plays a role. This is visible in the IPD measurements. It is anticipated that the EV-DO Revision A will offer significant improvements for real-time applications. Second, the amount of time needed to establish link and IP connectivity, which itself depends on the transport characteristics of the link, is a crucial factor during handovers. This can be seen in the delays depicting PPP connection establishment. This delay, although long, appears to be nearly constant. This is important from a handover planning perspective. Third, multiaccess handovers involving a LAN and a WAN are not symmetric; a user will likely run out of WLAN coverage sooner than WWAN coverage. To add to this, it takes longer to establish WWAN connectivity. So, more careful design is necessary for LAN-WAN handovers. Finally, delay and packet loss of mobility signaling messages are also important. Although delayed signaling may be tolerable, lost messages have a significant impact on perceived quality.

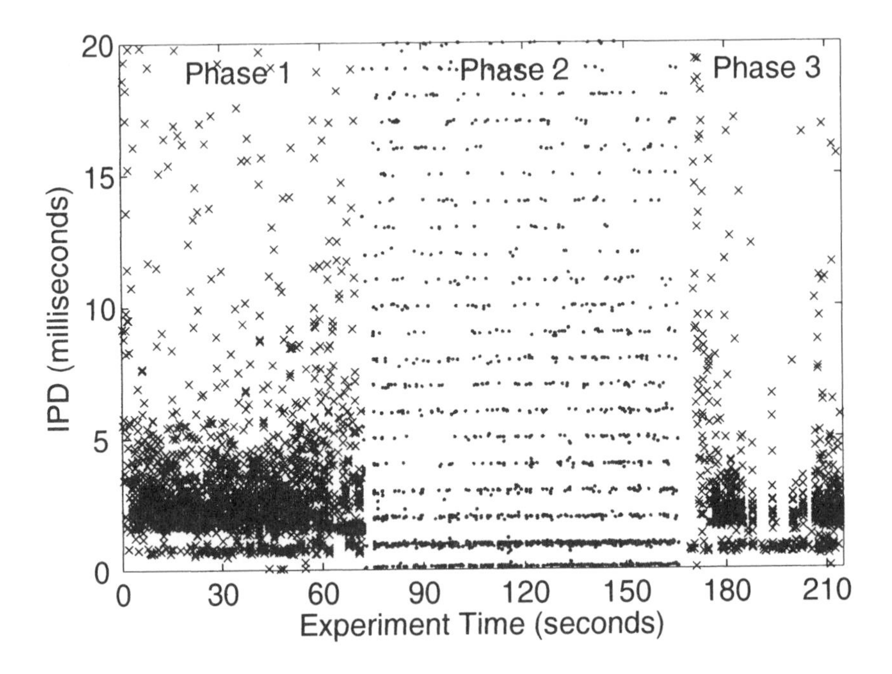

Fig. 21.4 Inter-packet Delay Time series

Our case study provides a few good insights into the dynamics of supporting real-time applications. Many issues remain. In the following section, we will identify some of these and suggest some research problems.

21.4 TAXONOMY OF RESEARCH PROBLEMS

Some of the open topics that need further research are described below.

1. Appropriate interface selection for each packet stream: This is a host-internal problem which determines the proper functioning of applications on the host. Each packet stream, in a sense, is bound to the interface used for initiating the session since the IP address used for communication is bound to the interface. Even when the Mobile IP home address is used as an application endpoint, the care-of address is still bound to a specific interface. So, for routing and forwarding treatment (QoS) purposes, the IP address is still bound to a specific interface and hence belongs to a "path." Mapping application characteristics and requirements for underlying interfaces that can best support them, as well as providing the ability to switch streams based on network events such as quality of connection and handover, need to be investigated.

2. Neighborhood discovery: Even if a single network is actively used for communication, knowledge about the neighborhood networks is important in order to be able to make handover decisions. With multiple network access, mobility is no longer based solely on physical movement. It also includes simply moving data from one network to another. In order to be able to do so, knowledge of the network neighborhood is crucial. Even though the presence of a network is easier to detect via radio signals, the IP layer needs information about the characteristics of the network itself. Having access to such information in an access-independent manner is crucial. Information that is necessary includes the type of authentication enforced and the amount of total bandwidth necessary. There is work in progress in the IEEE 802.21 to provide such an *Information Service*.

3. Access network selection: This typically follows the discovery process. Unlike the former, however, no standard solution is probably needed since each implementation can decide to use its own algorithm for selection. The devices may be preconfigured to automatically access networks with certain features, or user intervention may be necessary, or both modes of operation can be used. It is quite possible that the handover decision is arrived at in cooperation with a network server that has additional knowledge of available networks. Since such a server could easily be viewed as a control point, it is unclear whether the protocol exchange between the mobile node and such a server could also be standardized.

4. Target Network access authorization: Even if routing could be done efficiently, access control and resource authorization primitives could hold the packet streams in check until those operations are completed. When a user is offered seamless mobility experience, it should be an experience where network access and resource control operations are not visible to the user. At the same time, providers need to ensure that fair access and accounting are performed, which brings challenges to protocol design.

5. Switching flows from a high-speed link to a lower-speed link: This happens, for instance, when a mobile node has to switch from WLAN to CDMA EV-DO. Apart from the obvious drop in bandwidth, the new link establishment latency is a consideration, especially when the existing radio quality is quickly fading. The mobile node has to gain IP connectivity, which may involve access authentication, and complete location update signaling with *all* of its correspondents before it can switch the flows to the new interface. The forwarding treatment support (QoS) for all of those flows must also be set up on the new network. All of this incurs latency during an already time-critical period. Thus, a make-before-break design needs to account for scenarios where the existing link breaks away before all the operations can be completed on the link. We saw a good illustration of this problem in our case study earlier.

6. Switching from a lower-speed link to high-speed link: This happens when, for example, the mobile node has to switch from traditional cellular to WLAN. The link establishment latency is typically smaller in this scenario. However, the cost of regaining IP connectivity, context establishment and location update for all nodes may still weigh in favor of rerouting traffic to the new network so that connectivity, context establishment and location update operations can be performed at a less critical time.

7. Switching flows from one IP address to another A host may decide to move one or more of its packet streams from one network interface to another for a variety of reasons, including bandwidth and price considerations. A straightforward approach using Mobile IP is to initiate the protocol operations using an address valid on the target interface while still sending and receiving packets using the interface currently in use. Once the confirmation of new binding at the correspondent host reaches the mobile node, the packet stream can be switched over to the new interface. [5] This approach, however, affects *all* packet streams from the correspondent to the mobile node. A standard solution for selectively handing over some flows to a different interface is still not widely available.

8. Analytical modeling of multiple, powered on interfaces: For handhelds with limited processing, memory and power capacities, interface management is crucial. Well-established models that depict the advantages, such as the ability to exploit network diversity for running different sets of applications on different networks, as well as the disadvantages, do not exist. We expect such an analysis to be a function of usage models for multiaccess systems. Nevertheless, definitive studies in the field are lacking.

9. Mechanisms for fast wakeup for interfaces to provide "virtually on" support: A user may decide to switch to a dormant interface and expect it to be effectively "ON." Providing a fast response time when conserving network and host resources is crucial. For instance, in order to access a network, authentication is necessary. However, the authentication state may expire due to inactivity on an interface. Fast reauthentication techniques are an example of providing better response times for end users. This is even more crucial during handovers. New mechanisms must be explored to quickly establish the PPP connection for a previously authenticated mobile node so that real-time applications such as VoIP can be supported during handovers.

[5] Until the mobile node receives a Binding Acknowledgment, it will continue to use the old care-of address, which results in some packet loss.

21.5 CHAPTER NOTES

Tat Chan (formerly with the Nokia Research Center) conducted most of the experiments involving handovers. Amit Jardosh of the University of California at Santa Barbara analyzed the results and produced the graphs, some of which we have presented in this chapter. We are grateful to both of them.

REFERENCES

1. IS-856 Rev A - cdma2000 High Rate Packet Data Air Interface Specification (1xEVDO)

2. S. Baset., and H. Schulzrinne. "An Analysis of the Skype Peer-to-Peer Internet Telephony Protocol," in Proceedings of the INFOCOM '06, Barcelona, Spain, April 2006.

3. C. E. Perkins (Editor). "IP Mobility Support for IPv4," RFC 3344, Internet Engineering Task Force, August 2002.

4. H. Schulzrinne et al. "RTP: A Transport Protocol for Real-Time Applications," RFC 1889, Internet Engineering Task Force, January 1996.

22

Seamless Handovers

If your result needs a statistician, then you should design a better experiment. – Ernest Rutherford

—

22.1 INTRODUCTION

While bandwidth may be abundant at the Internet core, access networks are typically traffic bottlenecks requiring careful resource management. This is visible in dial-in, cable modem, DSL and wide-area wireless Internet access. The prevailing wisdom for Internet applications and services is a "simple core" and an "intelligent edge" network. As we discussed in Chapter 16, resource management at an intelligent edge network inevitably introduces state or context at the access network nodes in order to meet the needs of applications as well as those of the network itself. QoS and header compression are two good examples that require context management on the access networks due to application constraints and network constraints respectively. With node mobility, these contexts have to be carefully managed in order to support uninterrupted services to the user. In this chapter, we discuss the transfer of the header compression state synchronized with handovers. This is a reflection of experiments using the implementations of fast handover and context transfer, both of which we have discussed at length in previously.

The particular feature context of interest here is header compression [3, 4], which is compelling for the following reasons. First, header compression is a requirement in the lower-bandwidth links common in mobile networks, including even the higher-speed 3G air interfaces. Second, header compression is even more useful with IPv6 over wireless networks. Third, the operation of header compression requires the maintenance of quite a bit of state information. This necessitates tight context management on the access network, making the problem particularly challenging when mobility is involved. The process of establishing the compression context is time-consuming due to the error-prone, higher-RTT (Round-Trip Time) nature of some wireless (e.g., cellular) links. This delay is incurred often with frequent user mobility, requiring reestablishing compression contexts for *every* packet stream in use during each handover.

The emphasis here is not on specifying the details of the particular compression algorithms to be used. Instead, it is on investigating the dynamics of the handover system and on determining useful methods to reinstantiate the compression state of communications already in progress as the mobile node moves to a new point of attachment. So, the methods described here are useful with compression algorithms already specified [3, 6, 11].

22.2 BACKGROUND AND PROBLEM DESCRIPTION

22.2.1 Header Compression

In this section, we provide a simplified explanation of header compression operation. The detailed operations are specified elsewhere [3, 4, 6].

Header compression involves a compressor and a decompressor, both of which operate on unidirectional packet streams. Initially, the compressor sends normal *Full Headers* until it determines that the decompressor has established a context for the packet stream. Subsequently, the compressor only sends differences of those fields that change from packet to packet. Furthermore, if these *First Order* (FO) differences remain constant (i.e., the *Second Order* (SO) differences are zero) over a particular snapshot of the packet stream, the compressor sends only an indication as such, since the FO difference and hence the header field value itself can be extrapolated using a sequence number present in the compressed header. When the compressor operates in the SO state, it transmits the fewest header bits, and hence uses up the least amount of time over the air. See Figure 22.1; for common voice applications, the size of an IP header along with UDP and RTP [13], will exceed the size of the codec payload. This is true whether the payload is carried with IPv4 or IPv6.

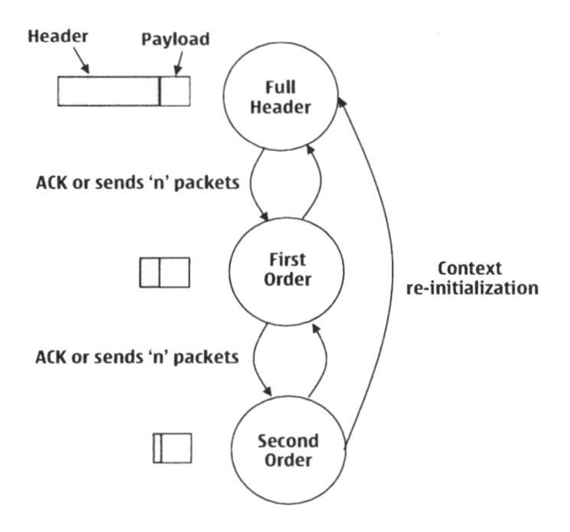

Fig. 22.1 Header Compression State Diagram

In order for the decompressor to correctly reconstruct headers, it needs to store state information (corresponding to static and dynamic header fields) and operate on that stored information in order to reconstruct the header fields. This operation includes retrieving static fields, such as IP addresses and port numbers, and extrapolating dynamic fields, such as an RTP timestamp, using the information received in the FO or SO headers. The compressor, on the other hand, has to make a decision on when to transition from Full Header (FH) to FO and then to SO. The compressor may, on the one hand, wait to receive a positive acknowledgement from the decompressor or, alternatively it may optimistically transition to a better state after sending a "sufficient number" of packets in the previous state. Either way, it has to send several packets over a low-speed, error-prone wireless link before it gains sufficient confidence to make the transition [3].

Independent of the method by which the compressor gains sufficient confidence, the compressor and the decompressor *must* have the same reference state with respect to which the differences in header values are represented at the compressor and appropriately reconstructed at the decompressor. In this approach, where the decompressor updates its context only when it receives packets that carry new reference state information, the packets following one or more lost packets can still be decompressed using the reference state information. We will use such a model [3, 11] for our reference here.

22.2.2 Seamless Mobility with Header Compression

With a brief but sufficient background on header compression, we can formulate the problem statement. First, all the compressed packets arriving at a router after handover must be *correctly* decompressed. No packet should have a wrong RTP timestamp or sequence number, for instance. Second, the number of compressed packets lost at a router due to lack of context must be almost zero in order to enable truly seamless handovers. Third, a compressor should be able to carry over its optimal state from one router to another. That is, if a compressor is in the SO state just before a handover, it should be able to retain that state, provided that the application pattern that determines how an RTP timestamp changes (for instance) does not change. The control variable in these experiments is the mean duration of the talk spurt which affects the number of packets generated and hence the number of packets of interest during handover.

In essence, we are interested in a smooth operation of header compression as the compressor engine is unplugged from one router and plugged back into another even as an application such as VoIP continues to generate packets during such a transition.

Before we proceed to the next section, we define two terms. *Uplink* refers to the direction from a mobile node towards a router, and *downlink* refers to the direction from a router towards the mobile node. So, when we say *uplink packets*, we mean packets sent by a mobile node to its correspondents or to the router itself, and *downlink packets* are those forwarded by the router to the mobile node.

22.3 DESIGN OF HEADER COMPRESSION CONTEXT TRANSFER

As we saw in an earlier chapter, the components of the design include defining the context, representing them in a packet format and synchronizing the transfer with handover. In addition, specific context processing and updating rules apply. We begin with the definition of a context.

22.3.1 Defining Context Structure

A *Compression Profile Type* (CPT) could, for example, represent IPv6/UDP/RTP, IPv4/TCP, or IPv6 only header compression contexts. A CPT could also be defined specifically for certain media types. For example, IPv6/UDP/RTP-G.723.1 could represent a profile type for a particular voice encoder. This could be important, because different media types will have different error characteristics, and fields within the RTP header may receive different handling from the applications that manage their packetization. A CPT indicates the context structure (i.e., various fields and their relative positions) and context semantics (i.e., how the fields should be interpreted to activate contexts).

A context includes values for the network layer (in our case, IPv6) header fields containing (among other values) the port numbers, the source, and the destination addresses associated with the packet stream. The context also includes the *most recently acknowledged* state information, such as the RTP timestamp and sequence numbers. Together with the IPv6 and UDP Full Header state variables, these RTP fields (those acknowledged by the decompressor) establish the *reference state*.

Furthermore, multiple contexts are each identified using a respective Context Identifier (CID). Thus, the [CID, CPT, Header Compression state variables] tuple defines the data structure for context transfer. Multiple such tuples are included for transfer when a number of contexts are present.

22.3.2 Relocation Timing and State Consistency

There are two times when the contexts can be relocated from the previous router to the new router. The timing of relocation is coupled to the process of forwarding of packets, because they need to use transferred contexts.

22.3.2.1 Context Transfer Prior to the Mobile Node's Attachment to the NAR: Suppose, for instance, that the PAR transfers the context at time T_0 in anticipation of mobile node's departure, whereas the mobile node remains connected to the PAR until $T_0 + \delta$. Any updates performed during δ will cause inconsistency between the contexts on the two routers. On the other hand, if the PAR stops performing header compression altogether after relocation (in order not to update the state), then the compressed packets sent by the mobile node will have to be dropped. See Figure 22.2-(a).

Closely related to the actual handover at the link layer is the time when the PAR starts forwarding downlink (i.e., those packets destined for the mobile node) packets towards the NAR. Such a switch in routing is necessary for the mobile node to receive packets still arriving for its previous IP address as it establishes new connectivity with the NAR. The PAR has to tunnel uncompressed packets towards the NAR in order to avoid potential updates to the downlink reference state when the mobile node is no longer present on its link. In general, determining whether the mobile node is still "lingering around" and then updating the reference state (the 'Yes' branch in Figure 22.2-(b)) is not a reliable operation, especially when handover is already in progress. However, tunneling packets immediately after relocation means that buffering is perhaps necessary at the NAR to avoid packet loss. In any case, relocation should precede tunneling of packets. These scenarios are characteristic of cases where predictive handover can be beneficial.

22.3.2.2 Context Transfer After the Mobile Node's Attachment to the NAR: In this case, the PAR could keep updating its context in the downlink, unaware of the mobile node's departure, causing the state on the mobile node to lag behind. In the uplink (i.e., in the direction towards the router from the mobile node), the state on the mobile node may jump ahead of that on the PAR. Hence, the state retrieved after the mobile node is attached the to NAR could be stale. Moreover, many compressed packets in uplink could be lost at the NAR until the context is retrieved. These scenarios reflect reactive handover.

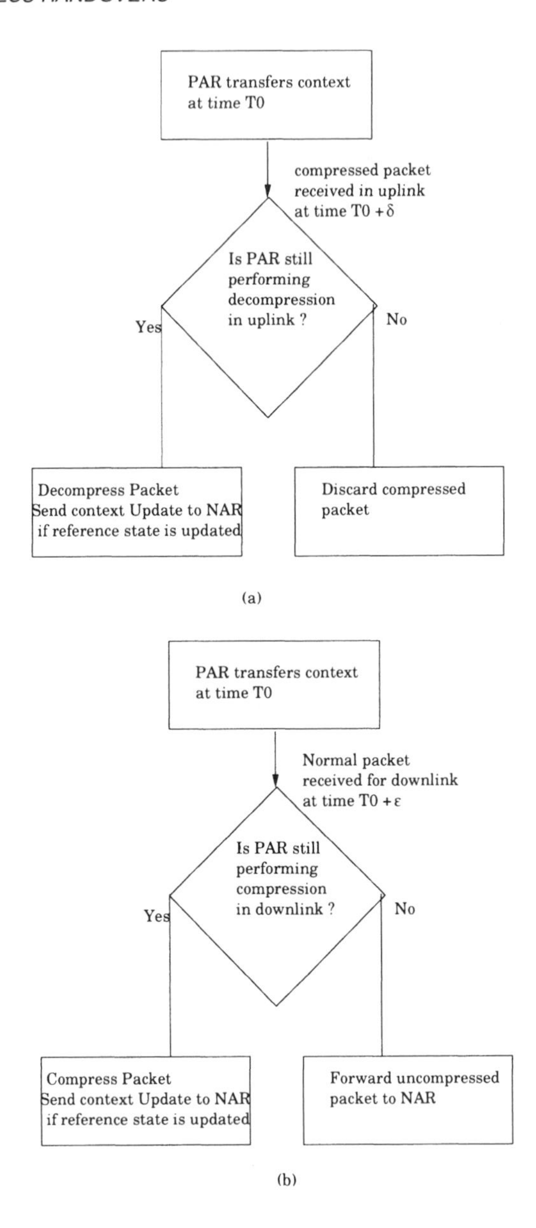

Fig. 22.2 Default Compression Behavior During Predictive Transfer

From the foregoing discussion, we see that for each time instance, we need to specify the header compression behavior to ensure state consistency. We also need a protocol to transfer the state closely synchronized with the handover.

22.3.3 State Update Rules

The following simple rules appear to be sufficient.

1. Sending Reference State Updates: The compressor on the mobile node sends any updates to the reference state until it receives a positive acknowledgment from the decompressor. If optimistic state transitioning is used (see Section 22.2), the compressor must switch to using acknowledgments. See Figure 22.3. The compressor on the PAR also needs to receive confirmation, after which it stops sending updates to the reference state. However, there is an exception to this rule depending on whether context has been transferred (see Rule 3 below).

2. Acknowledging Reference State Updates for the Uplink: The decompressor on the PAR does not acknowledge reference state updates once the context is relocated. This rule ensures that the compressor on the mobile node continues to send updates, i.e., PAR "enforces" the first rule. Hence, if the mobile node is still present on the PAR's link even after the context is relocated, the mobile node's compressor can only (correctly) update the reference state after attaching to the NAR. See Figure 22.3.

3. Avoiding Reference State Updates in the Downlink: The PAR must stop compressing packets in the downlink after the context is relocated. Instead, it must tunnel those packets to the NAR. See Figure 22.3 and Figure 22.4. This rule ensures that the compressor state on the NAR (and hence the decompressor state on the mobile node) is not inconsistent with respect to that on the PAR after relocation.

22.3.4 Synchronization with Handover Signaling

By synchronizing relocation with handover messages, the number of packets that need to be buffered is reduced. The number of packets that might be sent as updates to the reference state is also reduced. However, it is possible that the FBU message will not always reach the PAR reliably through the previous link. In addition, a standard WLAN card may reassociate without reliably providing a *link-going-down* trigger. In such cases, relocation takes place after the mobile node attaches to the new link. See Figure 22.4. While this might appear sub-optimal, the experiments indicate surprisingly beneficial results even in this reactive relocation. We describe the reasons in the following section.

22.4 PERFORMANCE STUDY

The experimental set up involves FreeBSD routers and Linux mobile nodes equipped with IPv6 and enhanced with a kernel implementation of handovers and context transfers. The access network is a WLAN (IEEE 802.11b) in which the mobile node changes its access point (each one connected to an access router) 100 times during each run of the experiment. The results are averages of multiple runs (at least three) corresponding to each parameter set.

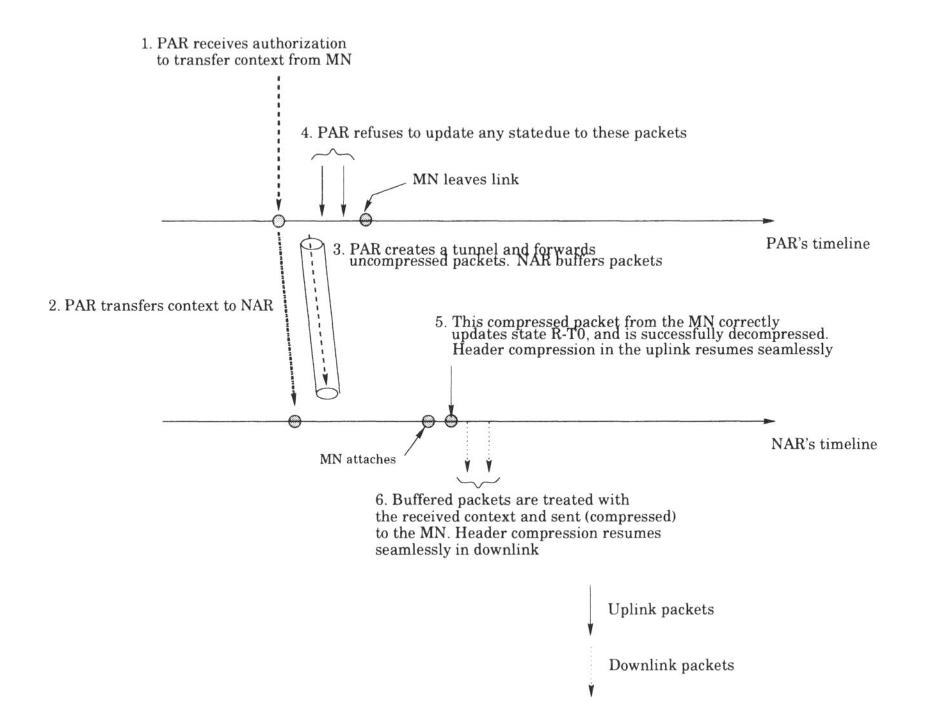

Fig. 22.3 Predictive Handover Timeline

In the setup, the mobile node sends voice packets to a distant correspondent. The packet generation is according to the ON-OFF source model. The ON period is geometrically distributed with mean λ^+, and the OFF period is exponentially distributed with mean λ^- [14]. During the ON period, packets containing 24 bytes of voice payload arrive at 20 ms intervals. The mean length of a talk spurt (ON period) is varied using probability p according to $\lambda^+ = (1 - p)/p$. This has the effect of changing the number of voice packets during each ON period, providing different data set for each experiment.

The results indicate the number of compressed packets lost at the decompressor running on the router due to lack of context. During a handover, the application may continue to generate voice packets. When a link is unavailable, these packets cannot be received. In order to measure the effectiveness of context transfer, we need to measure the packet loss when a link is available (i.e., when the mobile node has established link connectivity) but context is not in place. The router does not buffer incoming compressed packets if there is no corresponding state.

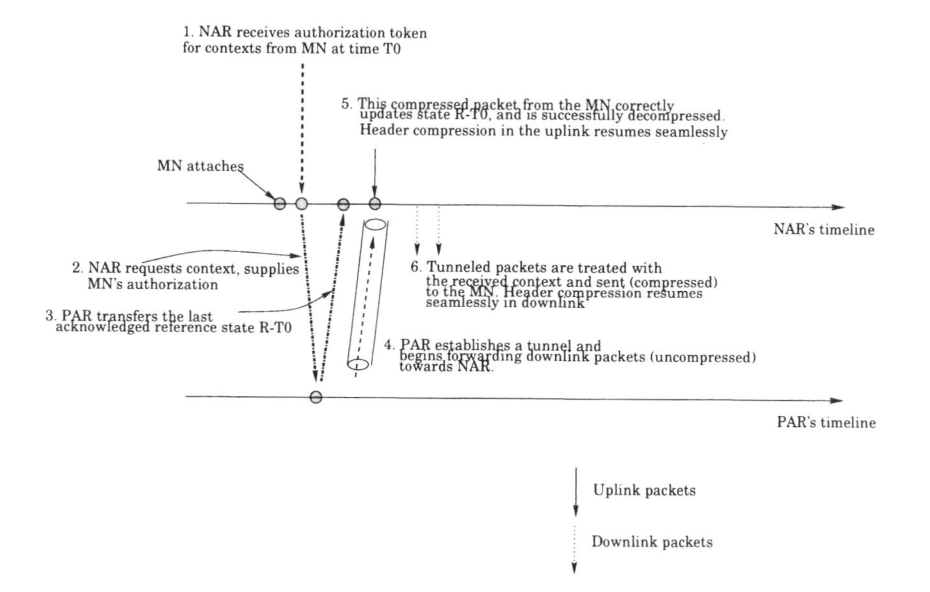

Fig. 22.4 Reactive Handover Timeline

In order to measure the correctness of the operation, it suffices to observe the monotonicity of the RTP timestamp and sequence numbers at the correspondent. Assume that the reference timestamp at the PAR (and the mobile node) is $Timestamp_i$ at the time of relocation. *If* this value changes on both the PAR and the mobile node to $Timestamp_j$, where $Timestamp_j > Timestamp_i$, then the mobile node sends increments assuming $Timestamp_j$ as the reference after attaching to the NAR. Using the received context, the NAR uses $Timestamp_i$ as the reference and hence breaks the monotonically increasing sequence of timestamps. This can be observed on the correspondent node.

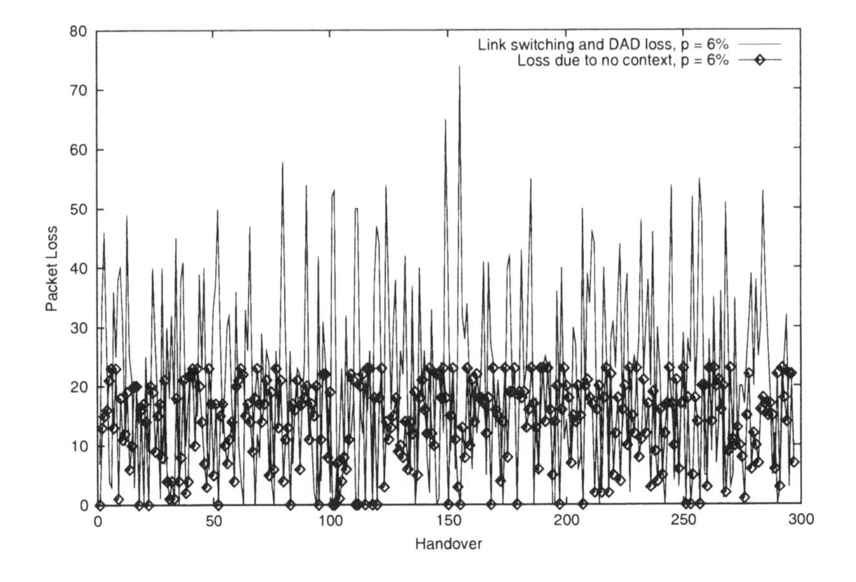

Fig. 22.5 Packet Loss with Basic Handover (No Context Transfer)

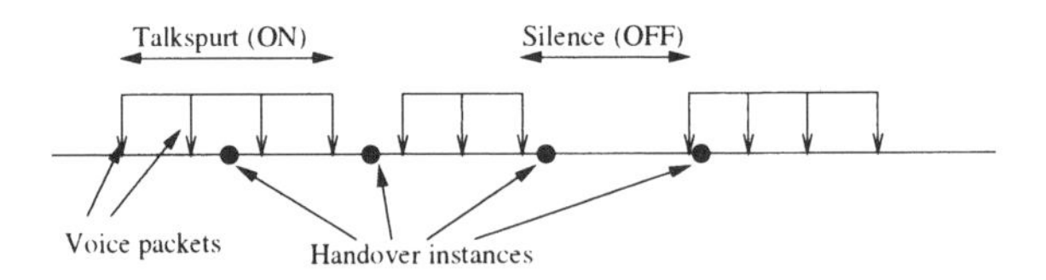

Fig. 22.6 Application State and Handover Epochs

22.4.1 Basic Handover

Figure 22.5 shows the number of packets lost when a basic handover takes place. That is, the mobile node has to perform movement detection, new IP address configuration and location update procedures as it undergoes a handover. All along, the application continues to generate payload at an average rate of 15 packets (corresponding to p = 0.06) per ON period, and the compressor on the mobile node sends compressed packets once the network interface is able to send IP packets. No context transfer is used. This scenario corresponds to a basic IP handover.

The packet loss observed is due to the following. During link switching, i.e., the process of terminating the existing WLAN link and re-establishing a new link, packets are lost on the mobile node since the application continues to transmit voice payload. This delay is about 100 ms with the interface card used and includes the host stack processing delays. [1] Apart from the link switching delay, the mobile node has to perform movement detection to determine whether it has moved to a new subnet by sending a Router Solicitation message and processing the Router Advertisement message in return. In addition to this round-trip delay, there is typically delay due to access routers staggering requests for advertisements [12]. On a WLAN with moderate load conditions, this delay is smaller than the interpacket separation delay (i.e., 20 ms) between voice payloads. The IP address configuration delay, however, is hardly trivial. The process of DAD, on average, introduces 1400 ms of delay, which is highly detrimental to voice quality. All voice packets generated during this process of assigning a unique address to the interface are dropped by the IP stack. [2] Finally, once the mobile node configures an IP address, it sends compressed packets which are dropped by the router since it has no context.

Let us focus on the number of packets lost due to the absence of context. If the compressor on the mobile node is in the FO or SO state, it can send many packets without waiting for acknowledgments to come back. In addition, acknowledgments can be sent sparsely [3], meaning that only one or two may be sent by the decompressor per RTT. Besides, even when the decompressor sends acknowledgments, the queuing delay causes them to be sent in bursts, rather than in lockstep with the received compressed packet. [3] This causes the compressor to jump ahead before it recognizes the decompressor's failure to reconstruct the header, and begins the process of context reinitialization (with a context initialization penalty on voice quality; see Section 22.2). A ceiling of four times the cellular link RTT (120 ms or six voice packets separated by 20 ms each) is used for the compressor to fall back to Full Header state. Hence, the maximum loss is capped at 24 packets, which is the worst-case. This loss can be made smaller by having the decompressor send a (negative) acknowledgment as soon as possible instead of rate limiting it (as recommended in [3], Section 5.7.6). Even so, in WLAN we see an average of four packets lost before context reinitialization, which additionally requires an average of four FH packets and four FO packets.

On a cellular link with 120 ms RTT where a Negative ACK (NACK) is sent immediately upon detection of unavailable context, up to six compressed packets sent before the NACK is processed by the compressor have to be dropped. Once the NACK is processed, context reinitialization takes place using Full Headers, as described in Section 22.2.

[1] The process of scanning for available access points takes much longer. In this implementation, the mobile node performs scanning ahead of handover, the first of which is done after establishing the link.

[2] A corresponding delay in IPv4 is due to DHCP address acquisition, which can take several seconds.

[3] This can be viewed as a flow-control problem between the compressor and the decompressor.

There is an instance when no packet loss occurs due to lack of context. This is when sufficient number of (FO or SO) packets have already been dropped by the IP stack during link switching and IP address configuration so that the very first packet reaching the decompressor is a FH packet which begins the process of context reinitialization. This is seen in Figure 22.5, where for each instance of zero packet loss due to absence of context, there is a corresponding non-zero loss due to link switching and DAD. There are also instances when no packet loss due to link switching delay and IP address configuration occurs, yet some packets are lost due to lack of context. This happens when handover coincides with an OFF period during which no packets are sent. However, when the new talk spurt begins, the compressed FO [4] packets sent are promptly dropped by the decompressor on the router. Until a Full Header packet establishes the context, all the compressed packets are lost.

In summary, the IP address configuration latency using DAD will often be quite detrimental to voice quality due to the 1400 ms average latency. The DHCP IPv4 address acquisition delay may be worse. The interplay between this latency and the application state (ON or OFF) during handover determines the actual effect on voice quality in terms of amount of packet loss. When handover coincides with an OFF period that is greater than the IP address configuration latency, no packets should be lost; however, packet loss still occurs due to lack of context. When handover coincides with an ON period, however, packet loss due to DAD dominates over packet loss due to lack of context. In either case, there is always a context reinitialization penalty since a new context needs to be established at the router. The extent of ON and OFF periods are parameters that control the overall packet loss. Figure 22.6 can be used to visualize this.

22.4.2 Predictive Context Transfer

In this section, we discuss improvements on the basic handover experiment described in the previous section. The bottlenecks identified in the previous section are IP address configuration latency and the lack of context at the NAR when the compressor on the mobile node starts sending packets. We evaluate fast handovers [9] to address movement detection and IP address configuration latencies.

By performing address configuration before moving to the new subnet, fast handover allows circumventing of DAD delay. Even when the address is not confirmed before the mobile node attaches to the new link, a single message (FNA) facilitates this confirmation. So, effectively, no packets are lost due to DAD. However, some packets are typically lost during link switching when link connectivity itself is not available.

[4]When a talk spurt ends, the new talk spurt starts in the FO state because the RTP timestamp jump is typically not on the slope corresponding to the previous talk spurt.

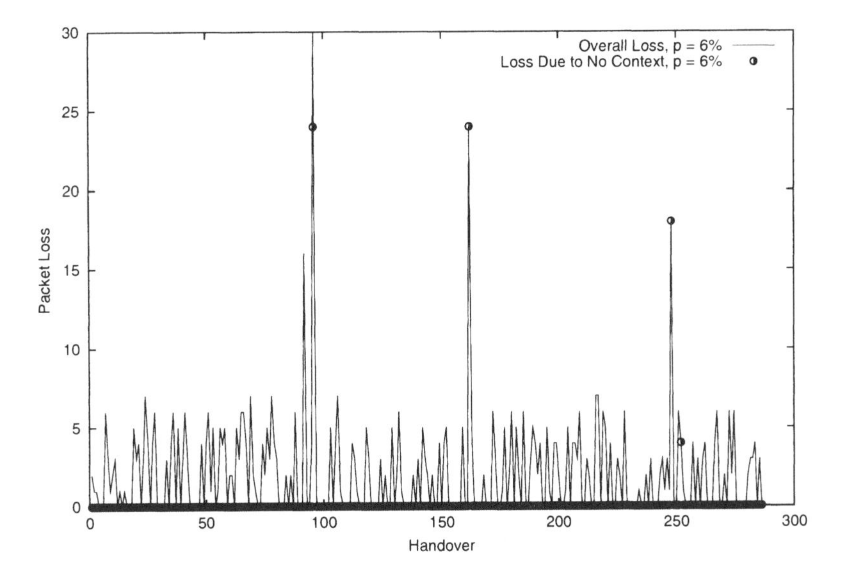

Fig. 22.7 Packet Loss with Predictive Context Transfer

In order to effect context transfer, we include a 'C' bit in the FBU to allow the PAR to transfer compression contexts along with the Handover Initiate (HI) message. Figure 22.7 shows the number of packets lost when the mean of the ON period, λ^+, is roughly 15 packets. The average number of packets lost is 2.108. In almost all handovers, no packets are lost due to lack of context. In a few cases, many packets are lost due to lack of context. This happens when the FBU packet is lost; no context transfer takes place in this case. The FBU packet can be lost if it does not exit the output queue before the link is terminated by the device driver. The mobile node still avoids IP address configuration latency, but the compressed packets sent immediately after link establishment are lost until a new context is established.

The results shown in Figure 22.8 reflect an increase in λ^+ to 24 packets during the ON period. The observed behavior is similar to that in Figure 22.7. No packets are lost due to lack of context unless signaling failure occurs. However, the average number of packets lost is 2.531 due to the higher average duty cycle of the ON period.

22.4.3 Reactive Context Transfer

In this case, context transfer initiation takes place after the mobile node attaches to the new router since the mobile node sends the FBU after it regains IP connectivity with the new router. The latencies identified in the basic handover are again avoided using fast handovers.

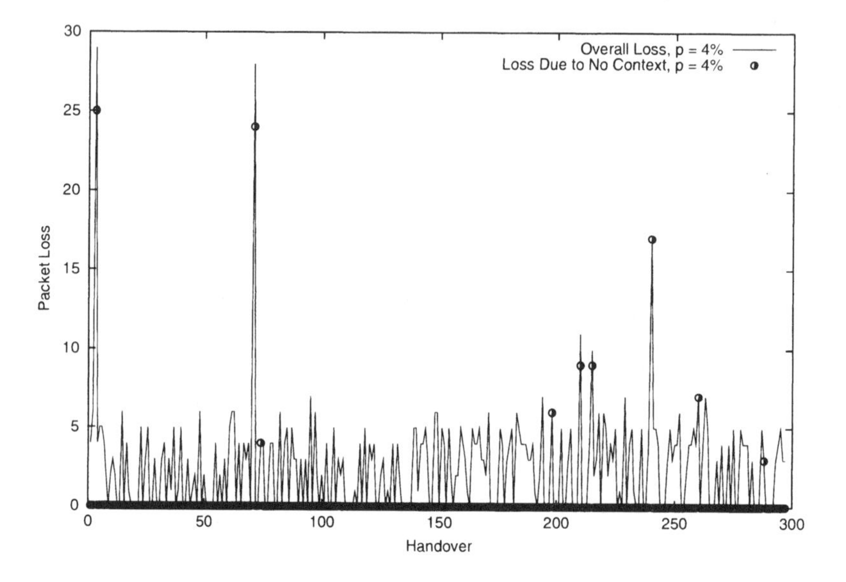

Fig. 22.8 Packet Loss with Predictive Context Transfer

The plot of packet loss when λ^+ is roughly 15 is shown in Figure 22.9. Contrary to what could be expected, there are hardly any packet losses due to lack of context. This can be attributed to the following reasons. First, the signaling is more reliable than that in the predictive scenario. It appears that the FBU always reaches the PAR unlike the situation in the predictive case, where it could be lost during link switching. Second, as long as the inter-packet separation between the FBU and the first compressed packet is longer than the time needed for context transfer, no packets are lost due to lack of context on the new router. This appears to be the case for almost all the handovers since the IP stack can transmit the FBU as soon as the link is established. In rare instances, at most one packet is lost when the FBU is immediately followed by a compressed packet. With reliable FBU transmission, the packet loss is never more than one because the voice packets are separated by 20 ms, which is ample time for context transfer in our testbed. In other words, the number of packets which can be lost is directly related to the time required for routers to communicate relative to the inter-packet separation between voice packets. The average number of packets lost in this case is 1.858.

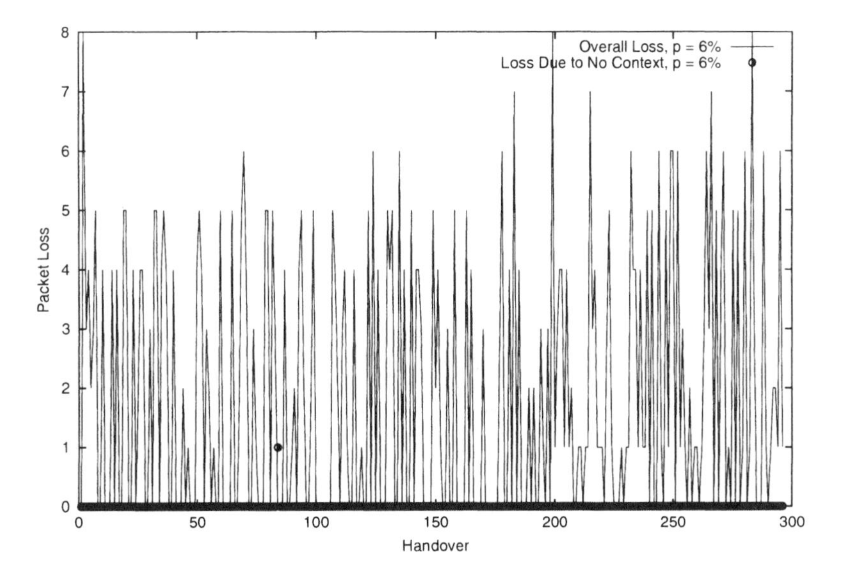

Fig. 22.9 Packet Loss with Reactive Context Transfer

This assessment is confirmed in Figure 22.10, where a mean of 24 packets for the ON period are sent. Again, no packet loss occurs due to lack of context. The overall packet loss, with an average of 2.667, increases since the average number of packets sent during an ON period is higher.

22.4.4 Behavior of State Transitions During Handover

The other performance metric of interest is how well a compressor is able to retain its favorable state (SO or FO) during handover. The compressor itself remains in one of the states immediately after handover, but whether it can carry over an optimal state depends on the expediency of handover and availability of context. In Figure 22.11, we plot the state transitions at the time of handover. State 1 refers to transition from SO state to SO state when handover occurs. State 2 refers to transition from SO to FO and so on. These states are shown in Table 22.1. In Figure 22.11, we see that without context transfer, all the state transitions occur towards the least efficient state (from SO to FH and from FO to FH). This is not surprising for the reasons we identified in Section 22.4.1. If the compressor is either in the SO state or the FO state immediately prior to handover, the only way it can retain that state is if there is a context available on the NAR.

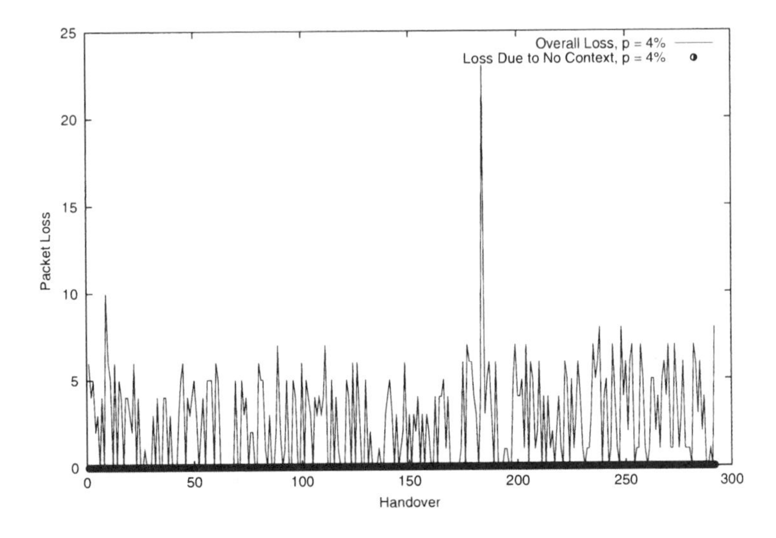

Fig. 22.10 Packet Loss with Reactive Context Transfer, Longer ON Times

State Value	Transition From	Transition To
1	SO	SO
2	SO	FO
3	FO	FO
4	FO	FH
5	SO	FH
6	FH	FH
7	FH	FO
8	FO	SO
9	FH	SO

Table 22.1 State Transition Representation

In comparison, Figure 22.12 shows the state transitions with reactive context transfer. The majority of the transitions are at level 3, level 2 and level 1, indicating movement from FO to FO, SO to FO and SO to SO. It is worthwhile to analyze this. If the compressor is in the SO state just before handover, it can remain in that state if the application does not transition to the OFF state during handover *and* the correct context is available on the NAR immediately after handover; correct context means that the reference state with respect to which the header differences are computed is consistent on the mobile node and on the NAR. The application transitioning to the OFF state depends on the ON period duty cycle, but when it does remain in the ON state longer than the link switching delay, the movement detection and IP address configuration delays should be small enough in order to make use of the context present on the router.

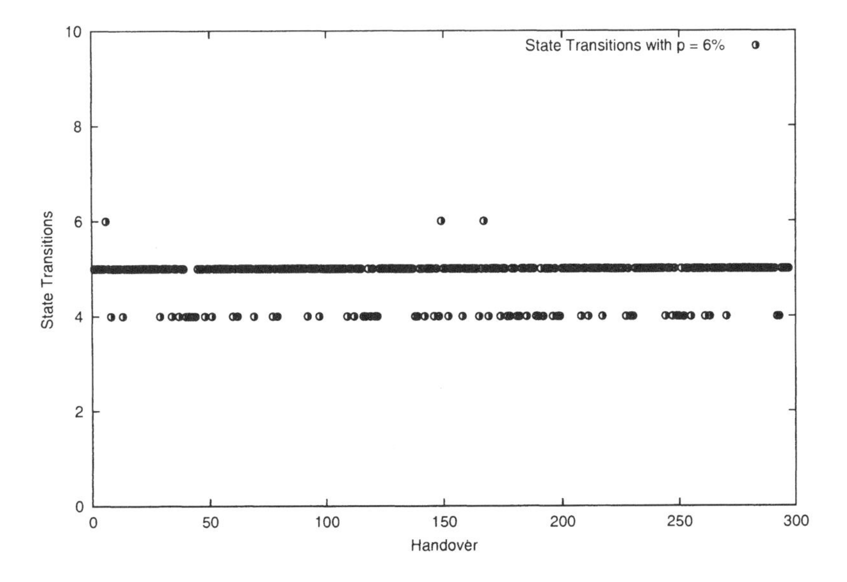

Fig. 22.11 Header Compression State Transition without Context Transfer

The transition from SO to FO occurs as soon as an ON period ends, regardless of link switching delay, fast handover and context transfer because the RTP timestamp will jump by an amount dependent on the duration of the OFF period, which varies from an instance to another. So, when a new ON period begins, the compressor sends packets in the FO state. If the handover is not fast enough or if the context is not present, packet losses occur. However, previous experiments demonstrate that this is unlikely with fast handovers and context transfers. The other possibility is that the compressor continues in FO state before and after handovers (State 3). A compressor leaves the previous link in the FO state because it did not receive positive acknowledgments to transition to SO prior to link switching. If the ON period continues, the FO state is maintained until an acknowledgment is received or the ceiling (set to 24 packets) for sending FO packets without waiting for an acknowledgment is reached. Since an acknowledgment cannot be received until the compressor synchronizes itself with the decompressor on the NAR, the compressor can at best retain the FO state if the combined delay of link switching, movement detection, IP configuration and context transfer is below the ceiling. [5] These results indicate that context transfer with fast handover helps maintain efficient state during handovers.

[5]Actually, the combined delay has to be less than the difference between the ceiling and the current count of unacknowledged FO packets, which is more stringent.

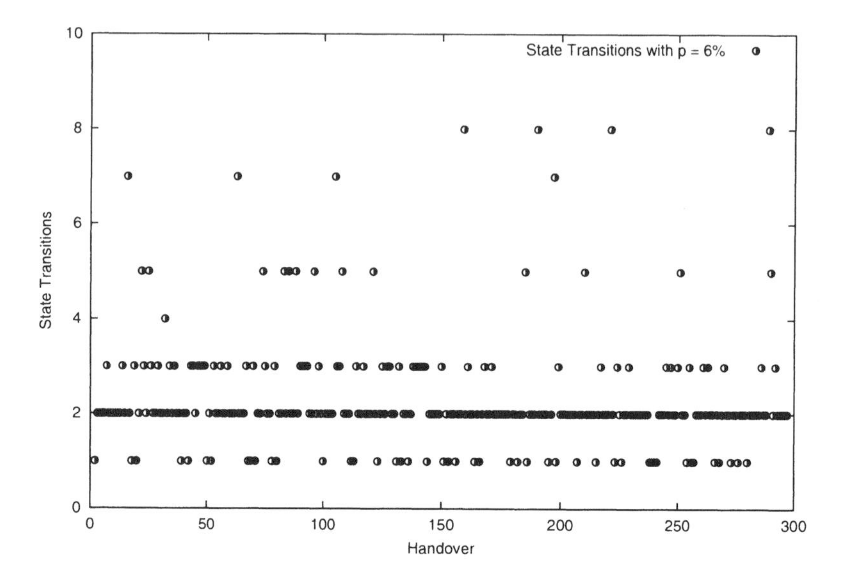

Fig. 22.12 Header Compression State Transition with Reactive Context Transfer

The other transitions shown in Figure 22.12 are left as exercises for the reader. Of particular interest are the relative possibilities of transitions from FH to FO and FO to SO, and why these transitions occur.

22.5 CHAPTER NOTES

Vijay Devarapalli (formerly with the Nokia Research Center in Mountain View, California) implemented the access router part of the fast handover protocol and conducted numerous experiments whose results are described here. Robert Chalmers from University of California at Santa Barbara implemented the original context transfer framework. Manish Tiwari and Sundar Jayaraman from Purdue University implemented the header compression and the original mobile node part of the fast handover protocol respectively. We are grateful to all of them.

Exercises

22.1 Define the header compression context for a IPv6/UDP/RTP stream where an RTP timestamp jumps by 160 every 20 ms. Which fields are static, which can be inferred and which need to be constructed on per-packet basis? How would you treat the checksum field?

22.2 There have been two general approaches to constructing FO and SO packets. One approach is to transmit the differences with respect to the previous packet [4], and the other is to use a fixed reference state of a previously acknowledged packet and transmit the differences with respect to that packet until the reference state changes [3]. What are the pros and cons of each approach? If a medium's Bit Error Rate (BER) is low, which is the preferred approach? If the RTT between a compressor and a decompressor is low, which is the preferred approach?

22.3 Analytical models of header compression are useful to understand state transitions. However, such models depend on the nature of the applications that use header compression. Assuming an ON-OFF source model for the application, develop an analytical model for the behavior of header compression during handovers. Assume a suitable model for the handover process itself.

22.4 Why does a compressor transition from FH state to FO state when using context transfers with fast handovers in Figure 22.12? Similarly, why does it transition from FO state to SO state?

22.5 If context transfer does not take place *in time*, packet loss will occur. Can this packet loss be avoided without context transfer, and if so, how? What can a decompressor do as soon as it recognizes that it cannot construct the header, and what is the overhead associated with its actions?

22.6 Consider that the handover stream also needs Differentiated Services [2] QoS support. This means there is a state corresponding to the packet classifier, the packet meter and the packet marker on the router. As a function of the load on the output queue at a router, analyze the effect of the absence of the QoS state on a packet stream. Extend the analysis to include multiple hops beyond the access router where at least the Differentiated Services Code Point needs to be marked for appropriate treatment.

REFERENCES

1. N. Asokan, P. Flykt, C. E. Perkins, and T. Eklund. "AAA for IPv6 Network Access" (work in progress), draft-perkins-aaav6-03.txt, March 2001.

2. S. Blake. et al. "An Architecture for Differentiated Services," RFC 2475, Internet Engineering Task Force, December 1998.

3. C. Bormann (Editor). "Robust Header Compression: Framework and Four Profiles: RTP, UDP, ESP, and Uncompressed," RFC 3095, Internet Engineering Task Force, July 2001.

4. S. Casner and V. Jacobson, "Compressing IP/UDP/RTP Headers for Low-Speed Serial Links," RFC 2508, Internet Engineering Task Force, February 1999.

5. S. Deering and R. Hinden, "Internet Protocol, Version 6 (IPv6) Specification," RFC 2460, Internet Engineering Task Force, December 1998.

6. M. Degermark, B. Nordgren, and S. Pink, "IP Header Compression," RFC 2507, Internet Engineering Task Force, February 1999.

7. D. Johnson, C. Perkins, and J. Arkko, " Mobility Support in IPv6," RFC 3775, Internet Engineering Task Force, 2004.

8. J. Kempf (Editor). "Requirements for Layer 2 Protocols to Support Optimized Handover for IP Mobility" (work in progress), draft-manyfolks-l2-mobilereq-00.txt, July 2001.

9. R. Koodli (Editor). "Fast Handovers for Mobile IPv6," RFC 4068, Internet Engineering Task Force, July 2005.

10. R. Koodli and C. E. Perkins. "Fast Handover and Context Transfer in Mobile Networks," ACM Computer Communication Review, Special issue on Wireless Extensions to the Internet, October 2001.

11. K. Le, C. Clanton, Z. Liu, and H. Zheng. "Efficient and Robust Header Compression for Real-Time Services," in Proceedings of IEEE Wireless Communications and Networking Conference (WCNC), September 2000.

12. T. Narten, E. Nordmark, and W. Simpson. "Neighbor Discovery for IP Version 6 (IPv6)," RFC 2461, Internet Engineering Task Force, December 1998.

13. H. Schulzrinne et al. "RTP: A Transport Protocol for Real-Time Applications," RFC 1889, Internet Engineering Task Force, January 1996.

14. K. Sriram and W. Whitt. "Characterizing Superposition Arrival Processes in Packet Multiplexers for Voice and Data," IEEE Journal on Selected Areas of Communication, September 1986.

15. S. Thomson and T. Narten. "IPv6 Stateless Address Autoconfiguration," RFC 2462, Internet Engineering Task Force, December 1998.

16. J. Wroclawski. "The Use of RSVP with IETF Integrated Services," RFC 2210, Internet Engineering Task Force.

23

Location Privacy and IP Mobility

Why do my callers hear a different ringtone when I am out of the country? My SMS still has my home timezone! - pondering over privacy quirks

—

23.1 INTRODUCTION

Location privacy in IP networks is an emerging concept. It is also a very broad topic, with different opinions and requirements. It is inherently difficult to capture the essence of an emerging topic with little common agreement on the problem itself, leaving solutions aside for now. So, we confine ourselves to privacy concerns in using IP addresses themselves and the role of mobility (roaming) in privacy due to change of IP addresses.

Very generally, the IP location privacy problem can be stated as follows: Given an IP address and a database that maps IP prefixes to their geographical sites, how can the current physical location of a host be made harder to determine? The very fact that a database exists that maps IP prefixes to geographical location provides a reference against which the possible solutions for this problem are compared. So, the key considerations here are the granularity with which a physical location can be determined and what mechanisms could be devised to hide the fact that a user has roamed to a new network.

The databases which exist on the Internet map the IP prefixes to the names of organizations under which the prefixes are registered. A typical query of an IP address results in attributes including the Internet Service Provider (ISP), the city, the region and the country. One can use this information to determine the exact location of a building whose address is used in registering the IP prefix. So, *if* a host is located in the same building whose physical address is returned by the Internet geography query service, a peer can map the host IP address to the geographical location. However, there is no way to be certain that the ISP physical address is the same as the location where the host is present. This is particularly true for wireless networks, where a provider's physical address has limited relevance to the physical location of a mobile host that uses the provider's IP prefix. In summary, existing solutions for mapping an IP address to a physical location appear to have the granularity of an ISP or an organization's physical location. This is understandable since the way an ISP or an organization subnets its IP network is orthogonal to registering ownership of network prefix chunks. When this address happens to be that of, for example, a small office building, the granularity can be considered to be small, with the caveat that knowledge of the size of the building is extraneous to the resolution of the IP address to the physical location. Furthermore, when IP tunnels are used, an IP address cannot provide mapping to a geographic area with any reasonable degree of accuracy.

To be sure, an IP address itself does not possess any geo-location properties. The network prefix has no relation to geographical coordinates and the host identifier is almost never constructed using them. Indeed, the idea of embedding geographical coordinates in the host or interface identifier is not only controversial but infeasible for mobile hosts whose location keeps changing. So, the only known approach to mapping an IP address to a physical location involves consulting databases that map IP prefixes delegated to ISPs or organizations to their physical addresses. However, protocols can be designed using IP assuming global positioning system (GPS) receivers in hosts. The Geopriv working group in the IETF is defining a framework considering privacy, authentication and authorization issues surrounding such protocols [12]. For instance, a new DHCP option to determine the geo-location of an IP address corresponding to a location (such as a wall socket) is defined. Using this option on supported systems, a host can determine its geo-location and use it for the purposes it deems fit. A malicious use of such an option could lead to a wireless host surveying various physical locations and building a detailed blueprint of the IP network at the granularity of subnets. This clearly causes more specific concerns than are raised by mapping a prefix to an ISP or some other organization. In our discussion, however, we will restrict ourselves to privacy pertaining to IP addresses.

Another notion of privacy involves profiling user data. In RFC 3041 [6], some extensions to IPv6 stateless address autoconfiguration are defined. The observation in [6] is that a constant Interface Identifier can be used to analyze the traffic pattern of a host, and hence of a user, even if the network prefix itself keeps changing. For instance, a network sniffer strategically placed on a link can record data which can subsequently be used to profile traffic patterns that could lead to information such as user activity during specific times of the day, when a user is in the office and so on. If such profiling is a concern, then changing the IID regularly using pseudorandom sequences generated by a hash (MD5) function is recommended. [1] Doing so would make it harder for an eavesdropper to correlate observed traffic with user activity.

User data profiling does involve IP addresses. Hence, mechanisms are needed to disrupt attempts that can correlate addresses to user activity. In this sense, RFC 3041 provides a mechanism that can be useful in some environments; a Mobile IPv6 node can configure its care-of address using the method defined in [6], and regularly change it. Using the same approach to home address does not appear to be as useful since a home address is used as a constant identifier which could be visible in the DNS. Changing the home address too often implies the DNS needs to be updated, most likely by the home agent, at each instance of change as well. In addition, changing home address requires re-negotiation of the security association between the mobile node and the home agent. For these reasons, changing home address often is not desirable. When a name or an IP address remains relatively constant in a system such as DNS, then there is limited applicability. So, even if a care-of address can be configured to thwart profiling, presence of home address in each data packet still serves as a constant identifier that can be exploited for traffic analysis purposes. This motivates the need to develop separate mechanisms to disrupt home address profiling.

Home address is not the only identifier which can compromise a user's location privacy. The MAC address, transport protocol port numbers, SIP URI [8] can all provide important information about a user's communication pattern. Indeed, any identifier with a global scope can reveal roaming. Like the home address, a SIP URI is an example of such a global identifier. Beyond these identifiers, if fields in IP communication (such as an IID) remain constant across network movements, such fields can also suggest roaming.

There is an important difference between inferring roaming based on profiling of fixed fields (such as an IID) and global identifiers. An attacker only needs to observe a home address or a SIP URI from a visited network to determine roaming, as opposed to profiling traffic across multiple network segments or collude with other attackers or do both [5]. It is arguably easier to simply monitor the usage of home address compared to traffic profiling of other fixed fields in IP communication. Also, inferring roaming based on profiling is independent of Mobile IP. In the remainder of this chapter, we focus on the location privacy problem that is applicable to Mobile IP.

[1] MD5 is no longer advocated for generating hash functions. HMAC_SHA1 is the preferred transform.

23.2 LOCATION PRIVACY PROBLEM WITH MOBILE IPV6

As we saw in the previous section, the location privacy problem is particularly applicable to Mobile IP. In this section, we define the problem more formally and illustrate it. As we do so, it may be useful to bear in mind the distinction between device roaming and user roaming. The former is primarily concerned with device identifiers (IP addresses) and the latter with user identifiers such as a SIP URI. Here, we focus on IP addresses. Together with a namespace service such as DNS, one could map device identifiers to user identifiers.

23.2.1 Revealing the Home Address

The home address of the mobile node is visible in all packets that the mobile node sends to a correspondent node beginning with the Binding Update message. The HoTI and HoT messages also contain the home address, but can be encrypted between the mobile node and its Home Agent. An onlooker can determine from the home address present in the destination option and in the routing header that a device has roamed to a new network. [2] When a binding between the device identifier and the user identifier is available, the onlooker can also determine that the user has roamed. This problem is independent of whether the mobile node uses route optimization (using the care-of address) or reverse-tunneling (using the home address) to communicate with the correspondent.

23.2.2 Disclosing the Care-of Address

A Binding Update from a mobile node can be interpreted as an indication of roaming to a correspondent node. This assumes that the correspondent is able to associate the care-of address to an home address, for instance by inspecting the Binding Cache entry. The home address is assumed to be obtainable by appropriate tools such as DNS.

23.2.3 Problem Illustration

Consider a Mobile Node present in its home network. Its correspondents can see an IP address valid on the home network. More specifically, the application endpoints are likely to see a user-friendly identifier such as a SIP URI when involved in a peer-peer communication such as VoIP. At the same time, the transport endpoints (i.e., RTP endpoints) will see the IP addresses. Hence, users may not notice any compromise in location privacy if they are uninterested in or unaware of any change of IP address as long as the communication continues as they normally see it. However, any user can capture packets belonging to a session and monitor the mobile node's IP address. Any change in IP address can provide sufficient information to a correspondent about roaming. Assessing the physical location based on IP addresses is similar to assessing the geographical location based on the area code of a telephone number. Just as in a telephone number, the granularity of the physical area corresponding to an IP address can vary, depending on how sophisticated the available tools are, how often an ISP conducts its network renumbering, and so on.

[2] We do not consider stolen mobile devices!

When the mobile node roams to another network, there are two parts to the location privacy problem: revealing information to its correspondents, and to onlookers.

The mobile node can communicate with its correspondents either directly or reverse tunnel its packets through the home agent. With reverse tunneling, the new IP address of the mobile node is not revealed, although the performance difference may not be acceptable to either parties. Hence, a user needs a policy on which mechanism to use with a particular correspondent. One could imagine having a policy in which those correspondents found in the *Contact List* are the only ones with which the user communicates directly. All others will first go through reverse-tunneling. Even with those correspondents with which route optimization is used, the transport protocol still sees the Home Address. Unless the correspondent runs some packet-capturing utility, the user cannot see which mode (reverse tunneling or route optimization) is being used, but knows that it is communicating with the same peer whose URI it knows. This is similar to conversing with a roaming cell phone user whose phone number, like the URI, remains unchanged.

On the other hand, independent of route optimization or reverse tunneling, the home address is revealed in data packets. When equipped with the ability to inspect packets "on the wire," an onlooker can determine that the mobile node has roamed and also determine that the user has roamed. This could compromise the location privacy even if the mobile node took steps to hide its roaming from correspondents.

When location privacy is compromised, it could lead to more targeted profiling of user data. An eavesdropper may specifically track the traffic containing the home address and monitor the movement of the mobile node with changing care-of address. Although the profiling problem is not specific to Mobile IPv6, it could be triggered by a compromise in location privacy due to revealing the home address. A correspondent may take advantage of the knowledge that a user has roamed when care-of address is revealed and act accordingly. This can be of concern to a mobile user especially when the correspondent is unacquainted or turns out to be untrustworthy.

Beyond location privacy concerns, both the home address and the care-of address could be subject to profiling, just like any other user traffic. However, applying existing techniques to thwart profiling may have implications for Mobile IPv6 signaling performance. For instance, changing the care-of address often causes additional Return Routability and binding management signaling, and changing the home address often has implications for IPsec SA management. We consider these issues in a later section.

Finally, the location privacy problem exists regardless of the way a home address (or a care-of address) is allocated. In the following sections, we provide some solutions to the location privacy problem. We recognize the importance of preserving the Return Routability protocol design principles. In particular, we are interested in creating no state at a correspondent node until a Binding Update succeeds and only relying on the parameters present in the BU message and the Return Routability protocol messages for the purposes of privacy support. We also recognize the importance of providing a solution that offers incremental deployment using the Mobile IPv6 protocol messages. We describe the solutions within the framework of the assumptions and risks involved in the Return Routability protocol itself. At the same time, we also describe a solution that requires minor additions to the Return Routability protocol but provides additional features of interest.

23.3 HIDING THE HOME ADDRESS FROM ONLOOKERS

The crucial problem in protecting location privacy in the presence of IP mobility is to conceal a constant identifier (home address) when the mobile node roams. A solution to this problem has to satisfy the following requirements:

- The field which replaces the constant identifier must be computed verifiably using the parameters, the identifier itself, and a shared secret all known to the two endpoints. A particular constraint is the computation and delivery of the replacing field when the shared secret itself is computed at the time of processing the replacing field using the constant identifier. In Mobile IPv6, this means that a *Privacy-Tag* that replaces the home address must be computable without preventing the MAC computation for the Binding Update which requires home address as a parameter.

- It must be possible to conceal the constant identifier beginning with the very first packet (the BU) sent from the visited network which needs to contain the constant identifier. Some packets (e.g., CoTI and CoT) sent from the visited network do not contain the constant identifier.

- The method should not introduce vulnerabilities in addition to those of the mobility signaling itself. For instance, the method should not create a state at a correspondent node prior to accepting the field which replaces the constant identifier.

- It should be possible to retrieve the original identifier from the mechanism used to conceal it. Such a retrieval should be secure and fast, since it needs to be applied on a per-packet basis.

Note that it is not feasible to simply omit the home address altogether once a Binding Update is completed since such packets would be indistinguishable from packets sent without any home address at all in the first place. This makes it necessary to include a field in place of home address.

In the following, we present Privacy-Tag to conceal the home address in Mobile IPv6.

23.3.1 Privacy-Tag Computation

Upon completing the Return Routability protocol (see Chapter 8), but before sending the BU message, the mobile node computes a Privacy-Tag as follows:

$$Privacy - Tag = String \oplus HoA \quad (23.1)$$

where,

$$String = First\ (128,\ HMAC\_SHA1(K_{bm}, Data)) \quad (23.2)$$

where,

$$Data = (HoA|CoA|HNI|CNI) \quad (23.3)$$

and,

$$K_{bm} = SHA1(HKT|CKT) \quad (23.4)$$

HNI is the Home Nonce Index present in the HoT message and CNI is the Care-of Nonce Index present in the CoT message. HKT is the Home Keygen Token present in HoT, which is calculated (as described in Chapter 8) as follows:

$$HKT = HMAC\_SHA1(K_{cn}, (HoA|nonce|0)) \quad (23.5)$$

CKT is the Care-of Keygen Token present in CoT and is calculated in a similar way using K_{cn} which is a key internal to a correspondent node. K_{bm} is the binding management key from Equation 23.4.

The mobile node places the Privacy-Tag in place of the home address in the destination option using a new Type value for the option so that the correspondent node can process the field accordingly. The mobile node then constructs a BU message shown in Figure 23.1.

The mobile node then uses the IPsec Encapsulating Security Payload [4] in tunnel mode to create a packet for its home agent as shown in Figure 23.2.

The mobile node subsequently reverse tunnels the packet to its home agent, which processes the IPsec tunnel, making use of the preexisting SA with the mobile node. The home agent forwards the inner Binding Update packet to its destination (i.e, the correspondent node).

When the correspondent node receives the Binding Update packet, it first computes the MAC for the Binding Update message using the home address from the source IP address field and care-of address from the Alternate CoA option. If the MAC matches what is present in the BU message, then the Binding Update is considered acceptable. The correspondent node then processes the new destination option containing the Privacy-Tag. It computes the $String$ using Equations 23.2, 23.3 and K_{bm}, and subsequently recovers the home address as:

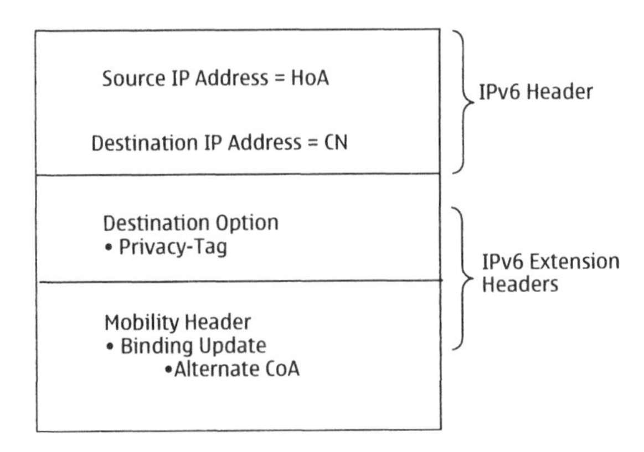

Fig. 23.1 The Binding Update Message with Privacy-Tag

$$HoA = Privacy - Tag \oplus String \tag{23.6}$$

If this home address matches the home address present in the source IP address, the correspondent node accepts the Privacy-Tag, and stores it together with the $String$ in its BCE. The correspondent node then sends Binding Acknowledgment to the mobile node's home address (according to Mobile IPv6 specification [3]) with a new *Status* value (2) indicating that Privacy-Tag is acceptable to use. Subsequently, the correspondent node itself begins using the Privacy-Tag in the routing header for packets sent to the mobile node's home address. In order to indicate the presence of a Privacy-Tag in the Type 2 routing header, a new 'P' bit in the existing Reserved field of the routing header could be used. This would allow easier firewall permeability, since the mobility-supporting firewalls must pass Type 2 routing headers already.

When data packets arrive, the correspondent node processes the destination option containing the Privacy-Tag. The correspondent node must be able to find a valid BCE for the care-of address (in the source IP address of the packet) and the Privacy-Tag. Only then does it swap the Privacy-Tag for the home address and resume normal Mobile IPv6 processing. In the reverse direction, the mobile node similarly replaces the Privacy-Tag in the routing header with its own home address before the routing header is processed. Instead of indexing the Privacy-Tag into the home address, we can also recover home address from a packet by computing $String$ on per-packet basis. We discuss this in Section 23.3.1.1.

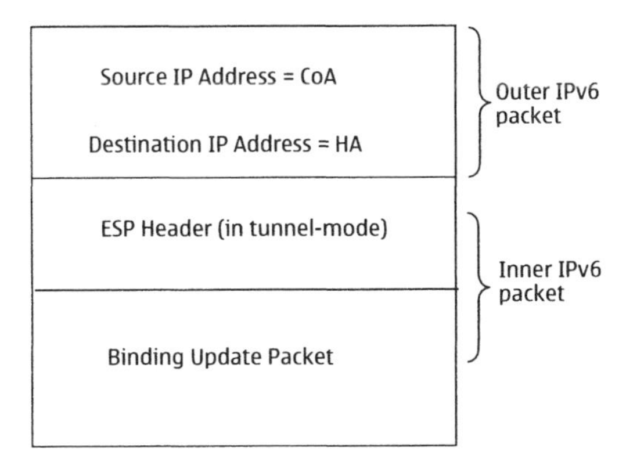

Fig. 23.2 The Reverse-tunneled Binding Update Message

The Privacy-Tag is recomputed every time the home address or the care-of address changes, for instance when the mobile node undergoes handover. And, when the lifetime of either the HoA cookie or the CoA cookie expires [3], the Privacy-Tag is recomputed. If the mobile node uses privacy extensions in [6] for care-of address, the care-of address may be recomputed even if the mobile node has not undergone a handover or a cookie lifetime has not expired. The Privacy-Tag is recomputed in this case as well. This allows it to offer the same kind of profiling protection that is available with any IPv6 address using RFC 3041.

23.3.1.1 Discussion The computation of the Privacy-Tag above carefully uses the same values which are available to a correspondent node at the time of processing the Binding Update. The $String$ computation uses parameters in HoT and CoT messages, which are sent along different paths. See Figure 8.1 in Chapter 8. Privacy-Tag by itself is not processed without verifying the MAC for the Binding Update. If the MAC for BU message fails, the packet is discarded. So, computation of bogus Privacy-Tags is avoided. If the MAC for the BU message matches, but the home address cannot be retrieved from the Privacy-Tag, then a normal Mobile IPv6 BCE with a care-of address and a home address is created, but the Privacy-Tag is not used. And an appropriate Status value (3) is returned in the Binding Acknowledgment message.

Spoofed Privacy-Tags are possible no more than spoofed home addresses themselves in the destination option in base Mobile IPv6. Both care-of address and Privacy-Tag must be spoofed, just like care-of address and home address spoofing, in order to fool a correspondent node to believe that the packet is originating from a genuine mobile node. A bogus node cannot replay the Privacy-Tag because the Privacy-Tag is only computed together with the Binding Update message, which contains replay protection. So, if a Binding Update is fresh, so is the Privacy-Tag. Subsequently, the Privacy-Tag is processed together with a care-of address for which a binding to a home address exists. In other words, a bogus node cannot replay a Privacy-Tag even if it passes the care-of address test, since both home address and care-of address need to be valid before a Binding Update can be sent. The computational load on the correspondent node includes a one-time calculation of the $String$, as in Equation 23.2, and an XOR operation to derive the home address.

Another useful property of the Privacy-Tag is recoverability of the home address from the Privacy-Tag itself. Since the home address is embedded in the Privacy-Tag, it is possible to recover it from each packet without using the Privacy-Tag as an index into the stored home address. This could be useful against intentional collisions where the Privacy-Tag is the same but not the parameters used to compute it. The possibility of such collisions is very small, however, given 128 bits for the field. Perhaps more interesting is the scenario in which the Privacy-Tag itself is changed from packet to packet. For instance, a sequence number field can be used as an additional parameter in Equation 23.2, which is incremented on each packet. Such a per-packet update can provide even more protection against profiling, and the home address can be recovered from the Privacy-Tag in each packet. The sequence number field itself needs to be accessible to the correspondent node in clear. This would require either a new destination option that carries both the sequence number and the Privacy-Tag or the use of sequence number from transport protocol headers. [3] Per-packet update also creates an additional issue: firewall treatment of the destination option and routing header. The present values in these fields are considered static. Accommodating dynamic values on a per-packet basis needs further consideration, and this is one of the topics for future work. With a Privacy-Tag that changes only when a BU message is sent, on the other hand, the firewalls can use the value in the destination option and the routing header just as they do with basic Mobile IPv6 itself.

[3]If the sequence number field is not present in one form or the other, then the correspondent node cannot determine its value in the event of packet loss.

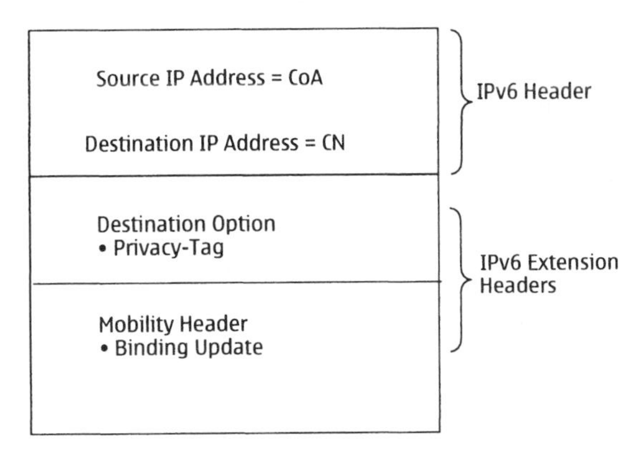

Fig. 23.3 The Route-Optimized Binding Update Message

Using reverse tunneling is necessary because the correspondent node needs the home address before it can calculate the MAC for the Binding Update, and subsequently calculate the $String$. By using IPsec encryption, the home address is made invisible for onlookers on the MN-HA path. The home address is visible in the HA-CN path. However, weaknesses arising along this path are considered to exist as part of today's Internet [7]. Nevertheless, we are interested in BU message transmission directly to a correspondent node without requiring reverse tunneling. This motivates us to apply the Privacy-Tag for route-optimized Binding Updates as well, and to discuss the ensuing security and privacy implications.

23.3.2 Privacy Tag in Route-Optimized Binding Update

When a Binding Update containing a Privacy-Tag is sent directly to a correspondent node, the packet format is as in Figure 23.3.

We can see that the home address is not present in the Binding Update. [4] This brings in some complexity, since a correspondent node needs a home address in order to verify the MAC for the Binding Update to validate the Privacy-Tag. The home address is recoverable from the Privacy-Tag, provided the correspondent node can compute the $String$, which itself must not include the home address. If it did, as in Equation 23.2, then the correspondent node has no access to it. Additionally, we cannot use K_{bm} to compute the $String$, since K_{bm} includes Home Keygen Token, which in turn includes the home address. See Equation 23.5. These requirements necessitate some small protocol extensions.

[4] Which is the whole idea of protecting it from onlookers.

The mobile node sets a 'P' bit in the existing Reserved field in the HoTI message. The correspondent node, if it supports the Privacy-Tag, computes an additional *Privacy Keygen Token* (PKT) as follows:

$$PKT = First(64, HMAC\_SHA1(K_{cn}, (Home\ Init\ Cookie|nonce|2))) \quad (23.7)$$

PKT computation is similar to HKT and CKT but uses the Home Init Cookie which the mobile node sends in HoTI. The correspondent node calculates HKT as usual, and returns it together with PKT to the mobile node in HoT message. Using PKT and CKT, the mobile node computes a new key:

$$K_{pm} = SHA1(PKT|CKT) \quad (23.8)$$

and then calculates

$$String = First\ (128,\ HMAC\_SHA1(K_{pm}, Data)) \quad (23.9)$$

where,

$$Data = (CoA|HNI|CNI) \quad (23.10)$$

As can be seen, the $String$ computation does not include the home address. Subsequently, the mobile node computes the Privacy-Tag using Equation 23.1.

The mobile node computes the binding management key K_{bm} as usual (using Equation 23.4), and computes the MAC for the Binding Update as it would for any BU message. It includes the Privacy-Tag as a new destination option and transmits the BU directly to the correspondent node. In addition, the mobile node includes the Home Init Cookie as a new Mobility Header option in the Binding Update message.

When a correspondent node receives a Binding Update without the Alternate CoA option, it uses the source IP address as the care-of address. It needs to recover the home address from the Privacy-Tag in the destination option. So, it first computes PKT as in Equation 23.7 using Home Init Cookie as one of the parameters, and CKT as specified in the Return Routability protocol. It computes K_{pm} using Equation 23.8. It then computes $String$ using Equation 23.9 and Equation 23.10. Subsequently, the correspondent node recovers the home address from the Privacy-Tag present in the packet using Equation 23.6. At this point, the correspondent node does not yet accept the Privacy-Tag.

With the home address and care-of address, the correspondent node computes the binding management key K_{bm}, computes the MAC and verifies it against that in the Binding Update. Only when the MAC matches, does the correspondent node accept both the Binding Update and the Privacy-Tag. It stores the Privacy-Tag and $String$ along with care-of address and home address in its Binding Cache. The processing of Privacy-Tag in packets sent to and from the mobile node is identical to that described earlier in Section 23.3.1.

23.3.2.1 Discussion The computation of the Privacy Keygen Token does not include the home address because the correspondent node cannot compute anything that includes the home address which is absent in the BU message. This means all the mobile nodes which are interested in hiding their home addresses when communicating with a correspondent node need to use some other unique field such as a Home Init Cookie, which is 64 bits long. The collisions for 64-bit-long field are very improbable, and the privacy management key K_{pm} has sufficient entropy for each such mobile node because it uses a hash of both PKT and CKT. Also each PKT and CKT is sent along different paths, which ensures protection against an on-path attacker (just as in the Return Routability protocol).

Furthermore, the MAC for the BU message is the same as in the Mobile IPv6 protocol. This MAC does include the home address in all computations. Although the Privacy-Tag computation is performed before this MAC is verified, the Privacy-Tag itself is not accepted until after the Binding Update is accepted. Hence, the method is secure by virtue of its synchrony with the BU signaling.

If we assume the presence of an eavesdropper in the HA-CN path, he could tamper with the 'P' bit, perhaps setting it to force a correspondent node to compute the PKT. The resulting DoS attack is limited to computation of a one-time token. The correspondent node does not store the PKT. The mobile node side will be surprised to see additional data bytes in the form of PKT. It must discard any data other than those it anticipates in the HoT message. The attacker can also unset a 'P' bit, in which case the mobile node may be led to believe that the correspondent node does not support the privacy extension for the home address. Even though we are enumerating these possibilities here, we emphasize that the basic Mobile IPv6 Return Routability protocol itself can be jeopardized by an attacker on the HA-CN path. For instance, the attacker can tamper the "Home Init Cookie" in the HoT message from the correspondent node to mobile node, forcing the mobile node to abandon route optimization. So, to be on fair grounds, we do not believe that HA-CN path attacks are justifiable only for the 'P' bit or the PKT.

23.3.3 Related Work

The primary choice when it comes to privacy in IPv6 addresses is the use of extensions defined in [6]. The IID in the IPv6 address is generated using a pseudo-random number generator, and the address itself is changed regularly in order to disrupt attempts against profiling. When this approach is applied for home addresses, the following considerations arise: Whenever a mobile node wishes to use a temporary home address, it must first send a Binding Update to its home agent and ensure its uniqueness by having the home agent perform DAD [10]. But the Binding Update itself must be secured using the IPsec security association between the mobile node and its home agent. This SA is established using the mobile node's home address as the identifier [1]. Since the mobile node needs to use a new temporary home address, *re-keying* the existing security association is necessary before the home agent accepts the Binding Update. Minimally, this implies more signaling, but more crucially the delay increases. There are additional considerations in maintaining multiple Home Addresses. Does a mobile node maintain a single permanent home address and then multiple temporary home addresses? This appears to be reasonable since a mobile node needs to be reachable at a permanent home address using the DNS. In order to facilitate this, it must follow the guidelines below:

- The mobile node must maintain the binding for old home addresses at the home agent for an appropriate amount of time. It must also assign the new temporary home address to its own interface once the Binding Update with its home agent succeeds, and maintain the old temporary home addresses for an appropriate period of time, in order to simultaneously use multiple temporary addresses with different correspondents. The list will also ensure that a previously used address is not used again.

- The mobile node must establish an IPsec SA each time it configures a new temporary address. The SAs based on the old home addresses should not be deleted immediately. Doing so will prevent the mobile node from using multiple temporary addresses simultaneously.

- The mobile node should not request any updates for DNS for its temporary home addresses.

In contrast, using the Privacy-Tag does not require additional signaling and managing multiple home addresses, their lifetimes and the lifetimes of their IPsec SAs.

The approach in [2] is perhaps one of the earliest proposals for hiding the home address. The method, which predates the Return Routability protocol, presents a Temporary Mobile Identifier (TMI) which a mobile node uses in place of home address. Since there was no Return Routability protocol when the method was presented, it does not follow the address ownership model of Return Routability (which is the basis of the Privacy-Tag derivation). More recently, there has been discussion in the IETF working groups about privacy general and privacy for Mobile IPv6 in particular.

What we have described thus far does not require changes to the Home Agent behavior. When a BU message is reverse tunneled, no protocol changes are required for Return Routability. With route-optimized Binding Updates, there are small extensions to the Return Routability and BU messages providing a road map for implementation. Along the same lines, we describe below another mechanism in which the home address is never disclosed in any of the messages, including the Return Routability messages. The mechanism, however, requires small changes to the home agent and the correspondent behavior.

23.3.4 Using a Pseudo Home Address

Since a home address is visible in all mobility signaling messages as well as in data packets, a mechanism using a pseudo value is worth considering. Essentially, the mobile node and the home agent derive a new home address as follows:

$$Pseudo\_HoA = HMAC\_SHA1(K_{ph}, PreviousPseudo\_HoA) \tag{23.11}$$

where K_{ph} is a symmetric key preestablished between the mobile node and its home agent (either manually or via IKE).

Pseudo_HoA is used in all messages including the HoT, HoTI, the Binding Update, and the prefix discovery messages. Hence, the real home address is never disclosed in any of the messages. See [11]. On the other hand, using Pseudo_HoA means that protocol changes are necessary. We consider this in the Exercises.

23.3.5 Protecting Mobile IPv6 Messages

Once a mobile node arrives on a new visited link, it sends Binding Update, Binding Acknowledgment, Mobile Prefix Solicitation and Mobile Prefix Advertisement messages, all of which contain the mobile node's home address. We have already discussed the message format for the Binding Update to conceal the home address from an onlooker. We leave the formats for these messages as an exercise for the reader. Following these formats is necessary for total protection against revealing the home address.

23.4 DISCLOSING THE CARE-OF ADDRESS

There are two fields which come into play when considering roaming indication to a correspondent: The home address and a Binding Cache Entry (BCE). When a mobile node visits a new network, it must obtain a topologically correct IPv6 address. When the mobile node wishes to communicate with a correspondent using route optimization, this address will be visible in the BCE. So, a BCE serves as a definite indication of roaming. Hence, a mobile node concerned about revealing roaming to a correspondent should not use route optimization. Instead, it must use reverse tunneling through the home agent. When doing so, it should use IPsec encryption to protect its home address from visibility to onlookers. The packet format is similar to the reverse-tunneled Binding Update format.

Even if a home address changes, for instance when the mobile node uses dynamic home address assignment, it does not provide a reliable indication of any roaming information to a correspondent node which could map it to obtain the rough geographic information. For instance, any ISP may have its prefix chunk subnetted over a large physical area, including geographically disjoint locations, and yet the mapping can only produce a single physical (postal) address of the ISP (used in registering the prefix chunk with an authority offering the address chunk). So, a correspondent node may not always be able to use a changing home address as a reliable indicator of roaming.

The latency of reverse tunneling through a distant home agent may be a concern for some nodes. When dynamic home agent assignment from the *visited* network is available, the mobile node can use a home address assigned by such a home agent without any perceptible difference is latency. Such a facility requires infrastructure support to ensure that the visiting mobile node can be properly authorized before it can be assigned a home agent from the visited network. Nevertheless, it is feasible and a mobile node may simply use reverse tunneling with all its correspondents to avoid revealing its roaming. However, a home address assigned from the visited network can provide a physical location information that confirms roaming. This is possible, in contrast to the scenario in the previous paragraph, when the visited network's ISP is not the same as the mobile node's home ISP.

As an existing solution, the Hierarchical Mobile IPv6 (HMIPv6) [9] provides limited protection against revealing a care-of address to a correspondent node. A mobile node using HMIPv6 would use a care-of address that remains constant as long as the mobile node is within the same jurisdiction of the Mobility Anchor Point (MAP). Depending on how large a subnet area a MAP covers, the movement of the mobile node within such an area is transparent to the correspondents. So, HMIPv6 hides the *subsequent* care-of address of the mobile node, but not roaming itself, from a correspondent, since the correspondent node knows about the care-of address. Nevertheless, Privacy-Tag can be used with HMIPv6 to hide the home address from onlookers.

23.5 CONCLUSION

The general problem of privacy when using IP communication is complex, mainly because it spans so many protocol layers where it is a problem, but also because there is no universally agreed-upon notion of what it means. One constituency, especially those involved in IETF, tends think that protection against profiling is critical for privacy. This is an important problem to address so that the privacy threats stemming from the predictability of communication fields are addressed. At first glance, this problem does not appear to be specific to any one protocol per se, but is a result of disparate yet constant identifiers in multiple protocol layers which can be profiled. So, if we were to guess at a solution, it would involve local host approaches to better manage its set of protocol layer identifiers. The results in this space should provide better privacy in general, not location privacy itself.

We have considered here the problem of maintaining location privacy for mobile users. Using Mobile IPv6 as the default mobility protocol, we have assessed the problem of revealing a home address to onlookers, which in effect means revealing roaming. We have described a method to conceal this identity in communication between a roaming mobile user and its correspondent while maintaining direct route-optimized communication between the two peers. The Privacy-Tag is simple to compute and does not require any state at a correspondent (just like the Return Routability protocol) before a binding is created. It minimally augments a BCE with the Privacy-Tag, which is used together with a care-of address to index the concealed home address, which is then swapped in the home address Option before regular Mobile IPv6 processing resumes. Even though the basic idea is the same, the way the Binding Update is sent affects the use of home address as a parameter in Privacy-Tag computation. One may choose the Privacy-Tag with reverse-tunneled Binding Updates for its zero changes to the Return Routability protocol, or send it directly using the small extension to the Return Routability protocol we have identified. In any event, the implementations can augment the Return Routability protocol to obtain location privacy. At the same time, we also mention that the specific techniques are likely to evolve as we better understand this space.

We have also identified the implications of disclosing a care-of address to a correspondent. While a perfect mechanism is not available to us without significantly changing the IP infrastructure itself, we believe a choice is available to a mobile node in selectively choosing the correspondents with which it wishes to engage in route optimization. We have also identified and discussed what constitutes roaming in the first place from a correspondent's perspective. In this discussion, we have found that the presence of a BCE can serve as a definitive indicator of roaming to a correspondent. Dynamic Home Agent assignment from the visited network appears to be one solution which can be useful, although its ubiquitous availability is still unassured.

Exercises

23.1 Often privacy is intertwined with confidentiality. However, privacy can also include profiling where exact identities and the user payload may not be known but the usage patterns may be derivable by monitoring. For instance, simply monitoring fields which remain fixed across movements can provide useful information. Location privacy is about protecting roaming information, as we saw in this chapter. Tabulate the differences between profiling and location privacy. Identify the interactions between them. Specifically, what is the role of *Sequence Number* in the Binding Update message in both profiling and location privacy?

23.2 What parameters, if any, have the potential to reveal roaming in addition to the IP address? How can they be obtained?

23.3 As we discussed, reverse tunneling seems to be the only mechanism available to a mobile node if it wishes to conceal its care-of address from its correspondents. Are there any other feasible mechanisms? If so, what is their deployment feasibility?

23.4 When a Pseudo_HoA is used in place of a globally routable HoA, what changes are necessary for the Return Routability protocol?

23.5 How is a Security Association established between the mobile node and its home agent for a Pseudo_HoA (in Section 23.3.4) and maintained?

REFERENCES

1. J. Arkko, V. Devarapalli, and F. Dupont. "Using IPsec to Protect Mobile IPv6 Signaling Between Mobile Nodes and Home Agents," RFC 3776, Internet Engineering Task Force, June 2004.

2. C. Castelluccia and F. Dupont. "A Simple Privacy Extension for Mobile IPv6 (work in progress)," Internet Engineering Task Force, February 2001.

3. D. Johnson, C. Perkins, and J. Arkko, "Mobility Support in IPv6," RFC 3775, Internet Engineering Task Force, 2004.

4. S. Kent and R. Atkinson. "IP Encapsulating Security Payload," RFC 2406, Internet Engineering Task Force, November 1998.

5. R. Koodli. "IP Address Location Privacy and Mobile IPv6: Problem Statement" (work in progress), draft-ietf-mip6-location-privacy-ps.txt, February 2007.

6. T. Narten and R. Draves. "Privacy Extensions for Stateless Address Autoconfiguration in IPv6," RFC 3041, Internet Engineering Task Force, January 2001.

7. P. Nikander, J. Arkko, T. Aura, G. Montenegro and E. Nordmark. "Mobile IP Version 6 Route Optimization Security Design Background," RFC 4225, Internet Engineering Task Force, December 2005.

8. J. Rosenberg et al. "SIP: Session Initiation Protocol," RFC 3261, Internet Engineering Task Force, June 2002.

9. H. Soliman, C. Casteluccia, K.El Malki, and L. Bellier. "Hierarchical Mobile IPv6 Mobility Management (HMIPv6)," RFC 4140, Internet Engineering Task Force, August 2005.

10. S. Thomson and T. Narten, "IPv6 Stateless Address Autoconfiguration," RFC 2462, Internet Engineering Task Force, December 1998.

11. Q. Ying, F. Zhao, and R. Koodli, "Mobile IPv6 Location Privacy Solutions" (work in progress), draft-irtf-mobopts-location-privacy-solutions, February 2006.

12. Geographic Location and Privacy IETF Working Group, http://ietf.org/html.charters/geopriv-charter.html

24

Mobile IPv4 Route Optimization using Return Routability

You become a writer by writing. It is a Yoga. – R. K. Narayan

24.1 INTRODUCTION

Return Routability mechanism for Mobile IPv6 [1] specifies a mechanism for establishing shared secret between a mobile node and an arbitrary correspondent node (CN) node on the Internet in order to enable route-optimized communication. It does this based on the assumption that the underlying routing infrastructure is able to route packets correctly based on their destination addresses. In other words, if a node can demonstrate that it is able to receive packets at an address by responding to messages sent to that address, assigning ownership of that address to the node cannot make the current Internet less secure. This is an important design point. The objective is not necessarily to design the best (or even better) possible security solution, but to design a mechanism that does *no more harm* to the current Internet than what might already be possible. Arguably, this position is moot. However, engineers are often faced with the challenge of balancing design with deployment considerations. Iron-clad security may be appealing but may turn out to be overkill in some scenarios. On the other hand, without provably secure tools, some applications may not be deployed at all (e.g., VPN). Keeping these concerns in mind, in spite of its documented weaknesses, Return Routability is generally accepted as the default mechanism for route optimization with Mobile IPv6.

When looking into the practical deployment of Return Routability, some interesting considerations arise. First, the routing infrastructure is based predominantly on IPv4. So, the routability of an IPv6 address is determined at least in part by the ability to correctly associate the IPv6 address with the IPv4 address of the tunnel endpoint that actually forwards the IPv6 packet encapsulated in an IPv4 packet. Any IPv6 transition method [3] should take this into account. Second, most terminals for which Return Routability is of interest are likely to be dual-stack. For instance, most mobile nodes with Mobile IPv6 are expected to be dual-stack, and those correspondent nodes which are not mobile nodes themselves but support route-optimized communication are likely to be dual-stack. In deployments where this is valid, Return Routability can be accomplished by the endpoints themselves without the need for additional support for IPv6 tunneling from the infrastructure elements. The use of IPv6 when routable IPv4 addresses are available could be questioned. However, the nodes cannot use a route-optimized path even with globally routable IPv4 addresses. Finally, even if the endpoints themselves are not able to tunnel Return Routability messages, route optimization for IPv4 traffic deserves consideration since most routing is based on IPv4. This motivates the need for a solution that allows direct route-optimized communication using both IPv6 and IPv4 addresses.

In this chapter, we investigate the following problem: how to enable route-optimized communication for IPv4 traffic using the Mobile IPv6 Return Routability mechanism. We identify the extensions to the existing Return Routability mechanism. The Mobile IPv6 working group in the IETF is working on a protocol for operation between a dual-stack mobile node and a dual-stack home agent for minimizing signaling and operational overheads associated with separate signaling for Mobile IPv4 and Mobile IPv6.

24.2 REFERENCE MODEL AND ASSUMPTIONS

In our model, a correspondent node is assumed to be capable of performing IP-in-IP encapsulation and decapsulation. Specifically, a correspondent node is able to perform IPv6-in-IPv4 encapsulation and decapsulation, as well as IPv4-in-IPv4 encapsulation and decapsulation. A correspondent is the endpoint for the IPv6-in-IPv4 tunnel. A home agent is also an endpoint for the IPv6-in-IPv4 tunnel. These tunnels are established only upon completion of the signaling specified in this document. A Mobile Node need not be an IPv6-in-IPv4 tunnel endpoint. The mobile nodes learn about a correspondent node's IPv4 address and IPv6 address via DNS or by other means.

A single node acts as a home agent for both IPv4 and IPv6. In other words, a mobile node's home agent has access to both IPv4 and IPv6 binding and security information. If a protocol is necessary to access the IPv4 bindings from an IPv6 home agent and vice versa, it is not described here. Finally, the home agent is able to process the new Mobility Options we describe.

24.3 PROTOCOL OPERATION

When a mobile node sends the Care-of Address Test Init (CoTI) message, it includes a new IPv4 CoA Mobility Header Option and a new IPv4 CN Address Mobility Header Option. The IPv4 CoA is what the mobile node uses on the visited network for its IPv4 traffic. The IPv4 CoA acts as the tunnel endpoint for the CoTI (and the ensuing CoT) message. When the address is co-located with the mobile node's interface, the CoTI message is encapsulated and the CoT message is decapsulated by the mobile node itself. When the IPv4 address is not co-located with the mobile node, the node offering the IPv4 address service for the mobile node must be able to perform encapsulation and decapsulation, and correctly forward the tunneled IPv4 and IPv6 packets to the mobile node. Furthermore, the node hosting the IPv4 address also needs to know the correspondent node's IPv4 address to which to tunnel the CoTI message. One way to do this is by having the tunnel endpoint inspect the packets for the Mobility Header type, and process the new options when present. Clearly, this operation is not necessary when the mobile node itself can act as a tunnel endpoint. In any case, the CoTI message with the new extension is encapsulated and sent to the correspondent node's IPv4 address. The source address of the tunnel is the IPv4 care-of address. As might be evident, the purpose of tunneling directly to a correspondent node's IPv4 address is to demonstrate to the correspondent node that the mobile node is reachable at the source IPv4 address of the tunnel.

When the mobile node sends the Home Address Test Init (HoTI) message, it includes a new IPv4 HoA Option, as well as a new IPv4 CN Address Option. There are no specific IPv4 tunneling requirements for the HoTI message. However, if the mobile node is the IPv4 tunnel endpoint, it may tunnel the HoTI message directly to the home agent's IPv4 address. In such a case, avoiding reverse tunneling of HoTI using IPv6 addresses may be considered; see Exercise 24.2 at the end of the chapter. When the home agent receives the HoTI message, it must process the new extensions. Note that the home agent must be able to inspect the Next Header field anyway in order to be able to apply IPsec.

The enhancement proposed here requires the home agent to also look for the new extension when the Next Header type is Mobility Header, and perform the following operations. It needs to act as the IPv6-in-IPv4 tunnel endpoint by including the mobile node's home address as the source IP address and the correspondent node's IPv4 address as the destination address for the tunnel. Both addresses are present in the new extension to HoTI. The home agent must ensure that a valid binding exists for the IPv4 home address before tunneling the packet to the correspondent node. Such a binding may be established using either Mobile IPv4 signaling or a single signaling mechanism for both Mobile IPv4 and Mobile IPv6, which is being specified by the MIP6 working group in the IETF.

When the correspondent node processes the new extension in the CoTI, it first determines whether the IPv4 care-of address in the extension is the same as the source address in the IPv4 tunnel. It then determines if the IPv4 correspondent node address in the extension is the same as its own IPv4 address present in the destination address of the IPv4 tunnel. If both verifications are true, it concatenates the IPv4 care-of address to the other parameters in generating the Care-of Keygen Token. The correspondent node then tunnels the CoT message to the mobile node's IPv4 care-of address. It does *not* establish a permanent IPv6-in-IPv4 tunnel for the care-of address. This is done only at the conclusion of successful Binding Update processing.

When the correspondent node processes the new extension in HoTI, it first determines whether the IPv4 home address in the extension is the same as the source address in the IPv4 tunnel. It then determines if the IPv4 correspondent node address in the extension is the same as its own IPv4 address present in the destination address of the IPv4 tunnel. If both verifications are true, it concatenates the IPv4 home address to the other parameters in generating the Home Keygen Token. The correspondent node then tunnels the HoT message to the mobile node's IPv4 home address. It does *not* establish a permanent IPv6-in-IPv4 tunnel for the home address. This is done only at the conclusion of successful Binding Update processing.

When the home agent receives the tunneled HoT message, it does the following. Since the destination address in the outer packet is the IPv4 home address, the IPv4 portion of the home agent operation requires verification of presence of a tunnel, according to RFC 3344 (Section 4.2.3) [2]. In this case, the destination addresses are the *same* in the tunneled and encapsulating packets. They belong to different IP versions, however. When the inner packet is of type IPv6, it is forwarded to Mobile IPv6 processing software, which in turn applies IPsec and returns the HoT packet to the Mobile IPv4 processing software. The home agent then changes the destination address of the outer IPv4 packet to the IPv4 care-of address corresponding to the IPv4 home address.

The mobile node receives both the CoT and HoT messages. The mobile node constructs the Binding Update message. The MAC calculation includes the IPv4 care-of address and the correspondent node's IPv4 addresses. The mobile node includes the IPv4 HoA and the IPv4 CoA as new Mobility Header Options. The Binding Update is sent using an IPv4 tunnel with IPv4 care-of address as the source address and the correspondent node's IPv4 address as the destination address just as in the CoTI message transmission.

Upon successful processing of the Binding Update message, the correspondent node performs the following bindings: IPv6 home address to IPv6 care-of address, as in Mobile IPv6, IPv4 home address to IPv4 care-of address, and finally, IPv6 care-of address to IPv4 care-of address. This allows the correspondent node to tunnel route-optimized IPv6 traffic using an IPv4 address. It also allows the correspondent node to tunnel the IPv4 traffic directly to the mobile node without having to traverse the home agent. In other words, it facilitates route optimization for IPv4 traffic.

24.4 MESSAGE FORMATS

A new IPv4 address option is defined as a Mobility Header option. This option is used to carry the mobile node's IPv4 care-of address and the correspondent node's IPv4 addresses in the Return Routability messages and Binding Update message.

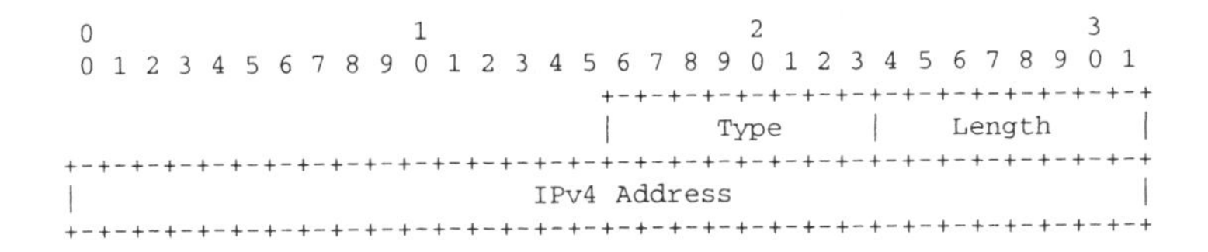
```
 0                   1                   2                   3
 0 1 2 3 4 5 6 7 8 9 0 1 2 3 4 5 6 7 8 9 0 1 2 3 4 5 6 7 8 9 0 1
                                +-+-+-+-+-+-+-+-+-+-+-+-+-+-+-+-+
                                |      Type     |    Length     |
+-+-+-+-+-+-+-+-+-+-+-+-+-+-+-+-+-+-+-+-+-+-+-+-+-+-+-+-+-+-+-+-+
|                          IPv4 Address                         |
+-+-+-+-+-+-+-+-+-+-+-+-+-+-+-+-+-+-+-+-+-+-+-+-+-+-+-+-+-+-+-+-+
```

Fig. 24.1 Mobility Header IPv4 Address Option

- Type: New Type denoting an IPv4 address
- Length: Length of the option in octets not including the Type and Length fields. MUST be set to 4.
- IPv4 Address: This field can include the MN's IPv4 HoA and CoA (in that order when both are present) and the CN's IPv4 address.

24.5 SECURITY IMPLICATIONS

With IPv6-in-IPv4 tunneling, a node could potentially include any IPv6 address inside an IPv4 packet. The solution in this chapter does not address this general problem. It does, however, require that the correspondent is a tunnel endpoint and the mobile node is either a tunnel endpoint or relies on a trustworthy node to provide the tunneling service. This consideration applies to tunneling of CoTI in an IPv4 packet. With HoTI, the home agent can ensure that the IPv6 home address and the IPv4 home address actually belong to the same mobile node. A correspondent node must ensure that the address present in the IPv4 Mobility Header option is the same as the source IP address of the encapsulating IPv4 packet.

Exercises

24.1 In our assumptions, we stated that the mobile node need not be a tunnel endpoint for IP-in-IP encapsulation. Why did we accept this assumption? Does it enable anything?

24.2 In the basic Return Routability mechanism, the mobile node reverse-tunnels the HoTI message to the home agent, which in turn forwards the packet to the correspondent. When the IPv4 address is co-located with the mobile node, the mobile node can reverse-tunnel the HoTI message using IPv4 addresses. What implications does this have on the security association and signaling? What are the savings, if any?

24.3 Prove that the approach described in this chapter is secure for IPv4 traffic. What are its weaknesses? Would it be sufficient considering the "do no more harm" principle?

24.4 Assume that the mobile node is behind a Network Address Translator (NAT) of IPv4 network. What extensions are necessary to support route optimization for private IPv4 addresses? What are the implications for the mobile node, the home agent and the Correspondent Node?

24.5 Now assume that the correspondent is behind a NAT. How would the protocol operate? In Exercise 24.4 and in this exercise, how do the nodes detect the presence of a NAT first?

REFERENCES

1. D. Johnson, C. E. Perkins, and J. Arkko. "Mobility Support in IPv6," RFC 3775, Internet Engineering Task Force, June 2004.

2. C. E. Perkins (Editor). "IP Mobility Support for IPv4," RFC 3344, Internet Engineering Task Force, August 2002.

3. The IPv6 Operations Working Group. http://www.ietf.org/html.charters/v6ops-charter.html

Glossary

Access Router The network element that provides IP connectivity. It is also the node that provides mobility services, including the home agent and fast handover functionalities.

Binding Cache A data structure maintained by the home agents and correspondent nodes which binds the home address to the care-of address. A **Binding Cache Entry (BCE)** refers to a specific binding of home address to care-of address along with other parameters such as Lifetime, Sequence Numbers, etc.

Binding Update List The list of peers with which a mobile node has sent a Binding Update.

Basic Service Set IDentifier or BSSID Typically the layer 2 address of the access point in WLAN.

Care-of Address The IP address acquired by the MN when visiting a network away from home. Sometimes denoted as CoA.

Context Transfer A protocol to reduce the overhead associated with reestablishing contexts as a function of handover, and hence supports the smooth operation of transport protocols.

Correspondent Node The node with which a mobile node communicates. Sometimes denoted as CN.

Fast Handovers A protocol to reduce delay and packet loss associated with handovers and hence support mobility of real-time applications.

Foreign Agent A node on the visited network that offers mobility services, such as its IP address and tunneling support, to visiting mobile nodes. Sometimes denoted as FA.

Handover A process in which a mobile node disconnects from its existing point of attachment and attaches itself to a new point of attachment. Handover includes operations at specific link layer as well as at the IP layer in order for the mobile node to be able to communicate again. One or more application streams typically accompany the mobile node as it undergoes handover.

Hierarchical Mobility The use of a local home agent in the visited domain to reduce signaling and provide some level of roaming transparency.

Home Network The network at which a mobile node is normally present at when it is not roaming.

Home Address The IP address of the mobile node valid on its home network. Sometimes denoted as HoA.

Home Network Prefix The prefix valid on the home network with which a home address can be derived.

Home Agent A router on the mobile node's home network which provides forwarding service to the home address when the mobile node is not present on its home network. Sometimes denoted as HA.

IPsec Suite of protocols used to secure IP traffic using a specified security association.

Internet Key Exchange The protocol used to dynamically establish security association between peers on the Internet. Often denoted as IKE.

Location Privacy Protecting the privacy of the location of roaming users. In Mobile IP, this means protecting the disclosure of home address and care-of address.

Medium Access Control (MAC) Protocol The protocol used by a particular link layer to communicate with neighbors and access points. A MAC layer is below the IP layer and helps transmit and receive IP datagrams.

Mobile Node A device with an IP software stack that freely roams from one network to another. Sometimes denoted as MN.

Mobility Anchor Point (MAP) A node that essentially acts as a home agent in the visited domain for mobile nodes.

Multiaccess The ability of a node to access multiple (typically) radio networks such as WLAN and cellular radio.

Neighbor Discovery (ND) A protocol used to resolve an IPv6 address to its link layer address. It is also used for router discovery.

Network Mobility (NEMO) Mobility involving an entire network of nodes in which a node takes responsibility for handling mobility for the network.

Prefix A bit string consisting of some number of bits which are common to addresses derived using it.

Return Routability A protocol that proves to a correspondent node that a mobile node owns the home address and that the mobile node is present at the care-of address. This helps the correspondent node to securely establish a Binding Cache.

Roaming A process in which a mobile node moves from one network to another, typically with no existing packet streams.

Routing Table A data structure where forwarding nodes such as routers keep the next hop information.

Security Association (SA) An abstraction used to define and implement IP security. Specifies what and how to secure IP traffic.

SSID The Service Set Identifier used in beacon advertisements sent by WLAN access points. It is generally expected to be a user-friendly text string.

Visited Network The network which a mobile node is visiting when away from its home network.

Index